Contemporary Phytomedicines

Contemporary Phytomedicines

Amritpal Singh Saroya
Herbal Consultant
Punjab
India

CRC Press
Taylor & Francis Group
Boca Raton London New York

CRC Press is an imprint of the
Taylor & Francis Group, an **informa** business
A SCIENCE PUBLISHERS BOOK

CRC Press
Taylor & Francis Group
6000 Broken Sound Parkway NW, Suite 300
Boca Raton, FL 33487-2742

Printed on acid-free paper
Version Date: 20161208

International Standard Book Number-13: 978-1-4987-7355-3 (Hardback)

Library of Congress Cataloging-in-Publication Data

Names: Amritpal Singh, 1971- author.
Title: Contemporary phytomedicines / Amritpal Singh Saroya.
Description: Boca Raton, FL : CRC Press, Taylor & Francis Group, [2016] | "A Science Publishers book." | Includes bibliographical references and index.
Identifiers: LCCN 2016050023| ISBN 9781498773553 (hardback : alk. paper) | ISBN 9781498773560 (e-book)
Subjects: | MESH: Plants, Medicinal | Phytotherapy | Plant Preparations--pharmacology | Drug Discovery
Classification: LCC RS160 | NLM QV 766 | DDC 615.3/21--dc23
LC record available at https://lccn.loc.gov/2016050023

Visit the Taylor & Francis Web site at
http://www.taylorandfrancis.com

and the CRC Press Web site at
http://www.crcpress.com

Preface

The term phytomedicine was coined by a French physician Henri Leclerc in 1913. Till recently phytomedicine has remained in the background. But due to emerging challenges to the conventional pharmaceutical industry (cost effectiveness and potency of the drugs), phytomedicine has made a dramatic comeback. Phytomedicine has witnessed several changes and several new concepts have been introduced. Phytomedicine, although, a separate discipline, has a strong linkage with Phytotherapy and Phytopharmacology.

As the title suggests the book is an attempt to bridge the gap between fundamental and emerging concepts in this field of medicine. The book has been divided into two parts.

Part A deals with core issues of the phyto-pharmaceutical drug industry. The book begins with an introductory chapter dealing with basic definitions with phytomedicine. Chapters 2–5 narrate emerging subjects such as *Phytopharmacovigilance, Phytopharmacoeconomics, Phytopharmacoepidemiology* and *Phytopharmacogenomics.* Chapter 6 discusses ethical issues in phytomedicine. Chapter 7 covers recent advances in drug delivery systems in phytomedicine whereas Chapter 8 is about application of nanotechnology in the field of phytomedicine. The further chapters cover metabolomics, regulatory and legal aspects of the phyto-pharmaceutical drug industry. The chapter on herbal bioavailability enhancing agents is the salient feature of Part-A.

Part B is related to applied research in the field of phytomedicine. Experimental findings on phyto-bioactive agents such as withanolides, steroidal alkaloids, phytosteroids and phytocannabinoids have been elaborated.

Nine annexures related to herbal drug registration are included.

I sincerely hope the book shall of genuine utility for academicians as well as phyto drug industry. Constructive criticism for improving the credibility of the book is welcome.

Contents

Acronyms and Abbreviations

CITES	=	The Convention on International Trade in Endangered Species of Wild Fauna and Flora
HPC	=	Health Professions Council
BHMA	=	The British Herbal Medicine Association
PLRs	=	*Product Licences of Right*
THRs	=	*Traditional Herbal Registrations*
HMAC	=	Herbal Medicines Advisory Committee
CSM	=	Committee on Safety of Medicines
BPC	=	British Pharmaceutical Codex
USP-NF	=	United States Pharmacopeia – National Formulary
ADR	=	Adverse Drug Reaction
AHN	=	Approved Herbal Name
AHP	=	The American Herbal Pharmacopoeia
AHPA	=	The American Herbal Products Association
AHS	=	Herbal Substance Name
Anvisa	=	Brazilian National Health Surveillance Agency
APC	=	Ayurvedic Pharmacopoeial Committee
API	=	Ayurvedic Pharmacopoeia of India
ARGCM	=	The Australian Regulatory Guidelines for Complementary Medicines
ARGCM	=	Australian regulatory guidelines for complementary medicines
ARTG	=	Australian Register of Therapeutic Goods
ASU	=	Ayurveda, Siddha and Unani
ATMs	=	African traditional medicines
AYUSH	=	Ayurveda, Yoga & Naturopathy, Unani, Siddha and Homoeopathy
BfArM	=	The Commissions of the German Federal Institute for Drugs and Medical Devices
BHP	=	British *Herbal Pharmacopoeia*
BNHUA	=	The Bangladesh National Formularies on Unani and Ayurvedic medicine
BP	=	The British *Pharmacopoeia*
CAM	=	Complementary and alternative medicine
CBER	=	Center for Biologics Evaluation and Research
CCH	=	Central Council for Homoeopathy

CCIM	=	Central Council for Indian Medicine
CCRAS	=	Central Council for Research in Ayurveda and Siddha
CCRH	=	Central Council for Research in Homeopathy
CCRUM	=	Central Council for Research in Unani medicine
CCRYN	=	Central Council for Research in Yoga and Naturopathy
CDER	=	Center for Drug Evaluation and Research
CDRH	=	Center for Devices and Radiological Health
CEP	=	Certificate of Suitability to the monographs of the European Pharmacopoeia
CFSAN	=	Center for Food Safety and Applied Nutrition
cGMP	=	Current Good Manufacturing Practice
CHM	=	Chinese herbal medicines
CMC	=	Chemistry manufacturing and controls
CPP	=	Certificate of Pharmaceutical Product
CRISM	=	Center for Research on Indian Systems of Medicine
CRS	=	Chemical Reference Substance
CTD	=	Common Technical Document
DCEP	=	Documentation to the Certification of Substances Division
DCP	=	Decentralised Procedures
DH	=	The Department of Health
DIN-HM	=	Homeopathic Medicine Number
DMF	=	Drug Master File
DSHEA	=	The Dietary Supplements Health and Education Act of 1994
DSU	=	Drug Safety Update
eCTD	=	Electronic Common Technical Document
EDQM	=	European Directorate for the Quality of Medicines and Healthcare
EFSA	=	European Food Safety Authority
EHTPA	=	European Herbal & Traditional Medicine Practitioners Association
EMA	=	European Medicines Agency
EP	=	European Pharmacopoeia
ESCOP	=	European Scientific Cooperative on Phytotherapy
FTC	=	The Federal Trade Commission
GACP	=	Good Agricultural and Collection Practices
GAP	=	Good Agricultural Practices
GHAFTRAM	=	Ghana Federation of Traditional Medicine Associations
GHP	=	Ghana *Herbal Pharmacopoeia*
GMP	=	Good Manufacturing Practice
GP	=	General practitioner
GRAS	=	Generally recognized as safe
HCN	=	Herbal Component Name
HINC	=	Herbal Ingredient Names Committee
HMP	=	Herbal Medicinal Product

HMPC	=	Committee for Herbal Medicinal Products
IDB	=	Investigator's Drug Brochure
IDMA	=	Indian Drug Manufacturers' Association
IMPD	=	The Investigational Medicinal Product Dossier
IND	=	Investigational New Drug
INN	=	International Nonproprietary Names
IRCH	=	International Regulatory Cooperation for Herbal Medicines
ISM&H	=	Indian System of Medicine and Homeopathy
ISM	=	Indian Systems of Medicine
KHP	=	The Korean Herbal *Pharmacopoeia*
LPC	=	Legal Procurement Certificate
LTR	=	Local Technical Representative
MA	=	A Full Marketing Authorisation
MD	=	Maximum dose
MDD	=	Maximum daily dose
MEB	=	Medicines Evaluation Board
MEDDRA (MedDRA)	=	Medical Dictionary for Drug Regulatory Affairs
MHRA	=	The Medicines and Healthcare products Regulatory Agency
MRP	=	Mutual Recognition Procedures
NAFDAC	=	The National Agency for Food Administration and Control
NCAHF	=	The National Council Against Health Fraud
NCCAM	=	The US National Center for Complementary and Alternative Medicine
NDA	=	The National Drug Authority
NDI	=	A new dietary ingredient
NHPD	=	The Natural Health Products Directorate
NHP	=	Natural Health Products
NIA	=	National Institute of Ayurveda
NIMPs	=	Non-investigational medicinal products
NIS	=	National Institute of Siddha
NMPB	=	National Medicinal Plants Board
NNHPD	=	The Natural and Non-prescription Health Products Directorate
NPN	=	Natural Product Number
NtA	=	Notice to Applicants
OAM	=	The Office of Alternative Medicine
OTC	=	Over The Counter products
Pas	=	Pyrrolizidine alkaloids
PHs Act	=	Public Health Service Act
PITAHC	=	Philippine Institute of Alternative Healthcare
PL	=	Product Licence
PLIM	=	Pharmaceutical Laboratory of Indian Medicine
PPRC	=	The *Pharmacopoeia* of the People's Republic of China

PSUR	=	Periodic Safety Update Reports
QA	=	Quality Assurance
QC	=	Quality Control
QPs	=	Qualified Persons
RAIG	=	The Regulatory Affairs Interest Group
RCT	=	Randomized Clinical Trial
RMS	=	Reference Member State
SATCM	=	State Administration of Traditional Chinese Medicine
SFDA	=	The China State Food and Drug Administration
SMEs	=	Small and medium-sized enterprises
SNVS	=	National Health Surveillance System
SPC	=	Summary of Product Characteristics
SR	=	Statutory Regulation
TAMA	=	Traditional and Alternative Medicines Act
TAMD	=	The Traditional and Alternative Medicine Directorate
TCM	=	Traditional Chinese Medicines
TGA	=	Therapeutic Goods Administration
TGAC	=	Therapeutic Goods Advertising Code
THMPD	=	Traditional Herbal Medicinal Products Directive
THMRS	=	Traditional Herbal Medicines Registration Scheme
THR	=	Traditional Herbal Registration
TKDL	=	Traditional Knowledge Digital Library
TM	=	Traditional medicine
TWHM	=	Traditional Western Herbal Medicine
UPC	=	Unani Pharmacopoeial Committee
URDs	=	Union Reference Dates
URHP	=	The Unified Register of Herbal Practitioners
USP RS	=	United States Pharmacopoeia Reference Substance
USP	=	United States Pharmacopoeia
WHO	=	The World Health Organization
AAMPS	=	Association of African Medicinal Plant Standards
TBAs	=	Traditional Birth Attendants

Definitions

Active Ingredient

Active ingredients refer to ingredients of herbal medicines with therapeutic activity. In herbal medicines where the active ingredients have been identified, the preparation of these medicines should be standardized to contain a defined amount of the active ingredients, if adequate analytical methods are available. In cases where it is not possible to identify the active ingredients, the whole herbal medicine may be considered as one active ingredient.

Botanical; Botanical Product

A finished, labeled product that contains vegetable matter, which may include plant materials (see below), algae, macroscopic fungi, or combinations of these. Depending in part on its intended use, a botanical product may be a food, drug, medical device, or cosmetic.

Botanical Drug Substance

A drug substance derived from one or more plants, algae, or macroscopic fungi. It is prepared from botanical raw materials by one or more of the following processes: pulverization, decoction, expression, aqueous extraction, ethanolic extraction, or other similar process. It may be available in a variety of physical forms, such as powder, paste, concentrated liquid, juice, gum, syrup, or oil. A botanical drug substance can be made from one or more botanical raw materials (see Single-Herb and Multi-Herb Botanical Drug Substance or Product). A botanical drug substance does not include a highly purified or chemically modified substance derived from natural sources.

Botanical Ingredient

A component of a botanical drug substance or product that originates from a botanical raw material.

Botanical Raw Material

Fresh or processed (e.g., cleaned, frozen, dried, or sliced) part of a single species of plant or a fresh or processed alga or macroscopic fungus.

Complementary/alternative medicine (CAM)

The terms "complementary medicine" or "alternative medicine" are used interchangeably with traditional medicine in some countries. They refer to a broad set of health care practices that are not part of that country's own tradition and are not integrated into the dominant health care system.

Ethnobotany

Ethnobotany is the study of how people of a particular culture and region make of use of indigenous plants.

Ethnomedicine

Ethnomedicine is a study or comparison of the traditional medicine practiced by various ethnic groups, and especially by indigenous peoples. The word ethnomedicine is sometimes used as a synonym for traditional medicine.

Ethnopharmacology

Ethnopharmacology is a related study of ethnic groups and their use of drugs. Ethnopharmacology is distinctly linked to plant use, ethnobotany, as this is the main delivery of pharmaceuticals.

Ethnopharmacy

Ethnopharmacy is the interdisciplinary science that investigates the perception and use of pharmaceuticals (especially traditional medicines, but not only), within a given human society.

Extraction

The commonly employed technique for removal of active substance from the crude drug is called extraction.

Herbal medicines

Herbal medicines include herbs, herbal materials, herbal preparations and finished herbal products, that contain as active ingredients parts of plants, or other plant materials, or combinations.

- **Herbs:** crude plant material such as leaves, flowers, fruit, seed, stems, wood, bark, roots, rhizomes or other plant parts, which may be entire, fragmented or powdered.
- **Herbal materials:** in addition to herbs, fresh juices, gums, fixed oils, essential oils, resins and dry powders of herbs. In some countries, these materials may be processed by various local procedures, such as steaming, roasting, or stir-baking with honey, alcoholic beverages or other materials.
- **Herbal preparations:** the basis for finished herbal products and may include comminuted or powdered herbal materials, or extracts, tinctures and fatty oils of herbal materials. They are produced by extraction, fractionation, purification, concentration, or other physical or biological processes. They also include preparations made by steeping or heating herbal materials in alcoholic beverages and/or honey, or in other materials.
- **Finished herbal products:** herbal preparations made from one or more herbs. If more than one herb is used, the term mixture herbal product can also be used. Finished herbal products and mixture herbal products may contain excipients in addition to the active ingredients. However, finished products or mixture products to which chemically defined active substances have been added, including synthetic compounds and/or isolated constituents from herbal materials, are not considered to be herbal.

Herbalism

See herbal medicine.

Marker Compound

A constituent of a medicinal herb, which is used for quality control and assurance of herbal product is known as marker compound.

Materia Medica proper

Materia Medica proper is defined as knowledge of natural history, physical characteristics, and chemical properties of drugs. It includes study of herbs, minerals and drugs from animal kingdom.

Multi-Herb (Botanical Drug) Substance or Product

A botanical drug substance or drug product that is derived from more than one botanical raw material, each of which is considered a botanical ingredient. A multi-herb botanical drug substance may be prepared by processing together two or more botanical raw materials, or by combining two or more single-herb botanical drug substances that have been individually processed from their corresponding raw materials. In the latter case, the individual single-herb botanical drug substances may be introduced simultaneously or at different stages during the manufacturing process of the dosage form.

Pharmacopoeia

Pharmacopoeia literally means "drug-making". Pharmacopoeia is a book describing drugs, chemicals, and medicinal preparations; especially: one issued by an officially recognized authority and serving as a standard.

Pharmacognosy

Pharmacognosy is the study of the physical, chemical, biochemical and biological properties of drugs, drug substances or potential drugs or drug substances of natural origin as well as the search for new drugs from natural sources.

Phytomedicine

Herbal-based traditional medical practice that uses various plant materials in modalities considered both preventive and therapeutic.

Phytopharmacology

Phytopharmacology is field of study of the effects of drugs on plants.

Phytopharmaceutical Science

Phytopharmaceutical Science is the development of drugs from plants and other natural compounds.

Plant Material

A plant or plant part (e.g., bark, wood, leaves, stems, roots, flowers, fruits, seeds, or parts thereof) as well as exudates thereof.

Regulatory affairs (RA)

Regulatory affairs is also known as government affairs. A regulatory affair is a profession within regulated industries, such as pharmaceuticals, medical devices, energy, banking, telecom, etc. Regulatory affairs have a very specific meaning within the healthcare industries including pharmaceuticals, medical devices, biologics and functional foods.

Single-Herb (Botanical Drug) Substance or Product

A botanical drug substance or drug product that is derived from one botanical raw material. Therefore, a single-herb substance or product generally contains only one botanical ingredient.

The Standardised Herbal Extract

The standardised herbal extract is a preparation, which contains a certain fixed proportion of the active constituent.

Therapeutic Activity

Therapeutic activity refers to the successful prevention, diagnosis and treatment of physical and mental illnesses; improvement of symptoms of illnesses; as well as beneficial alteration or regulation of the physical and mental status of the body.

Traditional Medicine

Traditional medicine is the sum total of the knowledge, skills, and practices based on the theories, beliefs, and experiences indigenous to different cultures, whether explicable or not, used in the maintenance of health as well as in the prevention, diagnosis, improvement or treatment of physical and mental illness.

Traditional Use of Herbal Medicines

Traditional use of herbal medicines refers to the long historical use of these medicines. Their use is well established and widely acknowledged to be safe and effective, and may be accepted by national authorities.

Herbal Glossary

Adaptogen : an agent that invigorates or strengthens the system.
Alterative : an agent used for purifying blood.
Anabolic : an agent having steroidal action.
Analeptic : an agent used to boost respiration and circulation.
Anodyne : an agent tahr relieves pain on local application.
Antiarrthymic : an agent used for treatment of heart disease.
Antibiotic : an agent used for killing microorganisms.
Anticoagulant : an agent used for preventing blood clotting.
Antidepressant : an agent used for counteracting depression.
Antihelmintic : an agent used to kill worms.
Antiperiodic : an agent used for preventing relapsing fever.
Antipruritic : an agent used to cure itching.
Antirheumatic : an agent used for curing arthritis and rheumatism.
Analgesic : an agent used for preventing pain.
Anti-inflammatory : an agent used for preventing inflammation.
Antipyretic : an agent used for lowering the fever.
Antiseptic : an agent used form preventing the growth of microorganisms.
Antispasmodic : an agent used for reliving the spasms of voluntary and involuntary muscles.
Aphrodisiac : an agent used to stimulate sex urge and maintain vitality.
Aperient : an agent used for mild laxation.
Ayurveda : the ancient healing system of India.
Bruising : a process of smashing of different parts of a medicinal herb in a pestle and mortar.
Cathartic : an agent used to relieve severe constipation.
Carminative : an agent used to dispel gas from the intestine and
prevent distension.
Cholagogue : an agent used to promote the flow of bile.
Choleretic : an agent that stimulates the formation of bile.
Convulsant : an agent which induces seizures.
Counterirritant : (See Rubefacient).
Crude drug : the form of the medicinal herb unchanged by processing other than separation of parts, drying or grinding.
Decoction : a process of boiling a coarsely bruised drug in water in tinned pots with covers for a definite period.

Diaphoretic	:	an agent used for increasing perspiration through the skin.
Diuretic	:	an agent used for increasing urine flow.
Ecbolic	:	an agent used for stimulating uterine musculature.
Elutriation	:	a process of separation of the coarser particles of a powder from the finer ones.
Emetic	:	an agent used for inducing vomiting.
Emollient	:	an agent, which softens the skin.
Expectorant	:	an agent used to promote the expulsion of mucus form the respiratory tract.
Expression	:	a process of pressing out juice or oil from plant products.
Extract	:	an process of manufacturing of concentrated preparations of the active principles of the vegetable drugs.
Fluid extract	:	a liquid extract of raw plant material, usually of a concentration ratio of 1 part raw herb to 1 part solvent.
Febrifuge	:	an agent used to reduce fever.
Hemostatic	:	an agent used to prevent flow of blood.
Hepatoprotective	:	an agent used for preventing injury to the liver.
Hygroscopic	:	a substance that readily attracts and retains water.
Hypnotic	:	an agent used to induce sleep.
Hypolipidemic	:	an agent which reduces high levels of cholesterol.
Incineration	:	a process of heating the organic substances with access of air, so that all the carbonaceous matter is burnt.
Infusion	:	a process of treating a moderately comminuted drug in a muslin bag soaked in cold or hot water.
Levigation	:	a process of grounding of solid substance with water to make a paste and dry.
Maceration	:	a process of soaking a ground up drug in a solvent and expression of fluid from it.
Medicated oil	:	oil preparation obtained by steeping the medicinal herb in oil for several days or months.
Sifting	:	a process of passing a powdered drug through a sieve to obtain powder of uniform strength.
Nervine	:	an agent used for improving the function of the nerves.
Nootropic	:	an agent having memory enhancing activity.
Pharmacy	:	the study of scope, the preparation of materials in suitable forms for use in Medicine.
Rubifacient	:	an agent used for increasing the blood supply to the skin, when applied locally.
Sedative	:	an agent used for calming the functional activity of the body.
Standardisation	:	a process of fixing the quantity of active constituent of a medicinal agent.
Stimulant	:	an agent used for boosting metabolism and circulation.
Stomachic	:	an agent used to promote stomach function.

Tincture : alcoholic solution of active constituents of vegetable drugs.

Trituration : the process of rubbing solid substances into finer ones with the help of a pestle and mortar.

Tonic : an agent used to increase energy and vigor in a specific part of the body.

Vasodilator : an agent used to dilate the blood vessels.

Vulnerary : an agent used to promote the healing of new cuts and wounds.

PART A
Industrial Aspects of Phytomedicine

Chapter 1

Introduction to Phytomedicine

1.1 Phytomedicine

The term phytomedicine (phyto+medicine) was coined by a French physician Henri Leclerc in 1913. Henri Leclerc published Precis de phytotherapie (a handbook of phytotherapy) in 1922. Henri Leclerc was credited with yet another publication, die pflanzenheikunde in der Arztlichen Praxis in 1944 (Plant based curative science in Medical Practice). Phytomedicine is defined as a herbal-based traditional medical practice that uses various plant materials in modalities (Farlex 2012). Phytomedicine is considered to be both preventive and therapeutic in nature and is also referred to as phytotherapy, herbal therapy or medical herbalism.

William Powel in the year of 1934 coined the term Phytotherapy. Phytomedicine should not be mixed with phytopharmacology, a term coined by David Macht, a scientist of Russian origin (Macht 1936).

1.2 Preliminary Definitions Related to Phytomedicine

1.2.1 Complementary/Alternative medicine

They refer to a broad set of health care practices that are not part of that country's own tradition and are not integrated into the dominant health care system. In fact, both the terms are interchangeable.

1.2.2 Traditional medicine

Traditional medicine is the sum total of the knowledge, skills, and practices based on the theories, beliefs, and experiences indigenous to different cultures, whether explicable or not, used in the maintenance of health as well as in the prevention, diagnosis, improvement, or treatment of physical and mental illness.

1.2.3 Medicinal plant/Medicinal herb

It includes crude plant material such as leaves, flowers, fruit, seed, stems, wood, bark, roots, rhizomes, or other plant parts, which may be entire, fragmented, or powdered.

1.2.4 Herbal medicines

They include herbs, herbal materials, herbal preparations, and finished herbal products, which contain as active ingredients parts of plants, or other plant materials, or combinations. Traditional use of herbal medicines refers to the long historical use of these medicines. Their use is well established and widely acknowledged to be safe and effective, and may be accepted by national authorities.

1.2.5 Herbal materials

They include, in addition to herbs, fresh juices, gums, fixed oils, essential oils, resins, and dry powders of herbs. In some countries, these materials may be processed by various local procedures, such as steaming, roasting, or stir-baking with honey, alcoholic beverages, or other materials.

1.2.6 Herbal preparations

They are the basis for finished herbal products and may include comminuted or powdered herbal materials, or extracts, tinctures, and fatty oils of herbal materials. They are produced by extraction, fractionation, purification, concentration, or other physical or biological processes. They also include preparations made by steeping or heating herbal materials in alcoholic beverages and/or honey, or in other materials.

1.2.7 Finished herbal products

They consist of herbal preparations made from one or more herbs. If more than one herb is used, the term mixture herbal product can also be used. Finished herbal products and mixture herbal products may contain excipients in addition to the active ingredients. However, finished products or mixture products to which chemically defined, active substances have been added, including synthetic compounds and/or isolated constituents from herbal materials, are not considered to be herbal.

1.2.8 Active ingredients

They refer to ingredients of herbal medicines with therapeutic activity. In herbal medicines where the active ingredients have been identified, the preparation of these medicines should be standardized to contain a defined amount of the active ingredients, if adequate analytical methods are available. In cases where it is not possible to identify the active ingredients, the whole herbal medicine may be considered as one active ingredient.

1.2.9 Marker compound

A constituent of a medicinal herb used for quality control and assurance of herbal product. A marker compound may or may not have therapeutic activity.

1.2.10 Pharmacopoeia

Pharmacopoeia is a book describing drugs, chemicals, and medicinal preparations; especially one issued by an officially recognized authority and serving as a standard. Pharmacopoeia is derived from ancient Greek world *pharmakopoiia* (*pharmakon+ poi–*). All the principle countries of the world have their national pharmacopoeias.

1.2.11 Extra pharmacopoeia

Extra pharmacopoeia contains information on drugs not included in the pharmacopoeia.

1.2.12 National Formulary

National Formulary (*NF*) includes formulations and unofficial preparations for widely sold products.

1.2.13 Addendum

Addendum is an addition required to be made to a document by its author subsequent to its printing or publication.

1.2.14 Monograph

A monograph is a paper on a single topic.

1.2.15 Herbal monograph

Herbal monographs normally include nomenclature, part used, constituents, range of application, contraindications, side effects, incompatibilities with other medications, dosage, use, and action of the herb.

1.3 History of Phytomedicine

It is not easy to trace the history of phytomedicine. Historical evidences, however do indicate that medicinal plants were used in ancient civilizations. Primitive man observed and appreciated the great diversity of plants available to him (Arturo 1941). The first evidence of use of medicinal plants in a health care system comes from China (2800 B.C.). Much of the medicinal use of plants seems to have been developed through observations of wild animals, and by trial and error (Logan 1942).

Shen Hung (3000 B.C.), the great Chinese emperor, wrote an account of 365 medicinal plants in his work, *Pen Tsao. ching* (Divine Husbandman's Materia Medica). The work is considered to be the earliest extant Chinese Pharmacopoeia. Shen Nong documented the use of Ma Huang (Ephedra) in the treatment of respiratory illness like bronchitis and asthma. Hammurabi, a king of Babylonia (1800 B.C.), wrote an account on usage of medicinal plants. He documented the use of peppermint in the treatment of digestive system ailments. Hammurabi prescribed the use of mint for digestive disorders (Charles 1976; Liu 1982).

Hippocrates (400 B.C.) wrote the first Greek herbal text. He explained the role of diet, exercise, and medicine in maintaining optimal health. Galen (200 A.D.), practitioner of herbal medicine, classified diseases according to the human anatomy. He further indicated specific remedies to cure diseases (Scarborough 1978; Kline 1997). Avicenna (1100 A.D.), the great Arabian physician, wrote the *Canon of Medicine*. Dioscorides, a Roman physician, wrote *De Materia Medica*, which described the medicinal use of plants ranging from almond to wormwood. *De Materia Medica* was the first systematic pharmacopoeia and was translated and preserved by the Arabs, and finally translated back into Latin by the 10th century (Ducourthial 2005).

Culpepper (1600 A.D.) wrote the principle and practice of herbal medicine in his work *The English Physician*. In his work, Culpepper has described 1653 drugs with information on the mode of preparation and dose. Many of his unpublished manuscripts were published after his death, but many more were lost in the Great Fire of London in 1666 (McCarl 1996). Marcus Aurelius (A.D. 161–180) explained the use of opium (*Papaver somniferum*) in the treatment of headache, epilepsy, asthma, and skin diseases. In fact, he documented the use of medicinal herbs in his work *Meditations* (Porter 1995).

Ayurveda originated from *Artharva Veda* and Vedic era is considered to be the time when *Ayurveda* flourished as a medical science. It is estimated that around 1000 B.C., two major texts of Ayurveda, *Charaka Samhita* and *Sushruta Samhita* were composed. *Charaka* and *Sushruta* are respected names in the fields of medicine and surgery respectively. Both the texts have dealt in detail with the use of medicinal plants (Indian Herbal Pharmacopoeia 1998). Chebulicmyrobalans (*Terminalia chebula*), *Arjuna* (*Terminalia arjuna*), *Guggul* (*Commiphoramukul*), *Shatavari* (*Asparagus officinalis*) and *Ashwagandha* (*Withania somnifera*) are popular medicinal plants targeted for application in modern science (Kapoor 1990).

The time period between 1488 to 1682 is known as the age of herbals (Gaebler 1964). Otto Brunfels wrote a herbal text in 1488, which was published in 1534. This period produced a number of distinguished herbalists like Gesner Conard, Leohard Fuchs, Hieronymus Boch, William Turner, and John Parkinson (Debus 1968). Friedrich Wilhelm Serturner (1783–1841) isolated morphine from *Papaver somniferum* in 1805 and showed the medical world that certain chemical constituents are responsible for curative actions of plant based remedies. The scientific community will always remain thankful to Serturner for his great service to the world of medicinal plants (Asimov 1982; Court 1985).

Felix Hoffman isolated aspirin from willow bark (*Salix* spp.). His work augmented the rational use of willow bark by ancient people. The bark was used in the treatment of arthritis and rheumatism. Aspirin is still prescribed in reducing pain and stiffness associated with joints (Dombrowski and Alfermann 1995). William Withering (1741–1799) reported to the scientific community about the separation of the cardiac glycoside, digoxin from foxglove (*Digitalis purpurea*). The discovery of digoxin proved to be a milestone in the history of medicine (particularly cardiology) as digoxin was once upon a time a first line drug in treating cardiac oedema.

Klie isolated reserpine from *Rauwolfia serpentina* and the alkaloid remained as the drug of choice for the treatment of hypertension for almost 50 years (Venkata Rao 2010). Jean Robiquet reported the isolation of antitussive alkaloid, codeine from the opium plant. Clark Noble did a great service for humanity by discovering Vinca alkaloids from the Madagascar periwinkle (*Catharanthus roseus* Linn). Vinca alkaloids (vinblastine and vincristine) are prized drugs for treating leukemia. Discovery of taxol from Pacific yew (*Taxus brevifolia*) by Mansukh C. Wani and silymarin from milk thistle (*Silybum marianum*) by Jack Masquelier are some recent examples of drug obtained from plants.

Before the discovery of antibiotics (penicillin and streptomycin), analgesics and steroids, man was completely dependent on medicinal plants as the healthcare system. With the discovery of phytochemicals, the whole interest of the scientific community shifted to organic synthesis and several drugs were synthesized. The growing popularity of the allopathic system of medicine was a major setback for herbal medicine. Emergency treatment and surgical advances are the gifts of the modern healthcare system to modern man (Asimov 1982).

Today we can see the renaissance of herbal system of medicine. Ayurveda and Traditional Chinese Medicine (TCM) are popular systems of healing in western countries. Recent studies have shown that an increasing number of patients are consulting doctors for alternative systems of healing. Relative safety and cost effectiveness may be factors responsible for the renaissance of the herbal system of medicine (Bone 1996; Shweta and Boaz 2015).

1.4 Pharmacopoeia and Formularies Related to Phytomedicine

1.4.1 Ayurvedic Pharmacopoeia of India (API)

The Ayurvedic Pharmacopoeia of India is a legal document of standards for the quality of Ayurvedic drugs and substances included therein (under the Drugs and Cosmetic Act 1940). The API is published by the Department of Ayurveda, Yoga & Naturopathy, Unani, Siddha and Homoeopathy (AYUSH), Ministry of Health and Family Welfare, Government of India (Anonymous, Indian Herbal Pharmacopoeia 1998).

1.4.2 Ayurvedic Formulary of India (AFI)

The scattered information on various formulations in classical Ayurvedic books has been compiled in such a way to make it suitable to develop pharmacopoeial standards and also to meet the requirements of Drugs and Cosmetics Act (Anonymous, Pharmacopoeial Laboratory for Indian Medicine 1978).

1.4.3 American Herbal Pharmacopoeia

The mission of the American Herbal Pharmacopoeia is to promote the responsible use of herbal products and herbal medicines. The American Herbal Pharmacopoeia has published monographs on 27 different botanicals including those used in Ayurveda (Anonymous, American Herbal Pharmacopoeia).

1.4.4 British Herbal Pharmacopoeia

The British Herbal Pharmacopoeia 1983, a classic maintained in print for its unique guidance on therapeutics, is very different to the British Herbal Pharmacopoeia 1996, which offers quality standards for 169 herbal raw materials. Taken together, Volume 1 (1992) and Volume 2 (2006) of the British Herbal Compendium cover almost all of the plant drugs for which specifications appear in the British Herbal Pharmacopoeia 1996 (Anonymous, British Herbal Pharmacopoeia 1996).

1.4.5 Ghana Herbal Pharmacopoeia

The Ghana Herbal Pharmacopoeia has been published by the Science and Technology Policy Research Institute, Council for Scientific and Industrial Research, Ghana in 2007.

1.4.6 Nigerian Herbal Pharmacopoeia

The Nigerian Herbal Pharmacopoeia was published in 2008.

1.4.7 European folk medicine

European folk medicine has roots in ancient Greek, Roman, and Arabic medical theories. The folk knowledge has been passed down generation to generation by written as well as oral modes over the centuries. Some of the medical traditions have survived the passage of time relatively intact. Some medical traditions have either changed or disappeared, whereas "novel" remedies and usage of plants have emerged (Bladt and Wagner 2007).

1.4.8 Bhutanese traditional medicine

The Bhutanese traditional medicine has evolved out of Tibetan medicine. The pharmacopoeia, ethnopharmacology, ethnobotany, and the ethnoquality aspects shares commonalities with the mainstream Tibetan medicine. Some practices are unique to Bhutanese traditional medicine; the current practices of Bhutanese traditional medicine includes traditional formulations, collection and use of medicinal plants, and above all, the inclusion of golden needle and water therapy in treatment procedures (Wangchuk et al. 2013).

1.4.9 Saami folk medicine

The Saami are the indigenous people of northern Sweden, Norway, Finland, and the Kola Peninsula. Medicine for animals came largely from the bear and reindeer. They use 120 medicinal plants in folk formulas. The dried Angelica root is called "urtas" in Lule Saami (DuBois and Lang 2013).

1.4.10 Ozarkian and Haitian folk medicine

There exist a similarity between Haiti and midwestern America Herbalism as far as collection and usage of herbs are concerned. Ozark people enjoy better access to doctors and hospitals as compared to Haitians. Sarsaparilla, catmint, and mints (peppermint, spearmint, lemon mint, and horsemint) are some common herbs used in practice. Senna is used as an efficacious febrifuge. Vervain is widely used in female diseases (Fournier and Dodard 1997).

1.4.11 Georgian traditional medicine

Georgian traditional medicine comprises ancient written classical documents and folk medicine. More than 500 manuscripts from the 10th to the 19th century have been found and described in Georgia and elsewhere (Waller and Killion 1972).

1.4.12 Sardinian folk medicine

Sardinia is a territory with one of the greatest floristic diversity in the Mediterranean area in Italy. In Sardinia, there is a folk pharmacopoeia that includes a large number of endemic species, and the number of pathologies that are still treated with natural active ingredients is also large. The use of *Scrophulariatrifoliata* L. for Basedow's disease and *Vincasardoa* (Stearn) Pign. against tuberculosis is noteworthy (Cugusi et al. 2015).

1.4.13 Traditional pharmacopoeia of Vulture-Alto Bradano, inland southern Italy

The traditional dermatological pharmacopoeia of Vulture-Alto Bradano has roots in folk remedies, spiritual illness and healing. These remedies find use in the treatment of 45 skin and soft tissue diseases in humans as well as animals. *Malvasylvestris* (Malvaceae), *Marrubiumvulgare* (Lamiaceae), and *Matricariarecutita* (Asteraceae) are commonly used herbal remedies (Quave et al. 2008).

1.4.14 Traditional Iberian pharmacopoeia

The Iberian Peninsula can be considered a small continent of around 600,000 km^2. It is separated from the rest of Europe by the Pyrenees, a mountainous barrier that has contributed to its relative isolation. The rich traditional lore of the Iberian Peninsula has attracted many folklorists, ethnographers, and medical anthropologists and ethnobotanists since the end of the nineteenth century. According to a recent review of medicinal plants popularly used in Spain, the number of species employed is around 1,200, more than 15% of the Iberian flora (Quave et al. 2012).

1.4.15 Izoceño-guaraní ethnomedicine and pharmacopoeia

Izoceño-guaraní people belong to the extended Chiriguano group. They are actually organized in independent communities settled down in Southeast Bolivia. Medicine is in the hands of PAYE who are recognized as specialists in their own group. Ethnopharmacological research throws light on over 306 species, 189 of them having medicinal uses. Animal products are also part of the treatment in addition to plants (Bourdy et al. 2009).

1.4.16 A Hausa herbal pharmacopoeia

Hausa practitioners practice a system of herbal medicine in Hausa (Northern Nigeria). Biomedical evaluation has proven efficacy of herbs mentioned in Hausa herbal pharmacopoeia for the treatment of malaria (Etkin 1981).

1.4.17 Bir pharmacopoeia

Bir Nighantu is also referred to as *Bir pharmacopoeia* and is a hand written herbal encyclopedia. The credit of compilation of *Bir Nighantu* goes to Pandit Ghana Nath Devkota under the instruction of Bir Samser, the former prime minister of Nepal (1885–1901 A.D.). Kosh Nath Devkota wrote an elaborated account found in *Nepali Nighantu*, published by the Royal Nepal Academy in the year of 1969. *Nepali Nighantu* provides information on 750 medicinal plants. Most probably, *Nepali Nighantu* is the pioneer written effort targeted towards the compilation of the traditional knowledge about medicinal plants of Nepal.

1.4.18 The bedianus codex

It is known as 'The Little Book of Herbs'. It was composed in 1552 at a college of Santa Cruz by Aztec Physician Marin de la Cruz. Painting of the plants is the salient feature. It contains information of 251 plant species. Ohuaxocoyolin was the drug used to treat glaucoma. It is considered to the *Begonia* species.

1.4.19 Mexican Herbal Pharmacopoeia

Mexican Herbal Pharmacopoeia is also known as the Aztec Herbal Pharmacopoeia. It describes medicinal uses of four hundred flowers (Waldstein 2006).

1.4.20 Phansomba

Phansomba is an Indian herbal folk medicine that consists of several species of the mushroom genus *Phellinus* (Vaidya et al. 2010).

Further Readings

Anonymous. American Herbal Pharmacopoeia. http://herbal-ahp.org.

Anonymous. Ayurvedic Formulary of India. http://plimism.nic.in/publications.html.

Anonymous. British Herbal Pharmacopoeia. http://bhma.info/index.php/british-herbal-pharmacopoeia-1996/.

Anonymous. Indian Herbal Pharmacopoeia Volume 1. Worli, Mumbai: Indian Drug Manufacturers Association, 1998.

Anonymous. Ghana Herbal Pharmacopoeia. Council for Scientific and Industrial Research, Ghana 2007. openlibrary.org/Ghana_herbal_pharmacopoeia.

Arturo C. A History of Medicine. New York: Alfred A. Knopf, 1941; 6.

Asimov I. Asimov's Biographical Encyclopedia of Science and Technology. 2nd Revised Edition. Garden City, New York: Doubleday, 1982.

Bladt S, Wagner H. From the Zulu medicine to the European phytomedicine Umckaloabo. *Phytomedicine* 2007; **14**: 2–4.

Bone K. Clinical Applications of Ayurvedic and Chinese Herbs: Monographs for the Western Herbal Practitioner. Warwick, Qld: Phytotherapy Press, 1996.

Bourdy G, Chāvez de Michel LR, Roca-Coulthard A. Pharmacopoeia in a shamanistic society: the Izoceño-Guaraní (Bolivian Chaco). *J Ethnopharmacol* 2004; **91**: 189–208.

Charles L. Asian Medical Systems: A Comparative Study. Berkeley: University of California Press, 1976.

Court WE. A history of herbal medicine. *Pharm Hist* (Lond). 1985; **15**: 6–8.

Cugusi L, Massidda M, Matta D, Garau E, Di Cesare R, Deidda M, Satta G, Chiappori P, Solla P, Mercuro G. A new type of physical activity from an ancient tradition: The Sardinian folk dance "Ballu Sardu". *J Dance Med Sci* 2015; **19**: 118–23.

Debus AG. World Who's Who In Science: A Biographical Dictionary of Notable Scientists from Antiquity to the Present. Chicago: Marquis, 1968.

Dhiman AK. History of Indian medicine and pharmacopoeia. *Aryavaidyan* 2003; **17**: 24–31.

Dombrowski K, Alfermann AW. Salicylic acid—the universal drug of human and herbal medicine? *Pharm Unserer Zeit* 1993; **22**: 275–85.

DuBois TA, Lang JF. Johan Turi's animal, mineral, vegetable cures and healing practices: an in-depth analysis of Sami (Saami) folk healing one hundred years ago. *J Ethnobiol Ethnomed* 2013; **13**: 57.

Ducourthial G. Dioscorides: in the beginning of the Materia Medica. *Rev Prat* 2005; **55**: 689–93.

Etkin NL. A Hausa herbal pharmacopoeia: biomedical evaluation of commonly used plant medicines. *J Ethnopharmacol* 1981; **4**: 75–98.

Farlex Partner Medical Dictionary © Farlex 2012.

Fournier AM, Dodard M. The health care delivery crisis in Haiti. *Fam Med* 1997; **29**: 666–9.

Gaebler H. 2000 years of herbals. *Ther Ggw* 1964; **103**: 242–56.

Kapoor LD. Handbook of Ayurvedic Medicinal Plants. Boca Raton: CRC Press, 1990.

Kline MD. Nicholas Culpeper Biography. Indiana University School of Medicine, December 19, 1997.

Liu ZZ. Shen Nong's Herbal, the earliest extant treatise on Chinese materia medica. *Zhong Yao Tong Bao* 1982; **7**: 43–5.

Logan C. Source Book of Medical History. New York: Dover Publications, 1942.

Macht D. Phytopharmacological reactions of normal, toxic and atoxic sera. Proceedings of the National Academy of Sciences (USA) 1936; **22**: 384–389.

McCarl MR. Publishing the works of Nicholas Culpeper, astrological herbalist and translator of Latin medical works in seventeenth-century London. *Can Bull Med Hist* 1996; **13**: 225–76.

Porter R, Mikulas T. Drugs and Narcotics in History. Cambridge: Cambridge University Press 1995; xii + 227 pp.

Quave CL, Pieroni A, Bennett BC. Dermatological remedies in the traditional pharmacopoeia of Vulture-Alto Bradano, inland southern Italy. *J Ethnobiol Ethnomed* 2008; **6**: 5.

Quave CL, Santayana MP-de, Pieroni A. Medical Ethnobotany in Europe: From Field Ethnography to a More Culturally Sensitive Evidence-Based CAM? *Evidence-Based Compl Altern Med* 2012, Article ID 156846, 17 pageshttp://dx.doi.org/10.1155/2012/156846.

Scarborough J. Theophrastus on herbals and herbal remedies. *J Hist Biol* 1978; **11**: 353–85.

Shweta S, Boaz RJ2. Phytomedicine and the Nobel Prize: Benefits of integrating traditional remedies into modern medicine. *Indian J Pharmacol* 2015 Nov–Dec; **47**(6): 698–9.

Vaidya JG, Bhosle SR, Bapat G, Garad SA, Sonawane HB, Ashghar S, Ali M, Prasad L, Sawant A, Varghese N. Phansomba: An Indian Herbal, Folk Medicine. pp. 289–305. *In*: Amani S Awaad, JN Govil and VK Singh (eds.). 2010.

Venkata Rao E. Reserpine and Indian traditional medicine. *Curr Sci* 2010; **98**: 465–466.

Waldstein A. Mexican migrant ethnopharmacology: pharmacopoeia, classification of medicines and explanations of efficacy. *J Ethnopharmacol* 2006; **108**: 299–310.

Waller T, Killion G. Georgia folk medicine. *South Folk Q* 1972; **36**: 71–92.

Wangchuk P, Pyne SG, Keller PA. An assessment of the Bhutanese traditional medicine for its ethnopharmacology, ethnobotany and ethnoquality: textual understanding and the current practices. *J Ethnopharmacol* 2013; **148**: 305–10.

Wujastyk D. Cannabis in traditional Indian herbal medicine. pp. 45–73. *In*: A. Salema (ed.). Ayurveda at the Crossroad of Care and Cure. Lisbon: Universidad Nova, 2000.

Zheng BC. Shennong's herbal-one of the world's earliest pharmacopoeia. *J Tradit Chin Med* 1985; **5**: 236.

Chapter 2

Phytopharmacovigilance

2.1 Introduction

Herbal drugs are a significant source of drugs used in synthetic, complementary, and alternative systems of medicine, including Ayurveda or Traditional Indian Medicine (TIM), Siddha, Unani, Homoeopathy, Western Medical Herbalism (WMH), Traditional Chinese Medicine (TCM), and Traditional African Medicine (TAM). The World Health Organization (WHO) has identified the growing importance of Complementary and Alternative systems of Medicine and issued guidelines for proper implementation of rules and regulations governing the phyto or herbal drug industry in the recent publication of articles on heavy metal content in ayurvedic formulations in the Journal of American Medical Association (JAMA).

2.2 What is Phytopharmacovigilance?

Pharmacovigilance is essential for developing reliable information on the safety of herbal medicines as used in Europe and the US. The existing systems were developed for synthetic medicines and require some modification to address the specific differences of medicinal herbs. Traditional medicine from many different cultures is used in Europe and the US which adds to the complexities and difficulties of even basic questions such as herb naming systems and chemical variability (Shaw et al. 2012).

> Allied to this is also the perception that a 'natural' or herbal product must be safe simply because it is not synthetic which means that the safety element of monitoring for such medicines can be overlooked because of the tag associated with such products. Cooperation between orthodox physicians and traditional practitioners is needed to bring together the full case details (Shetti et al. 2011).
>
> Independent scientific assistance in toxicological investigation and botanical verification can be invaluable for the full evaluation of any case report. Systematic pharmacovigilance is essential to build up reliable information on the safety of herbal medicines for the development of appropriate guidelines for safe effective use (Shawa et al. 2012).

Phytopharmacovigilance is an emerging issue addressing key factors related to the Phyto or herbal drug industry. Although Pharmacovigilance best suits the synthetic pharmaceutical drug industry and vigilance issues related to Phyto or herbal drugs is part of the conventional pharmaceutical drug industry. Globalization of the Complementary and Alternative System of Medicine (CAM) and challenges like drug resistance and cost effectiveness of synthetic drugs, have forced the emergence of Phytopharmacovigilance (Ernst 2004).

Phytopharmacovigilance targets several key factors related to the Phyto or herbal drug industry. Proper identification of medicinal plants, procurement of the raw material, process of extraction, the percentage of the active ingredients and marker compounds, microbial count, heavy metal limit, finished product, and trained manpower, all fall within the spectrum of Phytopharmacovigilance.

2.3 Pharmacovigilance of Herbal Medicines in the UK

In the UK at present, the Committee on Safety of Medicines/Medicines and Healthcare products Regulatory Agency's (CSM/MHRA) 'yellow card' scheme for Adverse Drug Reaction (ADR) reporting is the main method of monitoring the safety of herbal medicines. Despite recent initiatives to stimulate reporting of suspected ADRs associated with herbal medicines, such as extending the scheme to unlicensed herbal products, and including community pharmacists as recognised reporters, numbers of herbal ADR reports received by the CSM/MHRA remain relatively low. Proposed European Union legislation for traditional herbal medicinal products will require manufacturers of products registered under new national schemes to comply with the regulatory provisions on pharmacovigilance (Barnes 2003).

2.4 Phytopharmacovigilance of Ayurvedic Medicines

According to a survey conducted by the National Centre for Complementary and Alternative Medicine (NCCAM) in the USA, about 751,000 people in the United States had used Ayurveda and 154,000 people had used them within the past 12 months. Broadly speaking, two categories of medicines labelled as "Ayurvedic" are available in the market: firstly, classical Ayurvedic formulations, which are as per described in Ayurveda Samhitas and secondly patent and proprietary formulations made of extracts of herbs.

Classical Ayurveda prescribes metals and minerals as medicines given as Bhasma or in combination with plants as herbo-mineral formulations (e.g., Arogyavardhini). Manufacturing procedures for these medicines are stringent, and adverse reactions are described when precautions are not taken while manufacturing and administering these medicines. Although these medicines are widely used in India, doubts about their long-term safety come up due to the presence of toxic metals in them and there are reports related to adverse reactions (Rastogi 2011).

Nearly 21 percent of the Ayurvedic medicines tested were found to contain detectable levels of lead (most common), mercury, or arsenic. All metal-containing products exceeded one or more standards for acceptable daily metal intake. The prevalence of metal-containing products did not differ significantly by country of manufacture. Rasa Shastra products were more than twice as likely as non-Rasa Shastra products to contain metals, and several Rasa Shastra medicines manufactured in India could result in lead and/or mercury ingestion 100 to 10,000 times greater than acceptable limits (Saper et al. 2004).

A recent survey conducted among Ayurvedic physicians by our department examined their attitudes toward adverse reactions of Ayurvedic medicines and reporting these to the authorities. Of the 80 *vaidyas*, interviewed, 14 refused to accept that Ayurvedic drugs could produce adverse reactions and the rest felt that adverse reactions would occur only if Ayurvedic drugs were improperly manufactured and irrationally prescribed. Of these 66 doctors, 48 physicians said that they had seen "unexpected" reactions after administration of Ayurvedic drugs in their practice. Interestingly, only 14 of these 48 said that they had reported these reactions (Thatte and Bhalerao 2008).

The current model of pharmacovigilance and its associated tools have been developed in relation to synthetic drugs, and applying these methods to monitoring the safety of herbal medicines presents unique challenges in addition to those described for conventional medicines. Several problems relate to the ways in which herbal medicines are named, perceived, sourced, and utilized. This may be because of differences in the use of unorthodox drugs (e.g., herbal remedies) which may pose special toxicological problems, when used alone or in combination with other drugs (Wal et al. 2011).

2.5 Phytopharmacovigilance of Herb-Drug Interactions among Preoperative Patients

The Pharmacovigilance Center of the University of Florence conducted a survey across three hospitals in Tuscany on a sample population consisting of 478 patients admitted to the hospital for a preoperative assessment before surgical intervention. The aim of the study was to assess the concomitant use of herbal remedies and prescribe medications and to evaluate the most important potential interactions. Antihypertensive, antiplatelet, anticoagulant, and central nervous system agents were the main products involved. The use of herbal remedies is not devoid of risks and adverse effects due to potential interactions may be serious or even life threatening (Gallo et al. 2014).

2.6 Pharmacovigilance of Herbal Medicines in Nigeria

The study was carried out in Lagos West Senatorial District of Lagos State, Nigeria. Three categories of practitioners (378 respondents) were engaged and they included Traditional Herbal Sellers, Natural Health Practitioners, and Pharmacists. The results showed that herbal medicines are commonly recommended for malaria, typhoid, diabetes, and fever. Two hundred and eighty-one (74.3%) of the respondents claimed that herbal medicines have no adverse effects and only 91 (24.1%) of the respondents said there were some adverse effects reported by the users. The adverse effects reported include nausea, diarrhoea, and weight loss (Awodele et al. 2013).

2.7 Ethnobotany and Ethnopharmacology-contribution towards Phytopharmacovigilance

From the interweaving of EB/EP and pharmacovigilance arises a concept of ethnopharmacovigilance for traditional herbal medicines: the scope of EB/EP is extended to include exploration of the potential harmful effects of medicinal plants, and the incorporation of pharmacovigilance questions into EB/EP studies provides a new opportunity for collection of 'general' traditional knowledge on the safety of traditional herbal medicines and, importantly, a conduit for collection of spontaneous reports of suspected adverse effects (Rodrigues and Barnes 2013).

2.8 DNA Barcoading and Phytopharmacovigilance

The herbal drug industry should embrace DNA bar-coding for authenticating herbal products through testing of raw materials used in manufacturing products. The use of an SRM DNA herbal barcode library for testing bulk materials could provide a method for 'best practices' in the manufacturing of herbal products (Newmaster et al. 2013). A total of 4385 samples of 2431 species were collected, and these samples are from 61 commonly used herbs and their closely related species or adulterants. Based on assessments of the extent of genetic divergence, the DNA barcoding gap, and the ability for species discrimination, our results suggest that ITS2 is a powerful tool for distinguishing herbs (Pang et al. 2013).

Several studies have shown that substitution of plant species occurs in herbal medicines, and this in turn poses a challenge to herbal pharmacovigilance as adverse reactions might be due to adulterated or added ingredients. Recent developments in molecular plant identification using DNA sequence data enable accurate identification of plant species from herbal medicines using defined DNA markers. Identification of multiple constituent species from compound herbal medicines using ampliconmeta barcoding enables verification of labelled ingredients and detection of substituted, adulterated, and added species (de Boer et al. 2015).

Further Readings

Awodele O, Daniel A, Popoola TD, Salami EF. A study on pharmacovigilance of herbal medicines in Lagos West Senatorial District, Nigeria. *Int J Risk Saf Med* 2013; **25**: 205–17.

Barnes J. Pharmacovigilance of herbal medicines: A UK perspective. *Drug Saf* 2003; **26**: 829–51.

de Boer HJ, Ichim MC, Newmaster SG. DNA Barcoding and Pharmacovigilance of Herbal Medicines. *Drug Saf* 2015. [Epub ahead of print].

Ernst E. Challenges for phytopharmacovigilance. *Postgrad Med J* 2004; **80**: 249–250.

Gallo E, Pugi A, Lucenteforte E, Maggini V, Gori L, Mugelli A, Firenzuoli F, Vannacci A. Pharmacovigilance of herb-drug interactions among preoperative patients. *Altern Ther Health Med* 2014; **20**: 13–7.

Newmaster SG, Grguric M, Shanmughanandhan D, Ramalingam S, Ragupathy S. DNA barcoding detects contamination and substitution in North American herbal products. *BMC Med* 2013; **11**: 222.

Pang X, Shi L, Song J, Chen X, Chen S. Use of the potential DNA barcode ITS2 to identify herbal materials. *J Nat Med* 2013; **67**: 571–5.

Rastogi S. Why and how? Addressing to the two most pertinent questions about pharmacovigilance in Ayurveda. *Int J Ayurveda Res* 2011; **2**: 48–52.

Rodrigues E, Barnes J. Pharmacovigilance of herbal medicines: the potential contributions of ethnobotanical and ethnopharmacological studies. *Drug Saf* 2013; **36**: 1–12.

Saper RB, Kales SN, Paquin J, Burns MJ, Eisenberg DM, Davis RB. Heavy metal content of ayurvedic herbal medicine products. *JAMA* 2004; **292**: 2868–73.

Shaw D, Graeme L, Pierre D, Elizabeth W, Kelvin C. Pharmacovigilance of herbal medicine. *J Ethnopharmacol* 2012; **140**: 513–8.

Shetti S, Kumar CD, Sriwastava NK, Sharma IP. Pharmacovigilance of herbal medicines: current state and future directions. *Pharmacogn Mag* 2011; **7**: 69–73.

Thatte U, Bhalerao S. Pharmacovigilance of ayurvedic medicines in India. *Indian J Pharmacol* 2008; **40**: 10–2.

Wal P, Wal A, Gupta S, Sharma G, Rai A. Pharmacovigilance of herbal products in India. *J Young Pharm* 2011; **3**: 256–8.

Chapter 3

Phytopharmacoeconomics

3.1 Definition of Pharmacoeconomics

The term Pharmacoeconomics was coined in 1986 by Townsend. Pharmacoeconomics refers to the scientific discipline that compares the value of one pharmaceutical drug or drug therapy to another (Arnold and Ekins 2010). Pharmacoeconomic studies serve to guide optimal healthcare resource allocation in a standardized and scientifically grounded manner.

3.2 Objectives of Pharmacoeconomics

The objective of Pharmacoeconomics is that the outcome of the research must originate from within three dimensions when considering results and value of healthcare

- Acceptable clinical outcomes
- Acceptable humanistic outcomes
- Acceptable economic outcomes

3.3 Some Definitions (Napper and Newland 2003)

- Economic evaluation: Economic evaluation is the systematic appraisal of the costs and benefits of projects, or alternative ways of achieving the same outcomes, undertaken to determine the economic effectiveness of the alternatives.
- Cost-Effectiveness Analysis (CEA): A CEA is an economic evaluation in which the costs and consequences of alternative interventions are expressed as costs per unit of health outcome. CEA is used to determine technical efficiency, i.e., comparison of costs and consequences of competing interventions for a given patient group within a given budget.
- Cost-Minimization Analysis (CMA): A CMA is an economic evaluation in which consequences of competing interventions are the same and in which only inputs, that is, the costs are taken into consideration. The aim is to decide the least costly way of achieving the same outcome.

- Cost-Utility Analysis (CUA): A CUA is a form of economic study design in which interventions which produce different consequences, in terms of both quantity and quality of life, are expressed as "utilities". These are measures that comprise both length of life and subjective levels of well being.

3.4 Pharmacoeconomics of Herbal Medicines

There seems to be no compelling need for pharmacoeconomic analyses of herbal over-the-counter medicines, but such analyses are certainly warranted for herbal prescription medicines that have a high level of reimbursement. Such preparations are used in Germany, in particular, where physicians prescribed ginkgo, hawthorn, St. John's wort, horse-chestnut and saw palmetto to a value of more than DM50 million each in 1996 (De Smet et al. 2000).

> The published evidence suggests that ginkgo is of questionable use for memory loss and tinnitus but has some effect on dementia and intermittent claudication. St. John's wort is efficacious for mild to moderate depression, but serious concerns exist about its interactions with several conventional drugs. Well-conducted clinical trials do not support the efficacy of ginseng to treat any condition. Echinacea may be helpful in the treatment or prevention of upper respiratory tract infections, but trial data are not fully convincing. Saw palmetto has been shown in short-term trials to be efficacious in reducing the symptoms of benign prostatic hyperplasia. Kava is an efficacious short-term treatment for anxiety. None of these herbal medicines are free of adverse effects (Ernst 2002).

3.5 Herbal Drugs: Issues in Licensing and Economic Evaluation

In order to identify cost-effective care, reliable information is required about the costs as well as the efficacy and safety of the treatments being assessed. For most alternative therapies, such data are not available. Studies to gather such data are long overdue. The need of the hour is to take control of some herbal remedies and caution need to be exercised with the trend towards licensing of all herbal remedies. The licensing of those herbal remedies with equivocal benefits and few risks, as evidenced by a long history of safe use, increases barriers to entry and increases societal healthcare costs (Ashcroft and Po 1999).

3.6 Cost-effectiveness of St. John's Wort for treatment of Depression

St. John's wort (*Hypericum perforatum*) is most widely associated with the treatment of depression. Millions use it in the UK. A Cochrane review of its use for major depression has found it to be superior to placebo (Linde et al. 2008); a credential that few CAM treatments can claim. However, there remains a lack of evidence regarding the cost-effectiveness of the intervention.

A clear advantage over antidepressants has been demonstrated in terms of the reduced frequency of adverse effects and lower treatment withdrawal rates, low rates of side effects, and good compliance, key variables affecting the cost-effectiveness of a given form of therapy. The most important risk associated with use is potential interactions with other drugs, but this may be mitigated by using extracts with low hyperforin content. As the indirect costs of depression are greater than five times direct treatment costs, given the rising cost of pharmaceutical antidepressants, the comparatively low cost of *Hypericum perforatum* extract makes it worthy of consideration in the economic evaluation of mild to moderate depression treatments (Solomon et al. 2011).

A Markov model was constructed to estimate the health and economic impacts of St. John's wort versus antidepressants. Outcomes were treatment costs, quality-adjusted life years (QALYs), and Net Monetary Benefits (NMB). Probabilistic analyses were conducted on key model parameters. In this model, St. John's wort was shown to be a cost-effective alternative to generic antidepressants. Patients are more likely to receive treatment for a duration consistent with professional guidelines for treatment of major depression due to reduced incidence of adverse effects, improving outcomes. This represents an important option in the treatment of Major Depressive Disorder (Solomon et al. 2013).

Further Readings

Arnold RJG, Ekins S. Time for cooperation in health economics among the modelling community. *Pharmacoeconomics* 2010; **28**: 609–613.

Ashcroft DM, Po AL. Herbal remedies: issues in licensing and economic evaluation. *Pharmacoeconomics* 1999; **16**: 321–8.

De Smet PA, Bonsel G, Van der Kuy A, Hekster YA, Pronk MH, Brorens MJ, Lockefeer JH, Nuijten MJ. Introduction to the pharmacoeconomics of herbal medicines. *Pharmacoeconomics* 2000; **18**: 1–7.

Ernst E. The risk-benefit profile of commonly used herbal therapies: Ginkgo, St. John's Wort, Ginseng, Echinacea, Saw Palmetto, and Kava. *Ann Intern Med* 2002; **136**: 42–53.

Linde K, Berner MM, Kriston L. St. John's wort for major depression. Cochrane Database of Systematic Reviews 2008, Issue 4. Art. No.: CD000448. DOI: 10.1002/14651858.CD000448.pub3.

Napper M, Newland J. Health Economics Information Resources: A Self-Study Course. Bethesda: U.S. National Library of Medicine; 2003.

Solomon D, Ford E, Adams J, Graves N. Potential of St. John's Wort for the treatment of depression: the economic perspective. *Aust N Z J Psychiatry* 2011; **45**: 123–30.

Solomon D, Adams J, Graves N. Economic evaluation of St. John's wort (*Hypericum perforatum*) for the treatment of mild to moderate depression. *J Affect Disord* 2013; **148**: 228–34.

Chapter 4

Phytopharmacoepidemiology

4.1 What is Pharmacoepidemiology?

Pharmacoepidemiology is the study of the uses and effects of drugs in well defined populations (Strom 2006). With the growing availability of large healthcare databases, non-experimental studies of prescription medications are becoming increasingly common. However, the appropriate design and analysis of such studies can be challenging. The issues which are addressed in pharmacoepidemiology are

- Drug utilization research/quality of care
- Drug effects (effectiveness and safety)
- Analytic methods

4.2 Phytopharmacoepidemiology

Phytopharmacoepidemiology concerns itself with phyto or herbal drug uses by, and effects on, large numbers of people (De Smet Peter 1993).

4.3 Benefits of Phytopharmacoepidemiology

This new methodology can make substantial contributions to our understanding of herbal drug markets. Not only can it improve the timely detection and the quantitation of adverse reactions to botanical medicines, it can also help us to recognize the beneficial effects of herbal preparations. Moreover, it can increase our social and economic knowledge about herbal drugs by correlating the utilization patterns of these preparations to socially relevant determinants and prices.

Since such studies may require an internationally accepted herbal drug classification, a special set of herbal classification codes is proposed, which is fully compatible with the so-called ATC-classification (Anatomical, Therapeutic, Chemical classification). This latter system is endorsed by the Regional Office for Europe of the World Health Organization as the drug classification to be used in international pharmacoepidemiological studies (De Smet Peter 1993).

4.4 Phytopharmacoepidemiology in Addis Ababa, Ethiopia

A study was undertaken to study the utilization patterns of herbal drug use in urban Ethiopia. A cross-sectional community-based survey was conducted in Addis Ababa, the capital city of Ethiopia, using a pre-tested semi-structured questionnaire. The questionnaire was administered to 600 heads of households, largely mothers, selected using a multi-stage systematic random sampling technique, where the final sampling units were households.

> The prevalence of herbal drug use was found to be 37%. The main reasons given for choosing herbal medicine as the first line medication option were: dissatisfaction with the services of modern health institutions due to their time consuming nature, cost considerations, and perceived efficacy (Gedif and Hahn 2002).

4.5 Herbal ATC Classification

The Uppsala Monitoring Centre has published Guidelines for a Herbal ATC Classification and a Herbal ATC Index. However, these documents are no longer being supplied to external customers.

> Some herbal remedies have a longstanding use in medicine and their actions are well-defined. For various reasons it has been deemed impractical to incorporate hundreds of herbal remedies in the regular ATC classification. However, the experience from the ATC system-particularly in connection with the monitoring of adverse effects of drugs-has shown that such a system would also be suitable for herbal remedies.
>
> In 1998, De Smet proposed a system for ATC classification of herbal remedies which is fully compatible with the regular system. With a few modifications this system has been adopted and given guidelines. The ATC Index lists ATC codes per substance, while the herbals guideline is a help to assign ATC codes to herbal remedies.
>
> In both the ATC and the Herbal ATC systems remedies are divided into groups according to their therapeutic use. Wherever possible the level 1–4 codes in the herbal system are equal to the levels in the regular ATC system.

4.6 Ayurpharmacoepidemiology

Ayurpharmacoepidemiology is a new field developed by the synergy of the fields of clinical pharmacology, epidemiology, and Ayurveda. It will use the effects of Ayurvedic medicinal products on large populations to describe and analyze the practices, evaluate the safety and efficacy, and carry out medicoeconomic evaluations. Good pharmacoepidemiology practices in Ayurveda are projected to assist with issues of ayurpharmacoepidemiologic research.

The good pharmacoepidemiology practice guideline in the viewpoint of Ayurveda will be able to provide valuable evidence about the health effects of Ayurvedic herbs/drugs. The coupling of different fields like Pharmacovigilance, Pharmacoeconomics, and drug discovery with the Ayurvedic reverse pharmacology approach is the need of the hour for discovering potent drugs from nature. Pharmacoepidemiology provides solid platform for further basic sciences study in Ayurveda biology, Ayurgenomics, Ayurnutrigenomics, and Systems biology.

Several unanswered questions about the Ayurvedic drug use and informed interventions or policies can be addressed by informatics database. This can have great impact on the credibility and rationality of Ayurceuticals in the future.

4.7 Phytopharmacoepidemiology in Australia

Herbal medicine use is common in Australia, but little is known about the use of individual herbs. Methods: A cross-sectional population survey conducted in 2007 with a sample of 2526, in the Australian state of Victoria. Almost a quarter (22.6%, 95% confidence interval (CI): 20.9–24.2%) of survey participants had used at least one medicinal herb in the preceding 12 months. Aloe vera, garlic, and green tea were the most popular, each used by about 10% of participants. Health enhancement was the most common reason for herbal medicine use (69.6% of users) but relatively high proportions of users sought relief of specific medical conditions.

Over 90% considered their herbal medicine to be very or somewhat helpful. Less than half (46.6%) the users were aware that there were potential risks associated with herbal medicine. Relatively high proportions of female users had taken herbal medicine whilst pregnant (14.4%) and/or whilst breast feeding (10.0%). Over half (50.9%) of herbal medicine users had also used Western medicine for the same medical condition in the 12-month period. Almost the same proportion (49.9%) had used both forms of medication on the same day. In deciding whether or not to use herbal medicine, the vast majority of survey participants indicated that they would accept the advice of their medical practitioner (Zhang et al. 2008).

4.8 Phytopharmacoepidemiology in Obstetrics

A study reviewed the literature on the safety and efficacy of the most commonly used herbs to enable midwives to give evidence-based information to pregnant women. The participants were a total of 578 expectant mothers at least 20-weeks pregnant. 7.8% of the participants used one or more herbal remedy. The most commonly used herbal preparations during pregnancy were ginger, cranberry, raspberry leaf, chamomile, peppermint, and Echinacea. Altogether, 14 studies focusing on the safety and/or efficacy of these herbals in human pregnancy were identified. Ten studies of ginger, one of cranberry, two of raspberry leaves, and one of Echinacea were located (Holst et al. 2011).

4.9 Cam in Epileptic Patients in a Tertiary Care Hospital in India

The purpose of the study was to establish the pattern of use of CAM in epileptic patients. 1000 patients with seizure disorder visiting the Neurology outpatient department were interviewed regarding the use of complementary/alternative medicine (CAM) in the past. The pattern of use, persons who recommended CAM, and the reasons for trying these therapies and sequence of seeking them was noted in these patients.

> Overall, 32% of patients had used CAM. Ayurvedic medicine was used most frequently, either alone (43%) or in combination (38%) with other CAM therapies followed by homeopathy (12.5%). Use of CAM was seen among all age groups and at all levels of education and was more frequent in the rural population (67%). Influence of family and friends (50%) was the most common reason for trying these therapies. Most patients (57%) sought CAM providers first before seeking the services of a medical doctor in our study (Tandon et al. 2002).

Further Readings

De Smet Peter AGM. An introduction to herbal pharmacoepidemiology. *J Ethnopharmacol* 1993; **38**: 189–195 (Special Issue Proceedings of the Second International Congress on Ethnopharmacology).

Debnath P, Banerjee S, Adhikari A, Debnath PK. Ayurpharmacoepidemiology an Route to Safeguarding Safety and Efficacy of Ayurvedic Drugs in Global Outlook. *J Evid Based Complem Altern Med* 2015. pii: 2156587215624032. [Epub ahead of print]

Gedif T, Hahn H-J. Epidemiology of herbal drugs use in Addis Ababa, Ethiopia. *Pharmacoepidemiol Drug Safety* 2002; **11**: 587–591.

Holst L, Wright D, Haavik S, Nordeng H. Safety and efficacy of herbal remedies in obstetrics-review and clinical implications. *Midwifery* 2011; **27**: 80–6.

http://www.who-mc.org/DynPage.aspx?id=105834&mn1=7347&mn2=7259&mn3=7297&mn4=7502.

Strom B. Textbook of Pharmacoepidemiology. West Sussex, England: John Wiley and Sons. 2006; p. 976.

Tandon M, Prabhakar S, Pandhi P. Pattern of use of complementary/alternative medicine (CAM) in epileptic patients in a tertiary care hospital in India. *Pharmacoepidemiol Drug Safety* 2002; **11**: 457–463.

Zhang A, Story D, Lin V, Vietta L, Xue C. A population survey on the use of 24 common medicinal herbs in Australia. *Pharmacoepidemiol Drug Safety* 2008; **17**: 1006–1013.

Chapter 5

Phytopharmacogenomics

5.1 What is Pharmacogenomics?

Pharmacogenomics is the study of how genes affect a person's response to drugs. This relatively new field combines pharmacology (the science of drugs) and genomics (the study of genes and their functions) to develop effective, safe medications and doses that will be tailored to a person's genetic makeup.

5.2 Pharmacogenomics and Herb-Drug Interactions

The worldwide using of herb products and the increasing potential herb-drug interaction issue has raised enthusiasm for discovering the underlying mechanisms. Previous reviews indicated that the interactions may be mediated by metabolism enzymes and transporters in pharmacokinetic pathways. On the other hand, an increasing number of studies found that genetic variations showed some influence on herb-drug interaction effects. However, these genetic factors were not given due attention in the history. Pharmacogenomics may involve the pharmacokinetic or pharmacodynamic pathways to affect herb-drug interaction (Liu et al. 2015).

5.3 Pharmacogenomics of Kampo in the Treatment of Cancer

Kampo is a Traditional Japanese Herbal Medicine. The main focus is on Pharmacogenomics of natural products derived from Kampo medicinal plants with special emphasis on cancer treatment. One of these natural products with profound cytotoxicity against tumor cell lines is shikonin from *Lithospermum erythrorhizon*. This compound has been selected to demonstrate how molecular determinants of response of tumor cells to Kampo-derived natural products can be investigated by microarray-based approaches. Synthetic or semi-synthetic derivatives of natural products from Kampo medicine may lead to novel drugs with improved features for cancer treatment (Efferth et al. 2007).

5.4 Pharmacogenomics of Cameroon Ethnomedicine in the treatment of Cancer

Up to 974 compounds isolated from 148 medicinal plants were used as keywords in the NCI database to establish a library of 27 cytotoxic compounds. Two of the 10 most cytotoxic compounds, plumericin from *Plumeria rubra* and plumbagin from *Diospyros crassiflora* and *Diospyros canaliculata*, were analyzed in more detail. The IC(50) values for plumericin and plumbagin of 60 NCI cell lines were associated with the microarray-based transcriptome-wide mRNA expression.

Genes products identified for plumericin activity are mainly involved in enzymatic activity, transcriptional processes, or are structural constituents of ribosomes. Products identified for plumbagin activity are involved in several processes, but they are mostly the strucrural constituents of ribosomes or involved in enzymatic activity (Kuete and Efferth 2011).

5.5 Pharmacogenetics, Pharmacogenomics, and Ayurgenomics

The genotypic experiments have laid valuable insights into the genetic underpinnings of diseases. However, it is being realized that identification of sub-groups within normal controls corresponding to contrasting disease susceptibility could lead to more effective discovery of predictive markers for diseases. However, there are no modern methods available to look at the inter-individual differences within ethnically matched healthy populations.

Ayurveda, an exquisitely elaborate system of predictive medicine, which has been practiced for over 3500 years in India, can help in bridging this gap. In contrast to the contemporary system of medicine, the therapeutic regimen in *Ayurveda* is implicated on *tridoshas* and *prakriti*. According to this system, every individual is born with his or her own basic constitution, which to a great extent regulates inter-individual variability in susceptibility to diseases and response to external environment, diet, and drugs (Gupta 2015).

5.6 Phytopharmacogenomics of Herbal Drugs in African Populations

Most African people accept herbal medicines as generally safe with no serious adverse effects. However, the overlap between conventional medicine and herbal medicine is a reality among countries facing a transition in health systems. Patients often simultaneously seek treatment from both conventional and traditional health systems for the same condition. Commonly encountered conditions/diseases include malaria, HIV/AIDS, hypertension, tuberculosis, and bleeding disorders.

It is therefore imperative to understand the modes of interaction between different drugs from conventional and traditional health care systems when used in treatment combinations. Both conventional and traditional drug entities are metabolized by the same enzyme systems in the human body, resulting in both pharmacokinetics and pharmacodynamics interactions, whose properties remain unknown/unquantified. Thus, it is important that profiles of interaction between different herbal and conventional medicines be evaluated (Thomford et al. 2015).

Further Readings

Efferth T, Miyachi H, Bartsch H. Pharmacogenomics of a traditional Japanese herbal medicine (Kampo) for cancer therapy. *Cancer Genomics Proteomics* 2007; **4**: 81–91.

Gupta PD. Pharmacogenetics, pharmacogenomics and ayurgenomics for personalized medicine: a paradigm shift. *Indian J Pharm Sci* 2015; **77**: 135–41.

Kuete V, Efferth T. Pharmacogenomics of Cameroonian traditional herbal medicine for cancer therapy. *J Ethnopharmacol* 2011; **137**: 752–66.

Liu MZ, Zhang YL, Zeng MZ, He FZ, Luo ZY, Luo JQ, Wen JG, Chen XP, Zhou HH, Zhang W. Pharmacogenomics and herb-drug interactions: merge of future and tradition. *Evid Based Complement Alternat Med* 2015; **2015**: 321091.

Thomford NE, Dzobo K, Chopera D, Wonkam A, Skelton M, Blackhurst D, Chirikure S, Dandara C. Pharmacogenomics implications of using herbal medicinal plants on African populations in health transition. *Pharmaceuticals* (Basel) 2015; **8**: 637–63.

Chapter 6

Ethics in Phytomedicine

6.1 Introduction

The use of Complementary and Alternative Medicine (CAM) has grown dramatically in recent years, as has research on the safety and efficacy of CAM treatments (Ernst 1996). Minimal attention, however, has been devoted to the ethical issues relating to research on CAM. We argue that public health and safety demand rigorous research evaluating CAM therapies, research on CAM should adhere to the same ethical requirements for all clinical research, and randomized, placebo-controlled clinical trials should be used for assessing the efficacy of CAM treatments whenever feasible and ethically justifiable (Miller et al. 2004; Smith et al. 2011).

6.2 The Ethics of Traditional Chinese and Westren Herbal Medicine Research

The objectives of this study were to examine the experiences of Traditional Chinese Medicine (TCM) and Westren Herbal Medicine (WHM) researchers and Human Research Ethics Committee (HRECs) with the evaluation of ethics applications. Two cross-sectional surveys were undertaken by HRECs and TCM and WHM researchers in Australia. Anonymous self-completion questionnaires were administered to 224 HRECs and 117 researchers. A response confirming involvement in TCM or WHM research applications was received from 20 HRECs and 42 researchers.

The most frequent ethical issues identified by HRECs related to herbal products including information gaps relating to the mode of action of herbal medicines and safety when combining herbal ingredients. Researchers concurred that they were frequently requested to provide additional information on multiple aspects including safety relating to the side effects of herbs and herb-drug interactions. Overall adherence with the principles of ethical conduct was high among TCM and WHM researchers, although our study did identify the need for additional information regarding assessment of risk and risk management (Smith et al. 2011).

6.3 The Ethics of Complementary Medicines in Pediatrics

Nonjudicious use of CAM therapies may cause either direct harm or, by creating an unwarranted financial and emotional burden, indirect harm. When advising patients concerning CAM therapies, pediatricians face two major legal risks: medical malpractice and professional discipline. Pediatricians can incorporate these considerations into advising and clinical decision-making about CAM therapies to address the best interest of the pediatric patient while helping to manage potential liability risk (Cohen and Kemper 2005).

> Inclusion of CAM therapies in pediatric oncology and hematology—as in any medical subspecialty—is not itself "unethical", clinically inadvisable, or legally risky; the danger comes from over-reliance on one or more CAM therapies (particularly those with evidence of danger and/or paltry evidence of success) to the exclusion of conventional care that is curative and imminently necessary. Pediatricians can help address potential malpractice liability issues by evaluating the level of clinical risk, engaging the patient in shared decision making and documenting this in the medical record, continuing to monitor conventionally, and being prepared to intervene conventionally when medically required (Cohen 2006).

6.4 Ethical Obstacles in African Traditional Medicine

Medicine in Africa is regarded as possessing its own "life force", not just using a system of prescribing. This is because health problems are not only attributed to pathological explanations alone, but also to other "forces". Hence, traditional healers utter incantations to take care of negative forces which militate against achieving cure. Treatment in African traditional medicine (ATM) is holistic.

> It seeks to strike a balance between the patients' body, soul, and spirit. The problems arise from the infiltration of charlatans in the field, the practice of using mystical explanations for ill-health, and inadequate knowledge of the properties and clinical use of herbal remedies. Despite the problems, ATM can work in synergy with orthodox medicine utilizing strength of the same. ATM has to be applied within a uniform ethical system. Practitioners of ATM must follow the principles of autonomy and confidentiality (Omonzejele 2003).

Further Readings

Cohen MH, Kemper KJ. Complementary therapies in pediatrics: a legal perspective. *Pediatrics* 2005; **115**: 774–80.

Cohen MH. Legal and ethical issues relating to use of complementary therapies in pediatric hematology/oncology. *J Pediatr Hematol Oncol* 2006; **28**: 190–3.

Ernst E. The ethics of complementary medicine. *J Med Ethics* 1996; **22**: 197–8.

Miller FG, Emanuel EJ, Rosenstein DL, Straus SE. Ethical issues concerning research in complementary and alternative medicine. *JAMA* 2004; **291**: 599–604.

Smith C, Priest R, Carmady B, Bourchier S, Bensoussan, A. The ethics of traditional Chinese and western herbal medicine research: views of researchers and human ethics committees in Australia. *Evid Based Complement Alternat Med* 2011; **2011**: 1741427X.

Omonzejele P. Current ethical and other problems in the practice of African traditional medicine. *Med Law* 2003; **22**: 29–38.

Chapter 7

Herbosomes

7.1 Introduction

Phytomedicine, or herbal medicine, or phytotherapy, is the science of using herbal remedies to treat the sick (Kapoor 1990). They include herbs, herbal materials, herbal preparations, and finished herbal products, which contain as active ingredients parts of plants, or other plant materials, or combinations (Karnik 1994). Today's healthcare systems rely largely on plant material. Much of the world's population depends on phytomedicine to meet daily health requirements, especially within developing countries. Use of plant-based remedies is also widespread in many industrialized countries and numerous pharmaceuticals are based on or derived from plant compounds (Bone 1996; Sivarajan 1994).

7.2 Drug Delivery Systems in Phytomedicine

Ayurveda, the traditional Indian System of healing, has fundamental concepts as far as drug formulation and delivery are concerned. *Bhaishajya Kaplaba Vijnana* is full-fledged subject dealing with preparation of Ayurvedic Medicines (Anonymous, 1998). In phytomedicine, the pharmacopoeial preparations are known as *Official Galenicals* after Galen (A.D. 130 to 230), the Father of Pharmacopeia (Evans 1996).

Herbal drugs are often in impure state. In order to make them fit for therapeutic administration, they are subjected to various treatments. *Decoction* is the process of boiling in water coarsely comminute vegetable drugs for a definite period. Decoctions are therapeutically more active as they vigorously extracts the virtues of medicinal plants, roots, twigs, barks, and seeds (Anonymous 1998; Evans 1996).

Infusion is obtained when boiling or cold distilled water is poured on the drugs in a covered vessel and kept for fifteen minutes and then strained. Sometimes boiling is done for hours to prepare strong infusion. *Powders* are mixtures of dry substances reduced to fine powder and intimately mixed together. Powders may be of a single substance and more often of several (compound powders).

Expression is the process of pressing juice or oil from a medicinal or aromatic plant. *Bruising* is process of smashing up the different parts of a medicinal plant either by a pounding machine or pestle and mortar. *Tinctures* are alcoholic solutions of active principles of vegetable products, prepared by *maceration* or *percolation*.

Extracts are prepared by separating the soluble matter from vegetable tissues by application of a suitable solvent like alcohol, water, or ether. The resultant liquid is concentrated by evaporation to obtain liquid extract or concentrated nearly to dryness to obtain solid extract. *The standardized herbal extract* is a preparation, which contains a certain fixed proportion of the active constituent (Samuelsson 1999).

7.3 Recent Trends in Herbal Drug Delivery Systems

Sales of plant derived drugs reached $30 billion worldwide in 2002. At present about 50% of the total plant-derived drug sales come from single entities, while the remaining 50% come from herbal remedies (World Health Organization 1978). Development of drug from plants is a long and arduous process which involves many disciplines. It has been estimated that only 5 to 15% of the approximately 250,000 species of higher plants have been systematically investigated for the presence of bioactive compounds (Tyler 1994). In the industrialized countries, substances derived from plants are in everyday use-digitalin, ephedrine, morphine, quinine, and many more. Less often used substances like reserpine, guggulipid, and artemisinin are equally well known (Hussain 1993).

In the past few decades, considerable attention has been focussed on the development of novel drug delivery systems for herbal drugs (Saurabh and Kesari 2011). Novel herbal drug carriers cure a particular disease by targeting exactly the affected zone inside a patient's body and transporting the drug to that area. Novel drug delivery system is advantageous for delivering the herbal drug at predetermined rate and delivery of drug at the site of action which minimizes the toxic effects with the increase in bioavailability of the drugs (Shukla et al. 2012).

Various novel drug delivery systems such as liposomes, niosomes, microspheres, and phytosomes have been reported for the delivery of herbal drugs. Incorporation of herbal drugs in the delivery system also aids to increase in solubility, enhanced stability, protection from toxicity, enhanced pharmacological activity, improved tissue macrophage distribution, sustained, delivery and protection from physical and chemical degradation (Goyal et al. 2011; Jain et al. 2010).

The present chapter deals with the development of a new concept in herbal delivery system, i.e., "herbosomes" which are better absorbed and utilized and as a result produce better results than conventional herbal extracts owing to the presence of phosphatidylcholine which likely pushes the phytoconstituent through the intestinal epithelial cell outer membrane, subsequently accessing the bloodstream.

7.4 Herbosomes or Phytosomes or Planterosomes

Definition

The term "herbo" means plant, while "some" means cell-like. Herbosomes are basically standardized extracts or purified fractions complexed with phospholipids for a better bioavailability and enhanced activities. Herbosomes have improved pharmacokinetic and pharmacological parameters, which in result can advantageously be used in the management of hepatic diseases of acute and chronic origin resulting from toxic metabolic or infective origin or of degenerative nature (Rathore and Swami 2012).

Background

The use of herbrosomes is a new advanced modern dosage formulation technology to deliver herbal products and drugs with improved better absorption and, as a consequence, produce better results than those obtained by conventional herbal extracts (Gupta et al. 2007).

Phytosomes are advanced herbal products produced by binding individual component of herbal extract to phytolipids, mainly phosphatidylcholine, resulting in a product that is better absorbed and produces better results than the conventional herbal extract (Bombardelli et al. 1989). Most of the herbal drug bioactive constituents are water soluble molecules. However, poor absorption of water soluble phytoconstituents, limits their effectiveness on oral consumption or topical application (Venkatesan et al. 2000).

Herbal medicines are poorly absorbed either due to their large moleculer size which cannot absorb by passive diffusion, or due to their poor lipid solubility, severely limiting their ability to pass across the lipid-rich biological membranes, resulting in poor bioavailability. Many phytoconstituents have multiple rings and therefore, cannot be absorbed from the intestine into the blood by simple diffusion. Also, some herbal phytomolecules are poorly miscible with oils and other lipids and often fail to pass through the small intestine because of its lipoidal nature (Manach et al. 2004).

Many approaches have been developed to improve the oral bioavailability, such as inclusion by solubility and bioavailability enhancers, structural modification, and entrapment with the lipophilic carriers (Venkatesan et al. 2000; Longer et al. 1985).

Over the past century, chemical and pharmacologic science established the compositions, biological activities, and health giving benefits of numerous plant extracts. But often when individual components were separated from the whole there was loss of activity—the natural ingredient synergy became lost (Bhattacharya and Ghosh 2009).

As standardized extracts became established, poor bioavailability often limited their clinical utility. Then it was discovered that complexation with certain other clinically useful nutrients substantially improved the bioavailability of such extracts (Patil et al. 2012).

Basis of herbosomes technology

Herbosomes technology is also useful in pharmaceutical formulations intended for treatment of oral cavity in which the contact times are very short because phospholipid allows a greater adhesion of the product itself to the surfaces it comes into contact with and so vastly improve their absorption and bioavailability (Bombardelli et al. 1989).

The Phytosomes process produces a little cell because of which the valuable components of the herbal extract are protected from destruction by digestive secretions and gut bacteria. Phytosomes are better able to transition from a hydrophilic environment into the lipid-friendly environment of the enterocyte cell membrane and from there into the cell finally reaching the blood. It can also be used in anti-inflammatory activity as well as in pharmaceutical and cosmetic compositions (Dang 2000).

Preparation of herbosomes

Herbosomes are created when the standardized extract and active ingredients of a herb are bound to the phospholipids (such as phosphatidylcholine, phosphatidylethanolamine, or phosphatidyiserine with a polyphenolic component like simple flavonoids) on a molecular level. Herbosome structures contain the active ingredients of the herb surrounded by the phospholipids.

Phospholipids and phosphatidylcholine

Phospholipids are complex molecules that are used in all known life forms to make cell membranes. They are cell membrane building blocks, making up the matrix into which fit a large variety of proteins that are enzymes, transport proteins, receptors, and other biological energy converters. The phospholipids are readily compatible with the entire range of vitamins, minerals, metabolites, and herbal preparations currently consumed as the dietary phospholipids and omega-3 fatty acid works in functional synergy in cell membranes.

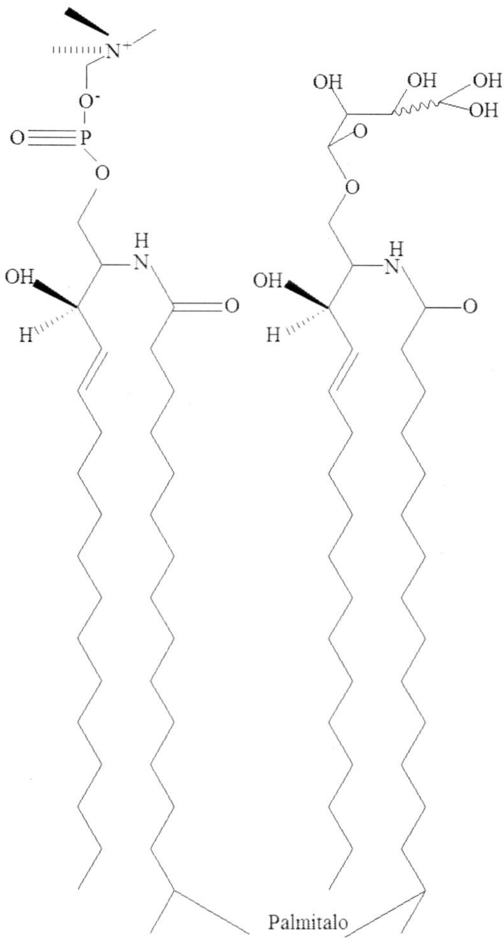

Fig. 7.1: Structure of phosphatidylcholine.

Phosphatidylcholine (Fig. 7.1) is a bifunctional compound miscible both in water and in oil environments, and is well absorbed when taken by mouth. Phosphatidylcholine is not merely a passive "carrier" for the bioactive compounds, but is itself a bioactive nutrient with documented clinical efficacy for liver disease, including alcoholic hepatitis.

Among these all phospholipids, phosphatidylcholine classes of phospholipids are very important in the drug delivery technology. The very first and most important advantage of phospholipids based vesicular system is the compatibility of phospholipids with membrane of human either internal membrane as well as skin.

Chemical analysis indicates that herbosome is usually a phytoconstituent molecule linked with at least one phosphatidylcholine molecule. Phosphatidylcholine is not merely a passive "carrier" for the bioactive phytoconstituent of the herbosomes but is itself a bioactive nutrient with documented clinical efficacy for liver disease, including alcoholic hepatic steatosis, drug-induced liver damage, and hepatitis (Bombardelli 1991).

The intakes of a herbosome preparation are sufficient to provide reliable clinical benefit, often leading to provide substantial phosphatidylcholine intakes. The herbosome process has been applied to many popular herbal extracts, including *Milk thistle*, *Ginkgo biloba*, *Grape seed*, *Green tea*, *Hawthorn*, *Ginseng*, etc.

The phytoconstituents lend themselves quite well for the direct binding to phosphatidylcholine, which means that the choline head binds to phytoconstituents while the fat-soluble phosphatidyl portion comprising the body and tail then envelopes the choline-bound material. This result is a little microsphere or cell being produced (Middleton and Kandaswami 1994).

Flavonoids

First recognized for their antioxidant properties, flavonoid is widely distributed in plants. To date, more than 4,000 naturally occurring flavonoid have been identified from plant sources having diverse biological activities (Bombardelli and Curri 1976).

The hypothesis of an interaction of flavonoid with phospholipids, which are ubiquitous in plant and animals, originated from the histo-chemical finding indicating that anthocyanosides from *Vaccinium myrtillus* L. show a strong affinity for specific cellular structure rich in phospholipids (Jose and Bombardelli 1987).

Advanced biochemical and pre-clinical studies have proved the potential of plant flavonoids and other hydrophilic natural compounds for the treatment of skin disorders, different types of carcinoma, anti-aging, and many other areas of therapeutics and preventive medicine. The hydrophilic nature and unique chemical structure of these compounds pose major challenge because of their poor bioavailability through the skin or gut. The use of herbosomes is novel formulation technology which helps to overcome these problems.

The flavonoid constituents of plant extracts lend themselves quite well for the direct binding to phosphatidylcholine. Herbosomes results from the reaction of a stoichiometric amount of the phospholipid (phosphatidylcholine) with the standardized extract or polyphenolic constituents (like simple flavonoids) in an aprotic solvent.

Phyto-phospholipid complex

In phosphatidylcholine, the phosphatidyl moiety is lipophilic and the choline moiety is hydrophilic in nature. Specifically the choline head of the phosphatidylcholine molecule binds to these compounds which then envelopes the choline bound material.

In the phytosome preparations, phospholipids are selected from the group consisting of soy lecithin, from bovine or swine brain or dermis, phosphatidylcholine, phosphatidylethanolamine, phosphatidyiserine in which acyl group may be same or different and mostly derived from palmitic, stearic, oleic, and linoleic acid (Yanyu et al. 2006).

Selection of flavonoids are done from the group consisting of quercetin, kaempferol, quercretin-3, rhamnoglucoside, quercetin-3-rhamnoside, hyperoside, vitexine, diosmine, 3-rhamnoside, (+) catechin, (–) epicatechin, apigenin-7-glucoside, luteolin, luteolinglucoside, ginkgonetine, isoginkgonetine, and bilobetine. On the basis of their physical-chemical and spectroscopic characteristics, these complexes can be considered as novel entities. In the complex formation of herbosomes the ratio between these two moieties is in the range from 0.5–2.0 moles. The most preferable ratio of phospholipid to flavonoids is 1:1 (Jain 2005).

Examples of phyto-phospholipid complex

Naringenin-phosphatidylcholine C complex: It was prepared by taking naringenin with an equimolar concentration of phosphatidylcholine. The equimolar concentration of phosphatidylcholine and naringenin were placed in a 100 mL round bottom flask and refluxed in dichloromethane for 3 h. On concentrating the solution to 5–10 mL, 30 mL of n-hexane was added to get the complex as a precipitate followed by filtration. The precipitate was collected and placed in vacuum desiccators (Jiang et al. 2001).

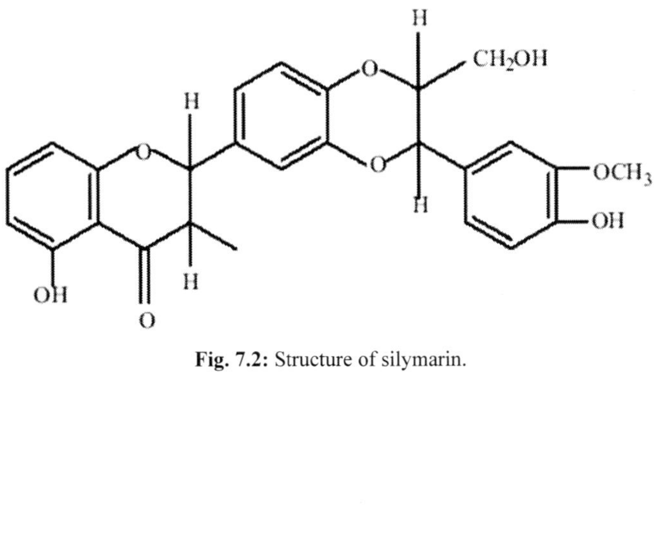

Fig. 7.2: Structure of silymarin.

Herba Epimedii total flavonoid phytosomes: Jiang et al., have optimized the preparation conditions using a uniform design and step regression and have prepared *Herba Epimedii* total flavonoid phytosomes (EFP) by means of solvent evaporation and investigated the cumulative dissolution of different ratios of EFP-PVP precipitates by means of dissolution release.

The oil/water apparent partition coefficient of icariin was enhanced more than 4-fold by phospholipids. The cumulative dissolution of *Herba Epimedii* flavonoid of the EFP-PVP precipitate was significantly higher than that of its physical mixture and a *Herba Epimedii* extract tablet (Jiang et al. 2001).

Silybin-phospholipids complex: A silybin-phospholipids complex was prepared using ethanol as a reaction medium. Silybin and phospholipids were resolved into the medium, after the organic solvent was removed under vacuum condition, and a silybin-phospholipids complex was formed (Yanyu et al. 2006).

7.5 Research on Phytosomes Preparation

Silybum marianum Linn. (Asteraceae)

Majority of the phytosomal studies have been focused to *S. marianum*, commonly known as milk-thistle, which contains well-known hepatoprotective flavonoid, silymarin.

Pre-clinical

1. Tedeco et al. reported that the Silymarin phytosomes show better anti-hepatotoxic activity than Silymarin alone and can provide protection against the toxic effects of Aflatoxin B1 on the performance of Broiler chicks (Tedeco et al. 2004).
2. Busby et al. reported that the use of a Silymarin phytosome showed a better fetoprotectant activity from ethanol-induced behavioral deficits than uncomplexed Silymarin (Busby et al. 2002).
3. Lee and Reyes conducted a series of studies on the Silymarin phytosome containing a standardized extract from the seeds of *Silybum marianum* administered orally to animals and found that it could protect the fetus from maternally ingested ethanol (Lee and Reyes 1999).
4. Yanyu et al. prepared the Silymarin phytosome and studied its pharmacokinetics in rats. In the study, the bioavailability of Silybin in rats was increased remarkably after oral administration of the prepared Silybin-phospholipid complex due to an impressive improvement of the lipophilic property of the Silybin-phospholipid complex and improvement of the biological effect of Silybin (Yanyu et al., 2006).
5. Bombardelli et al. reported that Silymarin phytosomes showed a much higher specific activity and a longer lasting action than the single components with respect to the percent reduction of edema, inhibition of myeloperoxidase activity, and the antioxidant and free radical scavenging properties (Bomberdelli et al. 1991).

Clinical

1. Barzaghi et al. conducted a human study designed to assess the absorption of Silybin when directly bound to phosphatidylcholine. The plasma Silybin levels were determined after administration of a single oral dose of Silybin phytosome and a similar amount of Silybin from *S. marianum* to healthy volunteers. The results indicated that the absorption of Silybin from the Silybin phytosome is approximately seven-times greater compared with the absorption of Silybin from the regular *S. marianum* extract containing 70–80% Silymarin (Barzaghi et al. 1990).
2. Moscarella et al. in one study of 232 patients with chronic hepatitis treated with the Silybin phytosome at a dose of 120 mg either twice daily or thrice daily for up to 120 days, investigated and found that the liver function returned to normal faster in patients taking the Silybin phytosome compared with a group of controls (49 treated with commercially available silymarin, 117 untreated or given placebo) (Moscarella et al. 1993).
3. Schandalik *et al.* used nine volunteer patients who had earlier undergone surgical gall bladder removal necessitated by gallstones. They received single oral doses of 120 mg silybin as silybin phytosomes, and bile was accessed for silybin levels. Silybin appeared in the bile after 48 hours accounted for 11 percent of the total dose. In the case of silymarin, approximately 3 percent of the silybin was recovered (Schandalik et al. 1994).

Thea sinensis **Linn. (Theaceae)**

Green tea extract generally contains a totally standardized polyphenolic fraction (not less than 66.5%, containing epigallocatechin and its derivatives) obtained from green tea leaves (*Thea sinensis*) and mainly characterized by the presence of epigallocatechin 3-O-gallate, the key compound.

Clinical

Francesco et al. studied on a recently developed oral formulation in the form of coated tablets containing highly bioavailable green tea extract was tested in obese subjects (n = 100) of both gender on a hypocaloric diet. Fifty subjects were assigned to the green tea extract plus hypocaloric diet, while the others 50 subjects followed the hupocaloric diet only. After 90 days of treatment, significant weight loss and decreased body mass index (BMI) were observed in the group taking the herbal extract (14 kg loss in the green tea group compared to a 5 kg loss in the diet-only group); waistline was reduced only in male subjects (Francesco et al. 2009).

Vitis vinifera **Linn. (Vitaceae-Grapeseed)**

In a randomized human trial, young healthy volunteers received grape seed phytosomes once daily for five days. The blood Total Radical-trapping Antioxidant Parameter was measured at several time intervals during 1st day, then also on 5th day. Already by 30 minutes after administration on 1st day, blood TRAP levels were significantly elevated over the control which received conventional standardized grape seed extract (Facina et al. 1994).

Ginkgo biloba Linn. (Ginkgoaceae)

In a bioavailability study conducted with healthy human volunteers the levels of *G. biloba* extract constituents (flavonoid and terpenes) from the phytosomal form peaked after three hours and persisted longer for at least five hours after oral administration. It was found that the phytosomal GBE produced 2–4 times greater plasma concentration of terpenes than did the non phytosomal GBE. Its improved oral bioavailability and good tolerability makes it the ideal ginkgo product even for long term treatment (Bombardelli and Mustich 1991).

Curcumin and naringenin

Maiti et al. developed phytosomes of curcumin and naringenin in two different studies. The antioxidant activity of the complex was significantly higher than pure curcumin in all dose levels tested. In the other study, the naringenin phytosome produced better antioxidant activity than the free compound with a prolonged duration of action, which may be helpful in reducing the fast elimination of the molecule from the body (Maiti et al. 2007).

Hesperetin

Hesperetin is a potent phytomolecule abundant in citrus fruits, such as grapefruit and oranges. In spite of several therapeutic benefits, viz., antioxidant, lipid-lowering, anti-carcinogenic activities their shorter half life and lower clearance from the body restricts its use.

> Recently Mukherjee et al. developed a novel hesperetin phytosome by complexing hesperitin with hydrogenated phosphatidyl choline. This complex was then evaluated for antioxidant activity in CCl_4 intoxicated rats along with pharmacokinetic study revealed that the phytosome had higher relative bioavailability than that of parent molecule at the same dose level (Mukherjee et al. 2008).

Quercetin phospholipid complex

Maiti et al. developed the quercetin phospholipid complex by a simple and reproducible method and also showed that the formulation exerted better therapeutic efficacy than the molecule in rat liver injury induced by carbon tetrachloride (Maiti et al. 2005).

Further Readings

Anonymous. Indian Herbal Pharmacopoeia Volume 1. Worli, Mumbai: Indian Drug Manufacturers Association, 1998.

Barzaghi N, Crema F, Gatti G, Pifferi G, Perucca E. Pharmacokinetic studies on Idb 1016, a silybin phosphatidylcholine complex in healthy human subjects. *Eur J Drug Metab Pharmacokinet* 1990; **15**: 333–38.

Bhattacharya S, Ghosh A. Phytosomes: the Emerging Technology for Enhancement of Bioavailability of Botanicals and Nutraceuticals. *The Internet Journal of Aesthetic and Antiaging Medicine* 2009; **2**: 1.

Bombardelli E, Curri SB. Anthologia Medica Santoriana. 1976; **5**: 177.

Bombardelli E, Curri SB, Loggia Della R, Del NP, Tubaro A, Gariboldi P. Complexes between phospholipids and vegetal derivatives of biological interest. *Fitoterapia* 1989; **60**: 1–9.

Bombardelli E. Phytosome: new cosmetic delivery system. *Boll Chim Farm* 1991; **130**: 431–38.

Bombardelli E, Mustich G. Bilobalide-phospholipid complex, their uses and formulation containing them. U. S. Patent No. EPO-275005, 1991.

Bomberdelli E, Spelta M, Della Loggia R, Sossa S, Tubaro A. Aging skin: protective effect of silymarin-phytosome. *Fitoteapia* 1991; **62**: 115–22.

Bone K. Clinical Applications of Ayurvedic and Chinese Herbs: Monographs for the Western Herbal Practitioner. Warwick, Qld: Phytotherapy Press, 1996.

Busby A, La Grange L, Edwards J, King J. The use of a siymarin/phospholipid compound as fetoprotectant from ethanol induced behavioral deficits. *J Herb Pharmacother* 2002; **2**: 39–44.

Dang Yi. New product concept. 2000, UPC code-0300540111783.

Evans WC. Trease and Evans' Pharmacognosy. London: WB Saunders Company Ltd, 1996.

Facina RM, Carini M, Aldini G, Bombardelli E, Morazzoni P, Morelli R. Free radicals scavenging action and antienzyme activities of procyanidins from *Vitis vinifera*. A mechanism for their capillary protective action. *Arzneim Forsch* 1994; **44**: 592–601.

Francesco DP, Anna BM, Angela B, Maurizio L, Andrea C. Green select phytosome as an adjunct to a low-calorie diet for treatment of obesity: a clinical trial. *Altern Med Rev* 2009; **14**: 154–160.

Goyal A, Kumar S, Nagpal M, Singh I, Arora S. Potential of novel drug delivery systems for herbal drugs. *Ind J Pharm Edu Res* 2011; **45**: 225–35.

Gupta A, Ashawal MS, Saraf S. Phytosomes: a novel approach towards functional cosmetics. *J Plant Sci* 2007; **2**: 644–9.

Hussain A. Status Report on Medicinal Plants for NAM countries. CSTDNADC, New Delhi, 1993.

Jain N, Gupta BP, Thakur N, Jain R, Banweer J, Jain DK, Jain S. Phytosome: a novel drug delivery system for herbal medicine. *Int J of Pharm Sci Drug Res* 2010; **2**: 224–8.

Jain NK. Liposomes as Drug Carriers, Controlled and Novel Drug Delivery, 1st Edition, CBS Publisher, New Delhi 2005; 308.

Jiang YN, Yu ZP, Yan ZM. Preparation of *herba epimedii* flavonoid and their pharmaceutics. *Zhongguo Zhong Yao* 2001; **26**: 105–8.

Jose MM, Bombardelli E. Pharmaceutical compositions containing flavanolignans and phospholipida active principles, 1987, U.S. Patent No-EPO209037.

Kapoor LD. Handbook of Ayurvedic Medicinal Plants. Boca Raton: CRC Press, 1990.

Karnik CR. Pharmacopoeial Standards of Herbal Plants, Vols. 1–2. Delhi: Sri Satguru Publications, 1994. Vol. 1: 189–92; Vol. 2: 125.

Lee C, Reyes E. Protective effects of the flavonoids mixture, Silymarin on fetal rat brain and liver. *J Ethnopharmacol* 1999; **65**: 53–6.

Longer MA, Ching HS, Robinson JR. Oral delivery of chlorthiazide using a bioadhesive polymer. *J Pharm Sci* 1985; **74**: 406–11.

Maiti K, Mukherjee K, Gantait A, Ahmed HN, Saha BP, Mukherjee PK. Enhanced therapeutic benefit of quercetin phospholipid: a comparative study on rats. *Iran J Pharmacol Ther* 2005; **4**: 84–90.

Maiti K, Mukherjee K, Gantait A, Saha BP, Mukherjee PK. Curcumin–phospholipid complex: preparation, therapeutic evaluation and pharmacokinetic study in rats. *Int J Pharm Pharmacol* 2007; **330**: 152–63.

Manach C, Scalbert A, Morand C. Polyphenols: food sources and bioavailability. *Am J Clin Nutr* 2004; **79**: 727–47.

Middleton E, Kandaswami C. The impact of plant flavonoid on mammalian biology: implications for immunity, inflammation, and cancer. pp. 619–652. *In*: JB Harborne (ed.). The flavonoids: Advances in Research Since 1986. 1st Ed. London: Chapman and Hall, 1994.

Moscarella S, Giusti A, Marra F, Marena C, Lampertico M, Relli P, Gentilini P, Buzzelli, G. Therapeutic and antilipoperoxidant effects of silybin phosphatidylcholine complex in chronic liver disease: preliminary results. *Curr Ther Res* 1993; **53**: 98–102.

Mukherjee K, Maiti K, Venkatesh M, Mukherjee PK. Phytosome of Hesperetin, A Value Added Formulation with Phytomolecules. 60th Indian Pharmaceutical Congress; New Delhi, 2008; p. 287.

Patil MS, Patil B, Chittam KP, Wagh RD. Phytosomes: novel approach in herbal medicines. *Asian J Pharm Sci Res* 2012; **2**: 1–18.

Rathore P, Swami G. Planterosomes: A potential phyto-phospholipid carrier for the bioavailability enhancement of herbal extracts. *Int J Pharm Sci Res* 2012; **3**: 737–55.

Samuelsson G. Drugs of Natural Origin: A Textbook of Pharmacognosy. Stockholm: Swedish Pharmaceutical Press, 1999.

Saurabh VK, Kesari A. Herbosome a novel carrier for herbal drug delivery. *Int J Curr Pharm Res* 2011; **3**: 36–41.

Schandalik R, Perucca E. Pharmacokinetics of silybin following oral administration of silipide in patients with extrahepatic biliary obstruction. *Drugs Exp Clin Res* 1994; **20**: 37–42.

Shukla A, Pandey V, Shukla R, Bhatnagar P, Jain S. Herbosomes: a current concept of herbal drug technology an overview. *J Med Pharm All Sci* 2012; **1**: 39–56.

Sivarajan VV. Ayurvedic Drugs and their Plant Sources. Lebanon, New Hampshire: International Science Publisher, 1994.

Tedeco D, Steidler S, Galletti S, Tameni M, Sonzogni O, Ravarotto L. Efficacy of Silymarin-phospholipid complex in reducing the toxicity of aflatoxin B1 in broiler chicks. *Poult Sci* 2004; **83**: 1839–43.

Tyler V. Herbs of Choice: The Therapeutic Use of Phytomedicinals. Binghamton, NY: Pharmaceutical Products Press, 1994: 119.

Venkatesan N, Babu BS, Vyas SP. Protected particulate drug carriers for prolonged systemic circulation—a review. *Indian J Pharm Sci* 2000; **62**: 327–33.

World Health Organisation. The promotion and development of traditional medicine. Geneva: World Health Organization, 1978. (Technical reports series no. 622).

Yanyu X, Yunmei S, Zhipeng C, Qineng P. The preparation of silybin–phospholipid complex and the study on its pharmacokinetics in rats. *Int J Pharmaceut* 2006; **307**: 77–82.

Chapter 8

Nanophytomedicine

8.1 Introduction

Over the past several years, great advances have been made on the development of novel drug delivery systems (NDDS) for plant actives and extracts. The variety of novel herbal formulations like polymeric nanoparticles, nanocapsules, liposomes, phytosomes, nanoemulsions, microsphere, transferosomes, and ethosomes has been reported using bioactive and plant extracts. The novel formulations are reported to have remarkable advantages over conventional formulations of plant actives and extracts which include enhancement of solubility, bioavailability, protection from toxicity, enhancement of pharmacological activity, enhancement of stability, improved tissue macrophages distribution, sustained delivery, and protection from physical and chemical degradation (Ajazuddin 2010).

Phytotherapeutics need a scientific approach to deliver the components in a sustained manner to increase patient compliance and avoid repeated administration. This can be achieved by designing novel drug delivery systems (NDDS) for herbal constituents. NDDSs not only reduce the repeated administration to overcome non-compliance, but also helps to increase the therapeutic value by reducing toxicity and increasing the bioavailability. One such novel approach is nanotechnology. Nano-sized drug delivery systems of herbal drugs have a potential future for enhancing the activity and overcoming problems associated with plant medicines (Ansari et al. 2012).

The use of nanotechnology for treatment, identification, monitoring, and managing biological systems has recently been called nanomedicine. In the herbal formulation research, incorporating the nano-based formulation has a great number of advantages for phytomedicine, including improvement of solubility and bioavailability, safeguard from toxicity, enhancement of pharmacological activity, improvement of stability, and increase in tissue macrophages distribution, sustained delivery, and protection from physical and chemical degradation (Patravale et al. 2015).

It has been widely proposed to combine herbal medicine with nanotechnology, because nanostructured systems might be able to potentiate the action of plant extracts, reducing the required dose and side effects, and improving activity. Nanosystems can deliver the active constituent at a sufficient concentration during the entire treatment period, directing it to the desired site of action. Conventional treatments do not meet these requirements.

8.2 Silymarin-loaded Solid Nanoparticle System

8.2.1 Prevention of paracetamol-induced hepatotoxicity

Silymarin nanoparticles (Smnps) were prepared by nanoprecipitation in polyvinyl alcohol stabilized Eudragit RS100(®) polymer. Process parameter optimization provided 67.39% entrapment efficiency and a Gaussian particle distribution of average size 120.37 nm. Silymarin release from the nanoparticles was considerably sustained for all formulations. Smnps were strongly protective against hepatic damage when tested in a paracetamol overdose hepatotoxicity model. Nanoparticles recorded no animal death even when administered after an established paracetamol-induced hepatic necrosis. Preventing progress of paracetamol hepatic damage was traced toan efficient glutathione regeneration to a level of 11.3 μmol/g in hepatic tissue due to Smnps (Das et al. 2011).

8.2.2 Hepatoprotective effect

Formulation of a silymarin-loaded nanoemulsion, comprising silymarin, castor oil, polyvinylpyrrolidone, Transcutol HP, Tween 80, and water at a weight ratio of 5/3/3/1.25/1.25/100 was accomplished using an SPG membrane emulsification technique at an agitator speed of 700 rpm, a feed pressure of 15 kPa, and a continuous phase temperature of 25°C. This resulted in generation of comparatively uniform emulsion globules with a narrow size distribution. Moreover, the silymarin-loaded solid nanoparticles, containing silymarin/castor oil/polyvinylpyrrolidone/Transcutol HP/Tween 80 at a weight ratio of 5/3/3/1.25/1.25, improved about 1,300-fold drug solubility and retained a mean size of about 210 nm.

Silymarin was located in the unaltered crystalline form in the nanoparticles. The drug dissolved rapidly from the nanoparticles, reaching nearly 80% within 15 minutes, indicating three-fold better dissolution than that of the commercial product. Further, the nanoparticles showed a considerably shorter time to peak concentration, a greater area under the concentration-time curve, and a higher maximum concentration of silymarin compared with the commercial product ($P < 0.05$). In particular, the area under the concentration-time curve of the drug provided by the nanoparticles was approximately 1.3-fold greater than that of the commercial product. In addition, the silymarin-loaded nanoparticles significantly reduced carbon tetrachloride-induced hepatotoxicity, indicating improved bioactivity compared with silymarin powder and the commercial product (Yang et al. 2013).

8.2.3 A comparative study on the hepatoprotective effect of silymarin and silymarin-nanoparticles

The present study was aimed to develop a novel silymarin-loaded solid lipid nanoparticle (Sm-loaded SLN) system with enhanced bioavailability and with an ability to provide excellent hepatic protection for poorly water-soluble drugs. Based upon the investigation results with apoptotic markers, PCNA and light microscopic findings, it was concluded that Sm-loaded SLN significantly reduced D-GaIN/TNF-α-induced hepatotoxicity, which suggested improved bioactivity compared to Sm (Cengiz et al. 2015).

8.3 Sesamol-loaded Solid Nanoparticle System

8.3.1 Role in carbon tetrachloride induced sub-chronic hepatotoxicity

Sesamol-loaded solid nanoparticle system (S-SLNs) prepared by microemulsification method were nearly spherical in shape with an average particle size of 120.30 nm and their oral administration at 8 mg/kg body weight (BW) showed significantly ($p < 0.001$) better hepatoprotection than free sesamol (FS) and a well established hepatoprotective antioxidant silymarin [SILY (25 mg/kg BW); $p < 0.05$) in CCl_4 induced sub-chronic liver injury in rats (Singh et al. 2014).

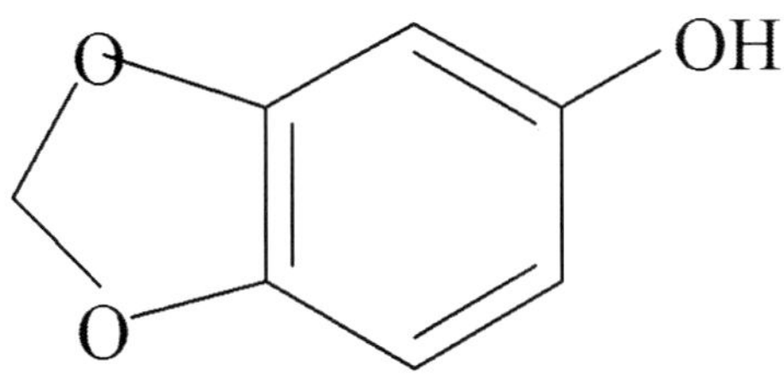

Fig. 8.1: Structure of sesamol.

8.3.2 Role in carbon tetrachloride induced hepatotoxicity

Sesamol, a component of sesame seed oil, exhibited significant antioxidant activity in a battery of *in vitro* and *ex vivo* tests including lipid peroxidation induced in rat liver homogenates. S-SLNs prepared by the microemulsification method were administered to rats post-treatment with CCl4 (1 ml/kg body weight (BW) twice weekly for two weeks, followed by 1.5 ml/kg BW twice weekly for the subsequent two weeks). S-SLNs (120.30 nm) at a dose of 8 mg/kg BW showed significantly better hepatoprotection than corresponding dose of free sesamol (FS; $p < 0.001$). Effects achieved with S-SLNs were comparable with silymarin (SILY), administered at a dose of 25 mg/kg BW (Singh et al. 2015).

8.4 Curcumin-loaded Solid Nanoparticle System

Curcumin, an established pleiotropic agent, has potential for hepatoprotection owing to its powerful antioxidant, anti-inflammatory, and antifibrogenic properties.

8.4.1 Attenuation of carbon tetrachloride induced hepatotoxicity

Curcumin solid lipid nanoparticles (C-SLNs) prepared using a microemulsification technique, were administered to rats post-treatment with CCl_4 (1 ml/kg body weight [BW] twice weekly for two weeks, followed by 1.5 ml/kg BW twice weekly for the subsequent two weeks). C-SLNs (12.5 mg/kg) significantly ($p < 0.001$–0.005) attenuated histopathological changes and oxidative stress, and also decreased induction of ALT, AST, and TNF-α in comparison with free curcumin (100 mg/kg), silymarin (25 mg/kg), and self-recovery groups (Singh et al. 2014a).

8.5 *Andrographis paniculata* Silver Nanoparticles

In vitro radical scavenging assay proved strong antioxidant effect of the AgNPs compared to 5% aqueous leaf extract. CCl(4) was used to induce hepatic injury in mice model. The biosynthesized AgNPs at three different doses (25, 50, 100 mg/kg BW of the animal) were used for treatment. Silymarin was used as a standard. Low dose (25 mg/kg BW) was effective in revival of all biological parameters to near normal in all intoxicated groups indicating the curing effects on CCl(4) induced liver injury (Suriyakalaa et al. 2013).

8.6 Application of Nanomedicine in Ayurvedic Research

A Bhasma means an ash obtained through incineration; the starter material undergoes an elaborate process of purification and this process is followed by the reaction phase, which involves incorporation of some other minerals and/or herbal extract. Thus bhasma can be defined as ancient Indian nanomedicine (Pal et al. 2014). Bhasma in accordance of classical expectation are *Swarna bhasma*, *Makshika bhasma*, *Abhrak bhasma*, *Tamra bhasma*, and *Louha bhasma*. X-ray diffraction, TEM, and particle size analysis revealed that these bhasma are in nanometer dimension. The bhasma may be considered as nanomedicine and are free from toxicity in therapeutic doses (Chaudhary 2014).

> To unravel the knowledge of bhasmas, an algorithm integrating ayurveda- and science-based experiments needs to connect the basic science of rasa shastra (Indian alchemy), the study of herbometallic medicines in ayurveda, to clinical data (Bhattacharya 2011). Bhasmas are generally safe drugs for human beings in spite of the presence of seemingly toxic elements and compounds as indicated by recent studies using modern analytical techniques. Nevertheless, more systematic nanomaterialistic investigations on Bhasmas are recommended for gaining the complete and reliable composition-processing-structure-effectiveness picture of these drugs (Adhikari 2014).

8.6.1 Nanoparticles of Jasada bhasma (zinc based bhasma)

Jasada bhasma is used in the treatment of diabetes mellitus in Ayurveda. Recent studies suggest that bhasmas comprise submicronic particles or nanoparticles. Thus a bhasma-inspired new drug discovery approach could emerge in which several metal based nanomedicines could be developed. Evidence suggests efficacy of zinc oxide nanoparticles in the treatment of diabetes mellitus as compared to *jasada bhasma* (Umrani and Paknikar 2015).

Further Readings

Adhikari R. Ayurvedic Bhasmas: overview on nanomaterialistic aspects, applications, and perspectives. *Adv Exp Med Biol* 2014; **807**: 23–32.

Ajazuddin Saraf S. Applications of novel drug delivery system for herbal formulations. *Fitoterapia* 2010; **81**: 680–9.

Ansari SH, Islam F, Sameem M. Influence of nanotechnology on herbal drugs: a Review. *J Adv Pharm Technol Res* 2012; **3**: 142–146.

Bhattacharya B. Elucidating the nanomaterialistic basis for ayurvedic bhasmas using physicochemical experimentation. *J Biomed Nanotechnol* 2011; **7**: 66–7.

Bonifácio BV, da Silva PB, dos Santos Ramos MA, Silveira Negri KM, Bauab TS, Chorilli M. Nanotechnology-based drug delivery systems and herbal medicines: a review. *Int J Nanomedicine* 2014; **9**: 1–15.

Cengiz M, Kutlu HM, Burukoglu DD, Ayhancı A. A comparative study on the therapeutic effects of silymarin and silymarin-loaded solid lipid nanoparticles on D-GalN/TNF-α-induced liver damage in Balb/c mice. *Food Chem Toxicol* 2015; **77**: 93–100.

Chaudhary A. Ayurvedic bhasma: nanomedicine of ancient India—its global contemporary perspective. *J Biomed Nanotechnol* 2011; **7**: 68–9.

Das S, Roy P, Auddy RG, Mukherjee A. Silymarin nanoparticle prevents paracetamol-induced hepatotoxicity. *Int J Nanomedicine* 2011; **6**: 1291–301.

Pal D, Sahu CK, Haldar A. Bhasma: the ancient Indian nanomedicine. *J Adv Pharm Technol Res* 2014; **5**: 4–12.

Patravale VB, Fernandes C, Pol A, Patel P, Parekh V. A special section on nanophytomedicine. *J Nanosci Nanotechnol* 2015; **15**: 4019–20.

Singh N, Khullar N, Kakkar V, Kaur IP. Hepatoprotective effects of sesamol loaded solid lipid nanoparticles in carbon tetrachloride induced sub-chronic hepatotoxicity in rats. *Environ Toxicol* 2014. doi: 10.1002/tox.22064. [Epub ahead of print].

Singh N, Khullar N, Kakkar V, Kaur IP. Attenuation of carbon tetrachloride-induced hepatic injury with curcumin-loaded solid lipid nanoparticles. *BioDrugs* 2014a; **28**: 297–312.

Singh N, Khullar N, Kakkar V, Kaur IP. Sesamol loaded solid lipid nanoparticles: a promising intervention for control of carbon tetrachloride induced hepatotoxicity. *BMC Complement Altern Med* 2015; **3**: 142.

Suriyakalaa U, Antony JJ, Suganya S, Siva D, Sukirtha R, Kamalakkannan S, Pichiah PB, Achiraman S. Hepatocurative activity of biosynthesized silver nanoparticles fabricated using *Andrographis paniculata. Colloids Surf B Biointerfaces* 2013; **102**: 189–94.

Umrani RD, Paknikar KM. Jasada bhasma, a zinc-based ayurvedic preparation: contemporary evidence of antidiabetic activity inspires development of a nanomedicine. *Evid Based Complement Alternat Med* 2015; **2015**: 193156.

Yang KY, Hwang du H, Yousaf AM, Kim DW, Shin YJ, Bae ON, Kim YI, Kim JO, Yong CS, Choi HG. Silymarin-loaded solid nanoparticles provide excellent hepatic protection: physicochemical characterization and *in vivo* evaluation. *Int J Nanomedicine* 2013; **8**: 3333–43.

Chapter 9

Metabolomics and Phytomedicine

9.1 Introduction

Metabolomics is defined as "systematic study of the unique chemical fingerprints that specific cellular processes leave behind" (Bennett 2005). Metabolism includes the final metabolites of a cell, or biological organisms (Jordan et al. 2009). Herbal metabolomics is unbiased, high-performance, and specific analysis of highly complex mixtures of plant extracts. This overall gain for metabolom analysis is at first a result of recent advances in mass spectrometry.

Metabolomics is rapidly evolving as the approach of choice across a broad range of sciences including phytomedicine. The metabolomes of herbal drugs are a particularly valuable natural resource for the evidence-based development of novel therapeutic applications in phytomedicine and nutraceuticals (Shyur and Yang 2008).

The ("-omic-" technologies) including metabolomics, genomics, and proteomics are basically high-throughput technologies. These technologies increase the number of genes which can be detected simultaneously. Further, they have potential in establishing the relationship between complex mixtures and complex effects in the form of gene expression. The "-omic-" technologies can play a significant role in chemical and pharmacological standardization of Phyto extracts and provide valuable information on toxicological aspects including acute, sub-acute and chronic toxicity. In the future, these technologies can play a significant role in giving an evidence based approach to phytomedicine, investigating the mode of action and allow to investigate herbal extracts without prominent active principle(s) (Ulrich-Merzenicha et al. 2007).

MS and NMR based metabolic fingerprinting or profiling is continuously being applied to drug discovery from natural resources, quality control of herbal material, and discovering of lead compounds (van der Kooy et al. 2009). Applying quick scanned time-of-flight (TOF) mass spectrometry *via* gas chromatography (GC) resulted in the numbers of detectable metabolites to

500–1000 in total extracts. Another attractive method would be direct injection of herbal extract to ultra-high-resolution Fourier Transform Ion Cyclotron Mass Spectroscopy, which is able to produce a significant fingerprint of metabolom (Hegeman 2010). Recently chromatographic fingerprinting and metabolomics have been used for quality control of traditional Chinese medicine or TCM (Liang et al. 2010).

9.2 Role of Metabolomics for the Classification and Quality Control of *Matricaria recutita* L. (Asteraceae)

NMR-based metabolomics, which combines high-resolution (1)H-NMR spectroscopy with chemometric analysis, is a novel analytical method for the standardisation of phytomedicines. *M. recutita* flowers from three different geographical regions (Egypt, Hungary, and Slovakia) were characterised using 1H-NMR spectroscopy followed by principal component analysis. It was found that the origin, purity, and preparation methods contributed to the differences observed in prepared chamomile extracts. In addition, this method also enabled the elucidation of the molecular information embedded in the spectra responsible for the observed variability (Wang et al. 2004).

9.3 Metabolomics and Microarray Gene Expression in the characterization of *Chelidonium majus* L. (Ranunculaceae)

Based on data from (1)H-NMR fingerprints and RP-HPLC analyses of the dichloromethane, water, ethanol, and ethanol 50% (V/V) extracts of *C. majus* showed a divergent composition of constituents depending on the solvent used. HepG2 liver cells responded differentially to the four extracts. Microarray analysis revealed a significant regulation of genes and signal cascades related to biotransformation. Liver-toxic signal cascades were also activated. Neither the activated genes nor the proliferation response could be clearly related to the differing alkaloid content of the extracts. A systems biology approach which combines a metabolomic plant analysis with a functional characterization by gene expression profiling in HepG2 cells is an appropriate strategy to characterize variations in plant extracts (Orland et al. 2014).

9.4 Metabolomics and Identification Issues of *Rheum palmatum* L. (Polygonaceae)

The comprehensive and unbiased information of 30 batches of *R. palmatum* covering raw and two general processing methods were given by metabolomic profiles. Using a molecular feature extraction algorithm, non-targeted compounds were analyzed in minutes. In total, 73 characteristic markers were extracted and identified by diagnostic ion filtering. They have been further analyzed by partial least squares-support vector machine-based pattern recognition. The comprehensive and rapid method for raw and processed pieces of *R. palmatum* classification shows good sensitivity, specificity, and prediction performance (Liu et al. 2016).

9.5 Integratation of Metabolomics and Pharmacokinetics for Evaluation of Phytomedicine

The pharmacokinetics of multi-component therapeutics is a great technical challenge, which has led to significant limitations in understanding the efficacies and toxicities of combination drugs and herbal medicines. Metabolomics has clearly demonstrated its value in elucidating the interaction of the biological systems, genome with its environments, and recently been used in the PK analysis of drugs, xenobiotics, and several nutrients. The metabolomics strategy can also be used in the pharmacokinetic study of plant derived agents to demonstrate their biological fates, thereby linking the plant metabolome to human/animal metabolome, and bridging the gap between multi-component agents and molecular pharmacology (Lan and Jia 2016).

9.6 DNA Sequencing and Metabolomics

Next generation DNA sequencing (NGS) demonstrates how the sophisticated analyses of samples, including complex mixtures unable to be evaluated effectively by previous techniques, can be rapidly screened (Coghlan et al. 2012). Modern DNA sequencing can rapidly provide a genetic audit of large numbers of cases that can be matched against established standards to focus the search for ingredients (Coghlan et al. 2012) such as toxic plants and/or potential allergens.

> By matching patterns of DNA sequences between the preparation under investigation and recognized reference material, the probability that a certain herb belongs to a particular family, genus, or species might ultimately be derived. Screening of preparations in metabolomics laboratories can also provide complementary information on other plant and pharmaceutical agents (Robertson et al. 2011).

Further Readings

Bennett D. Growing pains for metabolomics. *Scientist* 2005; **19**: 25–28.

Coghlan ML, Haile J, Houston J, Murray DC, White NE, Moolhuijzen P, Bellgard MI, Bunce M. Deep sequencing of plant and animal DNA contained within traditional Chinese medicines reveals legality issues and health safety concerns. *PLoS Genet* 2012; **8**: 1–11.

Hegeman AD. Plant metabolomics-meeting the analytical challenges of comprehensive metabolite analysis. *Brief Funct Genomics* 2010; **9**: 139–148.

Jordan KW, Nordenstam J, Lauwers GY, Rothenberger DA, Alavi K, Garwood M, Cheng LL. Metabolomic characterization of human rectal adenocarcinoma with intact tissue magnetic resonance spectroscopy. *Dis Colon Rect* 2009; **52**: 520–525.

Lan K, Jia W. An integrated metabolomics and pharmacokinetics strategy for multi-component drugs evaluation. *Curr Drug Metabol* 2016; **11**: 105–114.

Liang YZ, Xie PS, Chan K. Chromatographic fingerprinting and metabolomics for quality control of TCM. *Comb Chem High Throughput Screen* 2010; **13**: 943–53.

Liu Y, Li L, Xiao YQ, Yao JQ, Li PY, Yu DR, Ma YL. Global metabolite profiling and diagnostic ion filtering strategy by LC-QTOF MS for rapid identification of raw and processed pieces of *Rheum palmatum* L. *Food Chem* 2016; **192**: 531–40.

Orland A, Knapp K, König GM, Ulrich-Merzenich G, Knöß W. Combining metabolomic analysis and microarray gene expression analysis in the characterization of the medicinal plant *Chelidonium majus* L. *Phytomedicine* 2014; **21**: 1587–96.

Robertson DG, Watkins PB, Reily MD. Metabolomics in toxicology: preclinical and clinical applications. *Toxicol Sci* 2011; **120**: S146–70.

Shyur L-F, Yang N-S. Metabolomics for phytomedicine research and drug development. *Curr Opin Chem Biol* 2008; **12**: 66–71.

Ulrich-Merzenicha G, Zeitlera H, Jobstb D, Paneka D, Vettera HH. Application of the "-Omic-" technologies in phytomedicine. *Phytomedicine* 2007; **14**: 70–82.

van der Kooy F, Maltese F, Choi YH, Kim HK, Verpoorte R. Quality control of herbal material and phytopharmaceuticals with MS and NMR based metabolic fingerprinting. *Planta Med* 2009; **75**: 763–75.

Wang Y, Tang H, Nicholson JK, Hylands PJ, Sampson J, Whitcombe I, Stewart CG, Caiger S, Oru I, Holmes E. Metabolomic strategy for the classification and quality control of phytomedicine: a case study of chamomile flower (*Matricaria recutita* L.). *Planta Med* 2004; **70**: 250–5.

Chapter 10

Clinical Research in Ayurveda

10.1 Introduction

Ayurveda or Traditional Indian Medicine (TIM) is considered to be the oldest-practicing system of medicine. Recently, the herbal drug industry has witnessed explosive growth. CAM systems are in great demand, particularly Traditional Chinese Medicine (TCM) and Traditional Indian Medicine (TIM). The growing popularity of CAM among people, has led to the onset of research at molecular and clinical levels (Singh and Malhotra 2007).

With the development of new subjects like medicinal phytochemistry, phytophamacology, and phytopharmacotherapy, the importance of clinical research in TIM has become more significant. Analytical study of subjects like *Dravyaguna* (Medicinal Plant Pharmacology) and *Kayachikitasa* (Internal Medicine) is required for enhancing the practical utility of TIM (Singh and Malhotra 2001). Lack/documentation of clinical trials in TIM has triggered controversies, regarding therapeutic application of formulations used in TIM (Singh 2005).

Although, formulations of TIM have been used for centuries with success, testing at molecular levels is still a challenge (Kapoor 1990). Pharmacological intervening has opened a new age in CAM and TIM research. The concept of reverse pharmacology is rapidly catching up for developing cost-effective and potential drug candidates from medicinal plants (Vaidya 2006).

10.2 Classification of Clinical Studies in TIM

Clinical studies in TIM can be divided into two distinct groups:

1. Controlled studies
2. Uncontrolled studies

Recently, favorable clinical studies have appeared for single herb/polyherbal formulations used in TIM for varied ailments. The studies, seems to be appropriate with regard to several parameters like drug selection and standardization, design, patient participation, and results.

The present review is dedicated to rare clinical studies done on formulations used in TIM. The list of plants or formulations discussed in the review may be incomplete. The leading factor is lack of indexed publication dealing with clinical aspects of TIM. Moreover, the clinical knowledge documented by authors in Ayurvedic journals was done mostly around 1960 when pharmacological and clinical research were not in the limelight. Non-availability of full-length papers and English version also contributed to the incomplete list.

Materials and Methods

The keywords for the present review were clinical trials, clinical studies, TIM, single herb, polyherbal formulations, and Ayurveda. ABIM (Annotated Bibliography of Indian Medicine), a data bank on Indian Medicinal Plants provided by Central Council of Research in Ayurveda and Siddha (CCRAS) and journals related to clinical aspects of Complementary and Alternative Medicine (CAM) were used for searching data, updated until Feb 2007.

The references encountered in the search were later consulted. The data generated after a systemic literature study was documented according to human anatomy.

Integument system

Sookshma triphala in lipoma (Kulkarni 1995b), *Rudanti* (*Capparis moonii* Hook. f., Capparidaceae) in tubercular lymphadenitis (Sheth 1961), *Patoltriphaladi* and *Panchatiktaka kwatha* in scabies (Nair et al. 1991), and *Arogyavardhini rasa* and *Gandhaka rasayana* in leucoderma (Shetty et al. 2000).

Gastroenterology

Kutaja (*Holarrhena antidysenterica* Wall., Apocynaceae) in amoebiasis and giardiasis (Singh 1980), *Mustaka* (*Cyperus rotundus* Linn., Cyperaceae), and *Vibhituka* (*Terminalia belerica* Roxb., Combertaceae) in chronic diarrhea (Kumar 1982; Tripathi et al. 1983; Patwardhan et al. 1990), *Triphala* and *Haritaki* (*Terminalia chebula* Retz., Combertaceae) in constipation and worm infestation (Inamdar et al. 1962; Gaind et al. 1964; Tripathi et al. 1983), *Takrarishta* and *Sunthi* (*Zingiber officinale*) in malabsorption (Dixit 1976; Nanda et al. 1985), *Tulasi* (*Ocimum sanctum* Linn. Lamiaceae) and *Patola* (*Trichosanthes dioica* Linn., Cucurbitaceae) in peptic ulcer (Jall 1970; Ariyavansha et al. 1981).

Hepatology

Arogyavardhini Rasa in acute viral hepatitis, hepatic cirrhosis, and jaundice (Dange 1987; Wachasundar 2001d), *Phalatrikadi kwath* in jaundice (Dwivedi et al. 1984) and *Kalmegha* (*Andrographis paniculata* Nees., Acanthaceae) in jaundice (Tomar 1981).

Respiratory system

Haritakileha (Naik et al. 2001), *Pippalikshirapaka* (Upadhyaya et al. 1982), *Sirisatwakkvatha* (Sharma 1979) *Hardira* (*Curcuma longa* Linn., Zingiberaceae) (Jain et al. 1979), *Snuhi* (*Euphorbia prostrata* Linn. and *Euphorbia thymifolia* Linn., Euphorbiaceae) and *Pippali* (*Piper longum* Linn., Piperaceae) (Sharma and Sharma 1971b; Sharma et al. 1982) and *Pushkaramooladi choorna* (Saiprasad and Upadhyay 1998), in bronchial asthma, *trikatu* in allergic rhinitis (Sridhar 2001), and *Swarna basanta malti rasa* (Dube et al. 1978), *Pithecellobium dulce* Benth. (Fabaceae) (Kumar et al. 2004), and *Rudanti* (*Capparis moonii* Linn.) (Tandon et al. 1961) and *Krishna tulasi* (*Ocimum sanctum* Linn.) (Upasani and Mardikar 1989) in pulmonary tuberculosis.

Cardiovascular system

Lohasava and *Triphala mandoora* in anemia (Venkataraghavan et al. 1977).

Nervous system

Asthawarga kwatha and *Dhanvantara* yoga in paralysis (Rajagopalan et al. 1975), *Smrtisagara rasa* in amnesia (Tripathi and Singh 1999), *Ustookhudus churana* in migraine (Kumar et al. 2004a), and *Ashwagandha* (*Withania somnifera* Dunal. Solanaceae) (Rai 1979) in epilepsy.

Genitourinary system

Guduchi (*Tinospora cordifolia* (Willd.) Miers ex Hook. F., Menispermaceae) in uremia (Gupta et al. 1972), and *Sveta parpati* with *Kulatha kwatha* in urolithiasis (Kumar and Kumar 1995).

Musculoskeletal system

Sunthi guggulu (Kishore et al. 1982), *Vatari guggulu* and *Maharasnadi kwatha* (Swamy and Bhattathiri 1998) in rheumatoid arthritis, *Kanchanaragugulukwatha* (Rao 1982) in rheumatic diseases, *Goraksa* (*Dalbergialanceolaria* L.f., Fabaceae) in frozen shoulder and *Bhallataka* (*Semecarpus anacardium* Linn., Anacardiaceae) inosteoarthritis (Majumdar 1979c), *Guggul* (*Commiphoramukul* Engl., Burseraceae) (Singh 2004), and *Eranda veej ksheer paka* (Mohanty et al. 2000) in sciatica.

Endocrine system

Abhraka (mica) (Sankar and Aggarwal 1988), and *Chandraprabha vati* (Parimi et al. 1985) in diabetes mellitus and *Arogyavardhini Rasa* (Tewari and Jain 1982) in hypercholesterolemia/obesity.

ENT

Sharpunkha (*Tephrosia purpurea* Linn., Fabaceae) in adenoids and acute tonsillitis (Tewari and Jain 1983c).

Eye

Sookshma triphala in chalazion (Sudrik 1995).

Reproductive system

Ashokarishta and *Musalikhadiradi kwatha* in menorrhagia (Anonymous 2002d), and *Triphala kwath* in leucorrhoea (Singh and Londhe 1993).

Dental

Triphala in pyorrhea (Maurya et al. 1995; Maurya et al. 1997).

Results and Discussion

A systemic study afforded several single or poly-herbal, and herbo-mineral and purely mineral based formulations used in TIM. Much of the clinical research was related to respiratory and musculoskeletal system. Among polyherbal formulations, *guggul*-based formulations were the cornerstone for treating arthritis and rheumatism. Use of *Triphala* was highlighted in various clinical conditions.

The major drawback of these clinical studies is lack of control. The studies do emphasize clinical utility of formulations used in TIM, which may be the basis of reinitiating clinical trials. We also believe that instead of an expanding list of novel formulations, work should be initiated to evaluate the potential of already reported formulations to overcome the shortcomings encountered in earlier clinical studies.

Further Readings

Anonymous. Effect of *Ashokarishta* and *Musalikhadiradi kwatha* in menorrhagia. M.D. thesis, Government Ayurveda College, Trivandrum, 2002d.

Ariyavansha HAS, Gupta JP, Chaturvedi GN. Clinical studies on peptic ulcer and its treatment with Patola (*Trichosanthes dioica*). Thesis BHU, Varanasi, 1981.

Dange SV. Efficacy of *Arogyavardhini* an indigenous compound formulation, in acute viral hepatitis-a double-blind study. *Ind Pract* 1987; **55**: 1063–1069.

Dixit OP. Takrarishta in the management of secondary malabsorption (grahani). Role of takrarishta in the management of grahani roga: secondary malabsorption caused by Giardia lamblia. *J Res Indian Med* 1976; **11**: 50–59.

Dube CB, Sharma YK, Kansal CM. A comprehensive study of *Swarna basanta malti* in cases of *Rajayakshma* (pulmonary tuberculosis). *Nagarjun* 1978; **21**: 9–14.

Dwivedi ML, Tripathi SV, Dwivedi HS. Role of *Phalatrikadi kashaya* and *Arogyavardhini vati* in the management of jaundice (kamala). *Sac Ayur* 1984; **37**: 87–94.

Gaind KN, Mittal HC, Khanna SR. Anthelmintic activity of triphala. *Ind J Pharm* 1964; **26**: 106–107.

Gupta AS, Singh KP, Mahawar MM, Sharma VP. Preliminary report on phytochemical and clinical trials of *Tinospora cordifolia* Miers on six uraemia patients. *J Ind Med Prof* 1972; **18**: 8256–8257.
Inamdar MC, Rajarama Rao MR, Siddiqi HH. Purgative activity of triphala. *Ind. J. Pharm* 1962; **24**: 87–88.
Jain JP, Bhatnagar LC, Parsai MR. Clinical trials of Haridra (*Curcuma longa*) in cases of *tamakswasa* and *kasa*. *J Res Ind Med* 1979; **14**: 110–120.
Jall A. Clinical trials of *Ocimum sanctum* Linn. (*Tulsi*) in peptic ulcer and hyper-acidity patients. *J Res Ind Med* 1970; **4**: 238–239.
Kapoor LD. CRC Handbook of Ayurvedic Medicinal Plants. Boca Raton: CRC Press, 1990.
Karnick CR, Pathak NN. Clinical trials of crude drug *Tephrosia purpurea* (L). on adenoids and acute tonsillitis. *J Nat Integ Med Assoc* 1983c; **25**: 333–334.
Kishore P et al. Clinical studies on the treatment of amavata (rheumatoid arthritis) with *Sunthi guggulu*. *J Res Ayur Siddha* 1982; **3**: 133–146.
Kulkarni PH. Clinical assessment of effect of *sookshma triphala* in lipoma. pp. 66–71. *In*: PH Kulkarni (ed.). 1995b.
Kumar A, Kumar N. To evaluate the effect of Ayurvedic drugs (a herbomineral combination of *Sveta parpati* with *Kulatha kwatha*) in the management of *mutrasmari* (urolithiasis). *J Res Ayur Siddha* 1995; 16: 35–42.
Kumar S et al. Clinical effect of *Ustookhudus churana* on migraine (*vatic shirahshool*). *Sachitra Ayurved* 2004; **56**: 772–773.
Kumar S et al. *Manila tamarind* (*Pithecellobium dulce*) could help to treat tuberculosis. *J. Joseph Thas* (ed.). 83, 2004.
Kumar Y. Clinical studies on chronic diarrhea and its treatment with Ayurvedic drug *Mustaka* (*Cyperus rotundus*). Thesis, B.H.U., Varanasi, 1982.
Majumdar A. Clinical studies of drugs (*Bhallatak*, *Gourakh* and *Guggulu*) in osteoarthritis, frozen shoulder and sciatica. *Rheum* 1979c; **14**: 153–161.
Maurya DK et al. Role of *Triphala* in the management of pyorrhea. *Sac Ayur* 1995; **48**: 390–391.
Maurya DK et al. Role of *Triphala* in the management of periodontal disease. *Phytother Res* 1997; **16**: 91–93.
Mohanty J et al. *Eranda veej ksheer paka* in the management of *gridhrasi* (sciatica)—a clinical study. *Kayamaya's Siddha-Pani* 2000; **2**: 7.
Naik A, Nageswara Rao V, Mishra SK. Effect of *Haritaki leha* on *shwasa roga* (asthma): a clinical study. *Sac Ayur* 2001; **54**: 224–226.
Nair PRC, Menon TV, Vijayan NP, Prabhakaran VA. A comparative study of *Patoltriphaladi* and *Panchatiktaka kwatha yogas* in the treatment of *pama* (scabies). *J. Res Ayur Siddha* 1991; **12**: 151–162.
Nanda GC, Tewari NS, Kishore P. Clinical studies on the role of *Sunthi* in the treatment of *grahani roga*. *J Res Ayur Siddha* 1985; **6**: 78–87.
Parimi S et al. Anti-diabetic effect of *Chandraprabha vati*—a reappraisal (experimental study). *Sac Ayur* 1985; **48**: 395–399.
Pattan Shetty JK, Pushpalatha H, Bikshapathi T. Study of *Arogyavardhini* and *Gandhaka rasayana* in the treatment of leucoderma. *Sac Ayur* 2000; **53**: 438–440.
Patwardhan B et al. Clinical evaluation of *Terminalia belerica* in diarrhea. *Anc Sci Life* 1990; **10**: 94–97.
Rai NP. Clinical and experimental studies on epilepsy with special reference to its treatment with *Ashwagandha* (*Withania somnifera* Dunal). Thesis, B.H.U., Varanasi, 1979.
Rajagopalan K, Agnihotri RJ, Bhaskaran KP. A clinical trial with *Asthawarga kwatha* and *Dhanvantara yoga* in *pakshawadha*. *J Res Ind Med* 1975; **10**: 84–86.
Rao NH. *Kanchanara gugulu kwatha* in rheumatic diseases. A new dimension in kwatha preparations. *Rheum* 1982; **17**: 59–67.
Saiprasad AJV, Upadhyay BN. Management of *tamakashvasa* (bronchial asthma) with *Pushkaramooladi choorna*. *An Sci Life* 1998; **18**: 130–133.
Sankar VR, Aggarwal MP. Clinical studies of *Abhraga* (mica) chendooram in the management of diabetes mellitus (neerazhivu). *J Res Ayur Sidd* 1988; **9**: 38.
Sharma GP, Sharma PV. Effect of *Dugdhika on shwasa roga* (bronchial asthma)—{A} clinical study. *J Res Ind Med* 1971b; **6**: 118–124.
Sharma GD, Upadhyay BN, Tripathi SN. A clinical trial of *Euphorbia prostrata* and *Euphorbia thymifolia* in the treatment of bronchial asthma. *J Res Ayur Sidd* 1982; **3**: 109.

Sharma OD. Clinical and experimental studies on *tamaksvasa* and its management with *Albizia lebbeck*. Postgraduate Thesis (supervisor: S.N. Tripathi), B.H.U., Varanasi, 1979.
Sheth SC. Clinical trials of *Capparis moonii* Wight in tubercular meningitis and tubercular lymphadenitis. *Ind J Pharm* 1961; **24**: 116.
Singh AP, Malhotra S. Pharmacological considerations of ayurvedic herbs. *The U's Natural News* 2001; **2**: 37–40.
Singh AP. *Dravyaguna Vijnana*. Chaukhambha Orientalia, New Delhi. 699, 2005.
Singh AP, Malhotra S. A review of pharmacology of phytochemicals from Indian medicinal plants. *Int J Alt Med* 2007; **5**.
Singh KP. Some clinical studies on *Kutaja* (*Holarrhena antidysenterica* Wall.) in intestinal amoebiasis and giardiasis. Thesis B.H.U., Varanasi, 1980.
Singh RK, Londhe CS. Use of *triphala kwath* in *swet pradara* (leucorrhoea). *Deerghayu Int* 1993; **9**: 15–17.
Singh S. Ayurvedic therapies of sciatica (*gridhrasi*). pp. 185–201. *In*: Lakshmi Chandra Mishra (ed.). 2004.
Sridhar BN. The role of *Trikatu* yoga in the management of *pratisyaya*. *Aryavaidyan* 2001; **14**: 154–158.
Sudrik UV. Management of *Anjananamika* in *amavastha* with *swedana* and *Sookshma triphala*. pp. 11–14. *In*: PH Kulkarni (ed.). 1995.
Swamy GK, Bhattathiri PPN. *Vatari guggulu* and *Maharasnadi kwatha* in the management of *amavata*: a clinical study. *J Res Ayur Siddha* 1998; **19**: 41.
Tandon RN, Khanna BK, Bajpai RP. *Rudanti* in pulmonary tuberculosis. *J Ind Med Assoc* 1961; **36**: 143–145.
Tewari NS, Jain PC. Clinical evaluation of *Arogyavardhini* as a hypocholesterolaemic agent with special reference to obesity/corpulency. *J Res Ayur Siddha* 1982; **1**: 121–132.
Tomar GS. Clinical studies on liver diseases with special reference to jaundice (kamala) and its treatment with an indigenous drug—*Kalmegh* (*Andrographis paniculata* Nees). Thesis, Banaras Hindu University, Varanasi, 1981.
Tripathi JS, Singh RH. Views of the therapeutic activity of *rasa* preparations with special reference to some clinical studies on the *medhya rasausadhi smrti sagara rasa*. *Sac. Ayur* 1999; **52**: 236–239.
Tripathi VN, Tewari SK, Gupta JP, Chaturvedi GN. Clinical trial of *haritaki* (*Terminalia chebula*) in treatment of simple constipation. *Sac Ayur* 1983; **35**: 733–740.
Upadhyaya SD, Kansal CM, Pandey NN. Clinical evaluation of *pippali* (*Piper longum*) *kshira paka* in patients of bronchial asthma—a preliminary study. *Nagarjun* 1982; **25**: 256–258.
Upasani VV, Mardikar BR. A study report of effect of "*Krishna tulasi*" in *rajayakshma* (pulmonary tuberculosis). pp. 103–106. *In*: BR Mardikar et al. (eds.). 1989.
Vaidya ADB. Reverse pharmacological correlates of Ayurvedic drug actions. *Ind J Pharmacol* 2006; **38**: 311–315.
Venkataraghavan S, Bhattathiri PPN, Bhagavathy Amma KC, Chandrasekharan PP. Comparative study of *Lohasava* and *Triphala mandoora* in *panduroga*. *J Res Ind Med* 1977; **12**: 108–111.
Wachasundar N. Clinical evaluation of *Phalatrikadi kwath* and *Arogyavardhini* in early hepatic cirrhosis: a case report. pp. 79–82. *In*: PH Kulkarni (ed.). 2001d.

Chapter 11

Excepients for Phytodrugs

11.1 Introduction

An excipient is a natural or synthetic substance formulated along side the active ingredient of a medication (Bhattacharyya et al. 2006). Excipients are primarily used as diluents, binders, disintegrants, adhesives, glidants, and sweeteners in conventional dosage forms like tablets and capsules (USP 1992).

As the establishment of toxicity and approval from regulatory authorities poses a problem with synthetic excipients, of late more interest is being shown by researchers in herbal excipients.

11.2 Excipients for Phytodrugs (Shirwaikar et al. 2008)

For pills prepared from herbal or phytodrugs, an excipient can be solid or liquid. The basic purpose of using an excipient is to prevent the disintegration of the ingredients.

11.2.1 Gums

Gum Arabic in powder form is a good excipient. Gum Arabic imparts some hardness to the pills. Gum tragacanth is a good substitute for Gum Arabic. Wax, fat, oil, and creosote should be avoided with Gum Arabic. Acacia is mainly used in oral and topical pharmaceutical formulations as a suspending and emulsifying agent, often in combination with tragacanth. It is also used in the preparation of pastilles and lozenges and as a tablet binder.

11.2.2 Alcohol

Alcohol renders resinous substances soft. Pills where alcohol is used as an excipient should be prepared quickly otherwise they become fragile.

11.2.3 Glycerine

Pills where glycerine is used as an excipient remain soft. However, the hygroscopic character of glycerine is a major fault. This fault can be removed if water is added to 1/3rd weight of glycerine.

11.2.4 Licorice

Licorice is used as an excipient for pills containing oil or phenol.

11.2.5 Marshmallow

Marshmallow is used as an excipient for pills containing oil or phenol.

11.2.6 Proctor's paste

Proctor's paste contains a definite ratio of gum tragacanth, glycerine, and water. Proctor's paste is considered to be a useful excipient.

11.2.7 Soap powder

Soap powder is an excellent and appropriate excipient for pills containing herbal powders, extracts, and gum-resins. Soap powder renders pills soft and non-fragile.

11.2.8 Water

Water is an excellent excipient for preparing pills from powdered opium.

11.2.9 Wax

Wax is an excellent excipient for pills containing camphor, creosote, phenol, and volatile oils.

11.3 Excipients in Ghanaian Herbal Medicine

The dried seeds/fruits of *Aframomum melegueta*, *Piper guineense*, *Xylopia aethiopica*, and *Monodora myristica* are used as an excipients in Ghanaian herbal medicine. *A. melegueta* due to antioxidant properties can be used as a preservative in herbal preparations. Aromatic and pungent compounds have been identified in all the plants. Researchers are of the view that taste may play critical role in the use of the medicinal plants as excipients (Freiesleben et al. 2015).

11.4 Committee on Herbal Medicinal Products (HMPC) and Excipients in Herbal Teas

The pharmaceutical legislation does not provide the particulars for excipients' use in traditional herbal medicinal products as the quality aspect of the medicinal product is independent of its traditional use. The legislation does not impose any limitation to the number/percentage of excipients in herbal teas. However, the scientific opinion of the HMPC is that:

- usually no more than three excipients should be used in an herbal tea (more than three excipients imply technical obstacles in terms of quality testing) and excipients should not represent more than 30% of the total weight.
- more than three excipients or more than 30% of the total weight in a herbal tea would not raise concerns from a public health viewpoint provided that the marketing authorisation holder/traditional use registration holder can control the quality of the product and that appropriate justification on the need for more than three excipients is given by the marketing authorisation holder/traditional use registration holder (European Medical Agency 2013).

Further Readings

Bhattacharyya L, Schuber S, Sheehan C, William R. Excipients: background/introduction. *In*: A Katdare and M Chaubal (eds.). Excipient Development for Pharmaceutical, Biotechnology, and Drug Delivery Systems. USA CRC Press, 2006.

Freiesleben SH, Soelberg J, Jäger AK. Medicinal plants used as excipients in the history in Ghanaian herbal medicine. *J Ethnopharmacol* 2015; **174**: 561–8.

Shirwaikar A, Shirwaikar A, Prabhu SL, Kumar GA. Herbal excipients in novel drug delivery systems. *Indian J Pharm Sci* 2008; **70**: 415–22.

http://www.ema.europa.eu/docs/en_GB/document_library/Other/2011/03/WC500104038.pdf.

USP Subcommittee on excipients. *Pharm Forum* 1992; **18**: 4387.

Chapter 12

Certifications for Phytodrug Industry

12.1 Certifications

12.1.1 Hazard Analysis Critical Control Point (HACCP) Certification

Hazard Analysis Critical Control Point (HACCP) Certification is a prevention system that helps in analyzing food processes and determining its possible hazards (physical, chemical, and biological). Hazard Analysis Critical Control Point certification is applicable to Food, Herbal, and Agriculture Industry. As far as the herbal or phytodrug industry is concerned, HACCP is applicable to manufacturers of prescription and non-prescription drugs and remedies.

12.1.2 The Kosher certification

The Kosher symbol on a label represents more than a product that conforms to rigorous religious standards. It is valued as an independent verification mark of quality, integrity, and purity and is a powerful safeguard-likened by some in the industry to the famous Good. For all those requiring Kosher food products for religious or other reasons, Kosher certification is essential.

> The demand for Kosher certified products has burgeoned spectacularly over the years, with U.S. consumers forking over $195 billion (nearly 40% of all foods sold) on Kosher certified products, as compared to $1.65 billion in 1987.

12.1.3 Halal certification

Halal certification is vitally important to Muslim consumers interested in reputable and distinguishable differences in the packaging and presentation of Halal products, assuring that what they consume is handled within Islamic guidelines and requirements. In 2005, the Halal Journal reported a $150 billion market value for the global Halal food industry, embracing approximately 1.4 billion Muslims, in addition to millions of non-Muslims who choose to eat Halal certified products.

12.1.4 AYUSH Mark

Quality Council of India has been engaged in voluntary certification of quality of Ayurveda, Siddha and Unani (ASU) products. Through this scheme drug manufacturers are awarded the quality seal to the products on the basis of third party evaluation of the quality, subject to fulfillment of the regulatory requirements. AYUSH Standard and AYUSH Premium Marks are awarded for products moving into the domestic and international market respectively. 146 ASU products are reported to have been awarded the AYUSH Premium Mark and 97 products the AYUSH Standard Mark.

12.2 Botanical Drug Development

The Food and Drug Administration (FDA) has announced the availability of a draft guidance for industry entitled "Botanical Drug Development." This guidance describes FDA's current thinking on appropriate development plans for botanical drugs to be submitted in new drug applications (NDAs) and specific recommendations on submitting investigational new drug applications (INDs) in support of future NDA submissions for botanical drugs. In addition, this guidance provides general information on the over-the-counter (OTC) drug monograph system for botanical drugs. Although this guidance does not intend to provide recommendations specific to botanical drugs to be marketed under biologics license applications (BLAs), many scientific principles described in this guidance may also apply to these products. This draft guidance revises the guidance for industry entitled "Botanical Drug Products" issued in June 2004.

Chapter 13

Dietary Supplement Health and Education (DSHEA)

13.1 What is DSHEA?

The Dietary Supplement Health and Education Act of 1994, which spells out regulations regarding the manufacture and sale of dietary supplements, defines a dietary supplement as "a product (other than tobacco) intended to supplement the diet that bears or contains one of more of the following dietary ingredients: a vitamin, a mineral, an herb or other botanical, an amino acid, a dietary substance for use by man to supplement the diet by increasing the total dietary intake; or a concentrate, metabolite, constituent, extract, or combination of any ingredient noted in clause (A), (B), (C), (D), or (E)."

The act defines permissible labeling claims and places the burden of proof on the Food and Drug Administration to show that a product is unsafe. It also outlines safety requirements for new dietary ingredients.

13.2 DSHEA Act in Detail

An Act to amend the Federal Food, Drug, and Cosmetic Act to establish standards with respect to dietary supplements, and for other purposes.

§1. Short Title; Reference; Table Of Contents

(a) Short Title

This Act may be cited as the "Dietary Supplement Health and Education Act of 1994".

(b) Reference

Whenever in this Act an amendment or repeal is expressed in terms of an amendment to, or repeal of, a section or other provision, the reference shall be considered to be made to a section or other provision of the Federal Food, Drug, and Cosmetic Act.

(c) Table of Contents

The table of contents of this Act is as follows:

§2. Findings

Congress finds that –

(1) improving the health status of United States citizens ranks at the top of the national priorities of the Federal Government;

(2) the importance of nutrition and the benefits of dietary supplements to health promotion and disease prevention have been documented increasingly in scientific studies;

(3) (A) there is a link between the ingestion of certain nutrients or dietary supplements and the prevention of chronic diseases such as cancer, heart disease, and osteoporosis; and

(B) clinical research has shown that several chronic diseases can be prevented simply with a healthful diet, such as a diet that is low in fat, saturated fat, cholesterol, and sodium, with a high proportion of plant-based foods;

(4) healthful diets may mitigate the need for expensive medical procedures, such as coronary bypass surgery or angioplasty;

(5) preventive health measures, including education, good nutrition, and appropriate use of safe nutritional supplements will limit the incidence of chronic diseases, and reduce long-term health care expenditures;

(6) (A) promotion of good health and healthy lifestyles improves and extends lives while reducing health care expenditures; and

(B) reduction in health care expenditures is of paramount importance to the future of the country and the economic well-being of the country;

(7) there is a growing need for emphasis on the dissemination of information linking nutrition and long-term good health;

(8) consumers should be empowered to make choices about preventive health care programs based on data from scientific studies of health benefits related to particular dietary supplements;

(9) national surveys have revealed that almost 50 percent of the 260,000,000 Americans regularly consume dietary supplements of vitamins, minerals, or herbs as a means of improving their nutrition;

(10) studies indicate that consumers are placing increased reliance on the use of nontraditional health care providers to avoid the excessive costs of traditional medical services and to obtain more holistic consideration of their needs;

(11) the United States will spend over $1,000,000,000,000 on health care in 1994, which is about 12 percent of the Gross National Product of the United States, and this amount and percentage will continue to increase unless significant efforts are undertaken to reverse the increase;

(12) (A) the nutritional supplement industry is an integral part of the economy of the United States;
(B) the industry consistently projects a positive trade balance; and
(C) the estimated 600 dietary supplement manufacturers in the United States produce approximately 4,000 products, with total annual sales of such products alone reaching at least $4,000,000,000;

(13) although the Federal Government should take swift action against products that are unsafe or adulterated, the Federal Government should not take any actions to impose unreasonable regulatory barriers limiting or slowing the flow of safe products and accurate information to consumers;

(14) dietary supplements are safe within a broad range of intake, and safety problems with the supplements are relatively rare; and

(15) (A) legislative action that protects the right of access of consumers to safe dietary supplements is necessary in order to promote wellness; and
(B) a rational Federal framework must be established to supersede the current ad hoc, patchwork regulatory policy on dietary supplements.

13.3 BMPEA in Dietary Supplements

BMPEA is a substance that does not meet the statutory definition of a dietary ingredient. The Federal Food, Drug, and Cosmetic Act defines a dietary ingredient as a vitamin; mineral; herb or other botanical; amino acid; dietary substance for use by man to supplement the diet by increasing the total dietary intake; or a concentrate, metabolite, constituent, extract, or combination of the preceding substances. BMPEA is none of these, rendering misbranded any products that declare BMPEA as a dietary supplement.

The FDA issued warning letters to five companies regarding a total of eight products for which the product labeling lists BMPEA as a dietary ingredient. Two of the companies further identified the source of this stimulant as the botanical *Acacia rigidula*. Under existing law, including the Dietary Supplement

Health and Education Act passed by Congress in 1994, the FDA can take action to remove products from the market, but the agency must first establish that such products are adulterated (e.g., that the product is unsafe) or misbranded (e.g., that the labeling is false or misleading). While BMPEA was listed as a dietary ingredient on the product labels, the substance does not meet the statutory definition of a dietary ingredient.

The Federal Food, Drug, and Cosmetic Act defines a dietary ingredient as a vitamin; mineral; herb or other botanical; amino acid; dietary substance for use by man to supplement the diet by increasing the total dietary intake; or a concentrate, metabolite, constituent, extract, or combination of the preceding substances. BMPEA is none of these, rendering misbranded any products that declare BMPEA as a dietary supplement. Additionally, relating to the two companies that identified the botanical *Acacia rigidula* as the source of the BMPEA, research conducted by the FDA in 2013 established that BMPEA is not a constituent or extract of *Acacia rigidula*. FDA considers these specific products to be misbranded for this reason, as well.

13.4 DMBA in Dietary Supplements

DMBA is labeled as a dietary ingredient in some products marketed as dietary supplements. However, the FDA is not aware of any information demonstrating that DMBA was lawfully marketed as a dietary ingredient in the United States before October 15, 1994. As a result, for dietary supplements that contain DMBA to be lawfully marketed, one of the following must apply:

(1) the product containing the dietary ingredient must contain only dietary ingredients that have been present in the food supply as an article used in food in a form in which the food has not been chemically altered, or

(2) there must be a history of use or other evidence of safety establishing that the dietary ingredient, when used under the conditions recommended in the product labeling, will reasonably be expected to be safe; and prior to bringing the products to market, the manufacturer or distributor must notify FDA of the basis on which the manufacturer or distributor has concluded that a dietary supplement containing such dietary ingredient will reasonably expected to be safe.

Because neither of these conditions has been met by those marketing products that contain or are labeled as containing DMBA as a dietary ingredient, the FDA considers these dietary supplements to be adulterated.

Chapter 14

Acts Related to Banned or Restricted Phytoingredients

14.1 Introduction

Manufacturers, wholesalers, and sellers must ensure that any herbal products they place on the market do not contain banned ingredients. They must also ensure that restricted ingredients are used legally. Restrictions may be added or removed at any time. One must check that the ingredients are legal for a traditional herbal registration for a herbal medicine. The list of banned or restricted herbal ingredients for medicinal use has been prepared by the Medicines and Healthcare Products Regulatory Agency (MHRA).

14.2 The Medicines (Aristolochia and Mu Tong, etc.) (Prohibition) Order 2001 SI 1841

Not permissible to manufacture, import, sell, or supply any unlicensed medicine in the UK which contains the named herbal ingredients.

14.3 The Human Use Regulations 2012 No. 1916 Schedule 20 Part 1

Prohibits the sale or supply (including general retail or following a one-to-one consultation with a practitioner) of herbal medicines in the UK, if it contains one or more of the listed plants, except where sold in premises which are registered pharmacies and by or under the supervision of a pharmacist.

14.4 The Human Use Regulations 2012 No. 1916 Schedule 20 Part 2

Plants listed in Part 2 can only be sold in herbal medicines following a one-to-one consultation with a practitioner, at the dosages in column 2 or percentage of the substance in the product not exceeding that specified in column 3. If the dosage (or percentage of the substance in the product) specified is exceeded, the herbal medicines containing these plants can only be supplied in premises which are registered pharmacies and by or under the supervision of a pharmacist.

14.5 The Medicines for Human Use (Kava-kava) (Prohibition) Order 2002 SI 3170

Prohibits the sale, supply or importation of any medicine for human use which consists of or contains a plant (or part of a plant) belonging to the species Piper methysticum (known as Kava-kava) or an extract from such a plant, except those for external use only.

14.6 The Medicines for Human Use (Prohibition) (Senecio and Miscellaneous Amendments) Order 2008

Prohibits the sale, supply, or importation of any medicine for human use which consists of or contains a plant, or part of a plant, belonging to the species Senecio or an extract from such a plant.

Table 14.1: Banned and restricted herbal ingredients.

Botanical source	**Common name**	**Legal category**	**Maximum dose where permitted for internal use only. Max dose (MD)1, max daily dose (MDD)2**	**Maximum dose where permitted for external use only. Percentage (%)**
1	**2**	**3**	**4**	**5**
All *Aconitum* species*	Aconite	POM and SI 2130 – Parts II & III	No permitted dose unless made available by a prescription from a registered doctor or dentist	1.3% or below
Adonis vernalis	Adonis	SI 2130 – Parts II and III	100 mg (MD), 300 mg (MDD)	No dose permitted
Akebia quinata	Mu tong	SI 1841	Prohibited in all unlicensed medicines	Prohibited in all unlicensed medicines
Akebia trifoliata	Mu tong	SI 1841	Prohibited in all unlicensed medicines	Prohibited in all unlicensed medicines
Apocynum cannabinium	Canadian Hemp	SI 2130 – Part I	Can only be sold in premises which are registered pharmacies and by or under the supervision of a pharmacist	Can only be sold in premises which are registered pharmacies and by or under the supervision of a pharmacist
Areca catechu	Areca	SI 2130 – Part I	do	do

Table 14.1 contd....

Table 14.1 contd....

1	2	3	4	5
Aristolochia species**	Mu tong; Fangji; Birthwort; Long Birthwort; Indian Birthwort	SI 1841	Prohibited in all unlicensed medicines	Prohibited in all unlicensed medicines
Artemisia cina	Santonica	SI 2130 – Part I	Can only be sold in premises which are registered pharmacies and by or under the supervision of a pharmacist	Can only be sold in premises which are registered pharmacies and by or under the supervision of a pharmacist
Atropa belladonna	Belladonna herb	SI 2130 – Parts II and III	30 mg (MD), 90 mg (MDD)	do
Atropa acuminata	Belladonna herb	SI 2130 – Parts II and III	30 mg (MD), 90 mg (MDD)	do
Aspidosperma quebrachoblanco	Quebracho	SI 2130 – Part II and III	50 mg (MD), 150 mg (MDD)	do
Brayera anthelmintica	Kousso	SI 2130 – Part I	Can only be sold in premises which are registered pharmacies and by or under the supervision of a pharmacist	Can only be sold in premises which are registered pharmacies and by or under the supervision of a pharmacist
Catha edulis	Catha	SI 2130 – Part I	do	do
Chelidonium majus	Celandine	SI 2130 – Part II and III	2 g (MD), 6 g (MDD)	do
Chenopodium ambrosioides var *anthelminticum*	Chenopodium	SI 2130 – Part I	Can only be sold in premises which are registered pharmacies and by or under the supervision of a pharmacist	do
Cinchona species***	Cinchona bark	SI 2130 – Part II and III	250 mg (MD), 750 mg (MDD)	do
Clematis armandii	Mu tong	SI 1841	Not permitted in any unlicensed medicines	Not permitted in any unlicensed medicines
Clematis montana	Mu tong	SI 1841	do	do
Claviceps purpurea	Ergot	POM	Can only be made available by a prescription from a registered doctor or dentist	Can only be made available by a prescription from a registered doctor or dentist
Cocculus indicus		POM	do	do

Table 14.1 contd....

Table 14.1 contd....

1	2	3	4	5
Cocculus laurifolius	Fangji	SI 1841	Not permitted in any unlicensed medicines	Not permitted in any unlicensed medicines
Cocculus orbiculatus	Fangji	SI 1841	do	do
Cocculus trilobus	Fangji	SI 1841	do	do
Colchicum autumnale	Colchicum corm	SI 2130 – Part II and III	100 mg (MD), 300 mg (MDD)	Can only be sold in premises which are registered pharmacies and by or under the supervision of a pharmacist
Conium maculatum	Conium leaf and fruits	POM and SI 2130 – Parts II and III	No permitted dose unless made available by a prescription from a registered doctor or dentist	7.0% and below
Convallaria majalis	Lilly of the valley	SI 2130 – Parts II and III	150 mg (MD), 450 mg (MDD)	Can only be sold in premises which are registered pharmacies and by or under the supervision of a pharmacist
Crotalaria berberoana		SI 2130 – Part I	Can only be sold in premises which are registered pharmacies and by or under the supervision of a pharmacist	Can only be sold in premises which are registered pharmacies and by or under the supervision of a pharmacist
Crotalaria spectabilis		2130 – Part I	do	do
Curcurbita maxima	Cucurbita	SI 2130 – Part I	do	do
Datura stramonium	Stramonium	SI 2130 – Parts II and III	50 mg (MD), 150 mg (MDD)	do
Datura innoxia	Stramonium	SI 2130 – Parts II and III	50 mg (MD), 150 mg (MDD)	do
Digitalis leaf, Digitalis prepared	Foxglove	POM	Can only be sold in premises which are registered pharmacies and by or under the supervision of a pharmacist	Can only be sold in premises which are registered pharmacies and by or under the supervision of a pharmacist
Dryopteris filix-mas	Male fern	SI 2130 – Part I	do	do

Table 14.1 contd....

Table 14.1 contd....

1	2	3	4	5
Duboisia myoporoides	Duboisia	SI 2130 – Part I	do	do
Duboisia leichardtii	Duboisia	SI 2130 – Part I	do	do
Ecballium elaterium	Elaterium	SI 2130 – Part I	do	do
Embelia ribes	Embelia	SI 2130 – Part I	do	do
Embelia robusta	Embelia	SI 2130 – Part I	do	do
Ephedra species****	Ephedra	SI 2130 – Parts II and III	600 mg (MD), 1800 mg (MDD)	do
Erysimum canescens	Erysimum	SI 2130 – Part I	Can only be sold in premises which are registered pharmacies and by or under the supervision of a pharmacist	Can only be sold in premises which are registered pharmacies and by or under the supervision of a pharmacist
Gelsemium sempervirens	Gelsemium	SI 2130 – Parts II and III	25 mg (MD), 75 mg (MDD)	do
Holarrhena antidysenterica	Holarrhena	SI 2130 – Part I	Can only be sold in premises which are registered pharmacies and by or under the supervision of a pharmacist	Can only be sold in premises which are registered pharmacies and by or under the supervision of a pharmacist
Hyoscyamus niger	Hyoscyamus	SI 2130 – Parts II and III	100 mg (MD), 300 mg (MDD)	do
Hyoscyamus albus	Hyoscyamus	SI 2130 – Parts II and III	100 mg (MD), 300 mg (MDD)	do
Hyoscyamus muticus	Hyoscyamus	SI 2130 – Parts II and III	100 mg (MD), 300 mg (MDD)	do
Juniperus sabina	Savin	SI 2130 – Part I	Can only be sold in premises which are registered pharmacies and by or under the supervision of a pharmacist	do
Lobelia inflata	Lobelia	SI 2130 – Parts II and III	200 mg (MD), 600 (MDD)	do
Mallotus philippinensis	Kamala	SI 2130 – Part I	Can only be sold in premises which are registered pharmacies and by or under the supervision of a pharmacist	do

Table 14.1 contd....

Table 14.1 contd....

1	2	3	4	5
Mandragora autumnalis	Mandrake	POM	do	do
Papaver somniterum	Poppy capsule	POM	do	do
Pausinystalia yohimbe	Yohimbe bark	SI 2130 – Part 1	do	do
Pilocarpus jaborandi	Jaborandi	SI 2130 – Parts II and III	do	5.0% or below
Pilocarpus microphyllus	Jaborandi	SI 2130 – Parts II and III	do	5.0% or below
Piper methysticum	Kava-kava	SI 3170	Not permitted in unlicensed medicines, except those exclusively for external use	Not permitted in unlicensed medicines, except those exclusively for external use
Podophyllum hexandrum	Indian podophyllum	POM	Can only be made available via a prescription from a registered doctor or dentist	Can only be made available via a prescription from a registered doctor or dentist
Rauwolfia serpentina	Indian snakeroot	POM	do	do
Rauwolfia vomitoria	African serpentwood	POM	do	do
Rhus radicans	Poison Ivy	SI 2130 – Part I	do	do
Rhus toxicodendron	Poison Oak	SI 2130 – Parts II and II	do	do
Schoenocaulon officinale	Sabadilla; Cevadilla	POM	do	do
Scopolia carniolica	Scopolia	SI 2130 – Part I	do	do
Scopolia japonica	Scopolia	SI 2130 – Part I	do	do
Stephania tetrandra		SI 1841	Not permitted in any unlicensed medicines	Not permitted in any unlicensed medicines
Strophanthus species*****	Strophanthus	SI 2130 – Part I	Can only be sold in premises which are registered pharmacies and by or under the supervision of a pharmacist	Can only be sold in premises which are registered pharmacies and by or under the supervision of a pharmacist
Strychnos ignati	Ignatius bean	POM	Can only be made available via a prescription from a registered doctor or dentist	Can only be made available via a prescription from a registered doctor or dentist

Table 14.1 contd....

Table 14.1 contd....

1	2	3	4	5
Strychnos cuspida	Ignatius bean	POM	Can only be made available via a prescription from a registered doctor or dentist	Can only be made available via a prescription from a registered doctor or dentist
Strychnos nux vomica	Poison Nut	do	do	do
Ulmus fulva	Slippery Elm Bark (whole or unpowdered)	SI 2130 – Part I	do	do
Ulmus rubra	Slippery Elm Bark (whole or unpowdered)	SI 2130 – Part I	do	do
Veratrum viride	Indian Poke	POM	do	do
Veratrum album	White Hellebore	POM	do	do

**A. napellus, A. stoerkianum, A. uncinatum* var *japonicum, A. deinorrhizum, A. balfourii, A. chasmanthum, A. spicatum, A. lycoctonum*

** *A. clematis, A. contorta, A. debelis, A. fang-chi, A. manshuriensis, A. serpentaria*

****C. calisaya, C. ledgerana, C. officinalis, C. succirubra, C. micrantha*

*****E. sinica, E. equisetina, E. distachya, E. intermedia, E. gerardiana*

******S. kombe, S. courmonti, S. nicholsoni, S. gratus, S. emini, S. sarmentosus, S. hispidus*

Note:

1. Maximum dose or 'MD' means the maximum quantity of the substance contained in the amount of the medicinal product for internal use, which it is recommended should be taken or administered at any one time.
2. Maximum daily dose or 'MDD' means the maximum quantity of the substance contained in the amount of the medicinal product for internal use, which it is recommended should be taken or administered in any period of 24 hours.

Chapter 15

Comfrey Based Herbal Products

15.1 Introduction

Comfrey is a common name for plants in the genus *Symphytum* (Boraginaceae). Pyrrolizidine alkaloids (PAs) exist in many plants and many of them cause liver toxicity and/or cancer in humans and experimental animals. Comfrey contains as many as 14 pyrrolizidine alkaloids (PA), including 7-acetylintermedine, 7-acetyllycopsamine, echimidine, intermedine, lasiocarpine, lycopsamine, myoscorpine, symlandine, symphytine, and symviridine (Mei et al. 2010).

15.2 *Symphytum officinale* L. (Common Comfrey)

S. officinale is a perennial shrub that is native to Europe and some parts of Asia. Fond of moist soils, comfrey has a thick, hairy stem, and grows 2–5 feet tall. Its flowers are dull purple, blue or whitish, and densely arranged in clusters. The leaves are oblong, and often look different depending on where they are on the stem: Lower leaves are broad at the base and tapered at the ends while upper leaves are broad throughout and narrow only at the ends. The root has a black outside and fleshy whitish inside filled with juice.

15.3 *Symphytum asperum* Lepech. (Prickly Comfrey)

S. asperum is a coarse, hairy, rhizomatous perennial that is typically grown in shaded wildflower areas or naturalized areas for its attractive foliage and spring flowers. It grows to 3–4' tall. Ovate to elliptic leaves (to 2–8" long) are dark green and prickly hairy. Mature stems are not winged (leaf bases are not decurrent as is the case with *Symphytum officinale*. Small tubular flowers (each to 1/2" long) in scorpioid cymes open rose-pink in spring but mature to blue or purple. Flowers bloom May to August.

15.4 *Symphytum x uplandicum* Nyman (Russian Comfrey)

S. x uplandicum, is a naturally occurring hybrid (*S. officinale* x *S. asperum*), which typically grows in an upright clump to 18–24" tall with flower stems to 4–5'. It features large, oblong to elliptic-lanceolate, medium green leaves (to 14" long at the plant base) and bell-shaped, blue bell-like flowers that appear in drooping clusters (scorpiod cymes) from mid-spring to early summer. From pinkish buds, flowers open up rose but mature to purple.

15.5 Comfrey and FDA

Concern about comfrey-containing and particularly pyrrolizidine alkaloids-containing herbal supplements has been a foremost concern of the herbal products industry for many years and most manufacturers have already taken steps to remove these products from store shelves. In July 1996, the American Herbal Products Association Board of Trustees recommended that all botanical ingredients containing toxic pyrrolizidine alkaloids, including comfrey, display a cautionary statement that directs consumers to only use the products topically only when the skin is free of abrasions or cuts (No authors listed 2010).

The letter issued by FDA, advised herbal supplement manufacturers to remove products currently on the market that may contain comfrey and to ask customers who may be using them to stop immediately (FDA 2001). In addition to common comfrey (*Symphytum officinale*), the warning extends to other types of comfrey, including prickly comfrey (*Symphytum asperum*) and Russian comfrey (*Symphytum x uplandicum*). The FDA also ask manufacturers to report any adverse events that may be related to the consumption of comfrey products (Klepser and Klepser 1999).

15.6 The Efficacy and Safety of Comfrey

The mechanisms by which toxicity and mutagenicity are conveyed are still not fully understood, but seem to be mediated through a toxic mechanism related to the biotransformation of alkaloids by hepatic microsomal enzymes. This produces highly reactive pyrroles which act as powerful alkylating agents. The main liver injury caused by comfrey is veno-occlusive disease, a non-thrombotic obliteration of small hepatic veins leading to cirrhosis and eventually liver failure. Patients may present with either acute or chronic clinical signs with portal hypertension, hepatomegaly, and abdominal pain as the main features (Stickel and Seitz 2000).

15.7 Analysis of Herbal Teas made from the Leaves of Comfrey

Comfrey leaves were purchased from three commercial sources and used to prepare tea in a manner consistent with the methods used by consumers. The concentration of symphytine and echimidine varied considerably between teas prepared from leaves purchased from the different vendors of plant material. Moreover, a much higher concentration of symphytine was found in the tea when steps were included to reduce N-oxides prior to analysis. The treatment of pure symphytine with hot water did not generate the N-oxide derivative *de novo*. Since the pyrrolizidine alkaloids are known to be hepatotoxic, consumption of herbal teas made from comfrey leaves may be ill-advised (Oberlies et al. 2004).

15.8 Metabolism, Genotoxicity, and Carcinogenicity of Comfrey

The mechanisms underlying comfrey-induced genotoxicity and carcinogenicity are still not fully understood. The available evidence suggests that the active metabolites of PA in comfrey interact with DNA in liver endothelial cells and hepatocytes, resulting in DNA damage, mutation induction, and cancer development. Genotoxicities attributed to comfrey and riddelliine (a representative genotoxic PA and a proven rodent mutagen and carcinogen) are discussed in this review. Both of these compounds induced similar profiles of 6,7-dihydro-7-hydroxy-1-hydroxymethyl-5H-pyrrolizine (DHP)-derived DNA adducts and similar mutation spectra (Mei et al. 2010).

Further Readings

http://www.fda.gov/Food/RecallsOutbreaksEmergencies/SafetyAlertsAdvisories/ucm111219.htm, 2001.

Klepser TB, Klepser ME. Unsafe and potentially safe herbal therapies. *Am J Health Syst Pharm* 1999; **56**: 125–38; quiz 139–41.

Mei N, Guo L, Fu PP, Fuscoe JC, Luan Y, Chen T. Metabolism, genotoxicity, and carcinogenicity of comfrey. *J Toxicol Environ Health B Crit Rev* 2010; **13** :509–26.

No authors listed. Dangerous supplements: what you don't know about these 12 ingredients could hurt you. *Consum Rep* 2010; **75**: 16–20.

Oberlies NH, Kim NC, Brine DR, Collins BJ, Handy RW, Sparacino CM, Wani MC, Wall ME. Analysis of herbal teas made from the leaves of comfrey (*Symphytum officinale*): reduction of N-oxides results in order of magnitude increases in the measurable concentration of pyrrolizidine alkaloids. *Public Health Nutr* 2004; **7**: 919–24.

Stickel F, Seitz HK. The efficacy and safety of comfrey. *Public Health Nutr* 2000; **3**: 501–8.

Chapter 16

Herbal Bioenhancers

16.1 Introduction

The concept of bioavailability enhancing is well documented in Ayurvedic texts. Bioavailability enhancers have been described as *Yogvahidravya* in Ayurveda. Honey and *Trikatu* are typical examples of bioavailability enhancers agents. *Yogvahidravya* is described as an agent that enhances the medicinal activity of other drugs without losing its own potency.

Bose (1929) is credited with explaining the mechanism of action of bioenhancers. He noted that *Piper longum* increased the anti-asthmatic property of *Adhatodavasika* leaves. Bioenhancers or bioavailability enhancers are a novel chapter in medical sciences and it was first scientifically established in 1979 after the discovery of piperine. The term was first coined by CK Atal of Regional Research Laboratory, Jammu (India), while working on certain traditional drugs.

Bioenhancers of herbal origin have demonstrated a significant role in enhancing the bioavailability and bioefficacy of antihypertensives, anticancer, antiviral, antitubercular, and antifungal drugs at low doses (Ajazuddin et al. 2014).

16.2 Definition of Bioavailability Enhancers

Bioavailability enhancers are defined as substances that increase the bioavailability and bioefficacy of active substances with which they are combined without having any activity of their own at the dose used.

16.3 Trikatu and Bioavailability Enhancing

Trikatu is an Ayurvedic poly-herbal formulation consisting of equal amounts, Krshna Maricha (*Piper nigrum*), Pippali (*Piper longum*), and Sunthi (*Zingiber officinale*). Like Triphala, Trikatu is indicated in a number of diseases. Literal meaning of Trikatu is a collection of three acrid drugs. All the three ingredients of Trikatu have appetizer and digestive activity and according to Ayurvedic pharmacology, they enhance each other's activity.

16.3.1 Effect on the bioavailability of rifampicin

Coadministration of *Trikatu* does not influence the extent of bioavailability (AUC0-infinity) but reduces the rate of bioavailability (Cmax) of rifampicin. The latter effect may reduce the efficacy of rifampicin therapy (Karan et al. 1999b).

16.3.2 Effect on the bioavailability of carbamazepine

In the animals treated with a single dose of *Trikatu*, there was a significant decrease in Tmax of carbamazepine ($P < 0.05$). Multiple doses of Trikatu also shortened the Tmax of carbamazepine although not to a statistically significant level (Karan et al. 1999c).

16.3.3 Effect on the pharmacokinetic profile of isoniazid

Coadministration of *Trikatu* significantly reduced the Cmax (5.48 ñ 0.75 'g/ml vs 8.42 ñ 0.85 'g/ml; $P < 0.05$] and AUCo-([15.04 ñ 3.64 'g/ml. hr vs 24.76 (4.03 'g/ml. hr; $P < 0.05$] of isoniazid. *Trikatu* reduces the bioavailability of isoniazid in rabbits (Karan et al. 1998).

16.3.4 Effect on the pharmacokinetic profile of indomethacin

Trikatu reduces bioavailability of indomethacin (Karan et al. 1999a).

16.3.5 Effect on the pharmacokinetic profile of diclofenac sodium

In rabbits, it was observed that *Trikatu* significantly decreased the serum levels of diclofenac sodium. It was observed that the mean percent oedema inhibition shown by the combination of *Trikatu* and diclofenac was similar to that shown by *Trikatu* alone, but significantly less than that shown by diclofenac alone (Lala et al. 2004).

16.3.6 Effect on the pharmacokinetic profile of pefloxacin

A study revealed a decreased plasma concentration ($p > 0.05$) of pefloxacin following *Trikatu* administration during the absorption phase (10, 15, 20 min post pefloxacin administration). In contrast, the plasma concentrations of pefloxacin were significantly higher at 4, 6, 8 and 12 h (during the elimination phase) of the pefloxacin administration (Dama et al. 2008).

16.3.7 Effect on the bioavailability of ampicillin and norfloxacin

A study reported that coadministration of *Trikatu* and its components reduced bioavailability of ampicillin and norfloxacin in rabbits (Janakiraman and Manavalan 2009).

16.3 *Aloe vera* L. (Aloaceae) as Bioavailability Enhancer

Aloe vera gel and the whole leaf extracts of *A. vera* have been shown to increase plasma concentration and improved absorption of ascorbic acid (vitamin C) and (tocopherols) vitamin E. The *Aloe vera* gel extract has been reported to increase the absorption of vitamin C after 24 hours. *Aloe vera* gel and the whole leaf extracts of *A. vera* resulted in the improvement of the absorption of vitamin E and prolonged its plasma concentration, especially after eight hours (Vinson et al. 2005).

16.4 *Stevia rebaudiana* Bertoni (Asteraceae) as Bioavailability Enhancer

Extracts/fractions/pure isolated constituents of *S. rebaudiana* either alone or in combination with the alkaloid, piperine have a selective role as bioavailability enhancers of drugs, nutraceuticals, and herbal drugs/formulations (Gokaraju and Gokaraju 2010).

16.5 *Zingiber officinale* Rosc (Zingiberaceae) as Bioavailability Enhancer

Zingiber officinale is used in the range of 10–30 mg/kg body weight acts as a bioenhancer. It has shown to increase the bioavailability of Azithromycin (85%), Cephalexin (85%), Cefadroxil (65%), Amoxycillin (90%), Cloxacillin (90%), and Erythromycin (105%) (Drabu et al. 2011).

16.6 *Glycyrrhiza glabra* L. (Fabaceae) as Bioavailability Enhancer

Glycyrrhizin present in *G. glabra* has been reported to increase the bioavailability of taxol, the anticancer molecule in the breast cancer cell line. The inhibition of cancerous cell growth by taxol in the presence of glycyrrhizin was higher than treatment with taxol alone. Further, glycyrrhizin is reported to enhance the transport of antibiotics (ampicillin, nalidixic acid, rifampicin, and tetracycline) and vitamins (B_1 and B_{12}) across the gut membrane (Drabu et al. 2011).

16.7 *Allium sativum* L. (Alliaceae) as Bioavailability Enhancer

Allicin, the active bioenhancing agent in garlic enhances the fungicidal activity of the antifungal agent, Amphotericin B against *Candida albicans*, *Aspergillus fumigates*, and *Saccharomyces cerevisiae*. Amphotericin B exhibited enhanced antifungal activity against *S. cerevisiae* when given with allicin (Ogita et al. 2012).

16.8 *Sinomenium acutum* Thunb (Menispermaceae) as Bioavailability Enhancer

Co-administration of paeoniflorin (monoterpene glucoside having a poor absorption rate and a very low bioavailability (3–4%) when administered orally) with sinomenine (an alkaloid from *S. acutum*) dramatically altered the pharmacokinetic behaviors of paeoniflorin in rats (Liu et al. 2005).

16.9 *Cuminum cyminum* L. (Apiaceae) as Bioavailability Enhancer

Bioavailability/bioefficacy enhancing activity of *C. cyminum* has been revealed toward a number of drugs (Qazi et al. 2009). Volatile oil and flavonoids including luteolin are supposed to bioavailability enhancers. Luteolin especially has been demonstrated to be a potent P-glycoprotein inhibitor in the literature (Boumendjel et al. 2002).

16.10 *Moringaoleifera* L. (Moringaceae) as Bioavailability Enhancer

Niaziridin, the nitrile glycoside enhanced the absorption of antibiotics (rifampicin, tetracycline, and ampicillin), vitamins, and nutrients (Khanuja et al. 2006).

Further Readings

Ajazuddin, Alexander A, Qureshi A, Kumari L, Vaishnav P, Sharma M, Saraf S, Saraf S. Role of herbal bioactives as a potential bioavailability enhancer for Active Pharmaceutical Ingredients. *Fitoterapia* 2014; **97**: 1–14.

Boumendjel A, Di Pietro A, Dumontet C, Barron D. Recent advances in the discovery of flavonoids and analogs with high-affinity binding to P-glycoprotein responsible for cancer cell multidrug resistance. *Med Res Rev* 2002; **22**: 512–529.

Dama MS, Varshneya C, Dardi MS, Katoch VC. Effect of *trikatu* pre-treatment on the pharmacokinetics of pefloxacin administered orally in mountain Gaddi goats. *J Vet Sci* 2008; **9**: 25–9.

Drabu S, Khatri S, Babu S, Lohani P. Use of herbal bioenhancers to increase the bioavailability of drugs. *Res J Pharm Biol Chem Sci* 2011; **2**: 107.

Gokaraju GR, Gokaraju RR. Bioavailability/bioefficacy enhancing activity of Stevia rebaudiana and extracts and fractions and compounds thereof. Date of Publication: 06/05/2010. United States Patent Number, US20100112101A1, 2010.

Janakiraman K, Manavalan R. Studies on effect of coadministration of Trikatu and its components on oral bioavailability of Ampicillin and Norfloxacin in rabbits. *J Pharm Res* 2009; **2**: 27–30.

Karan RS, Bhargava VK, Garg SK. Effect of *trikatu* (piperine) on the pharmacokinetic profile of isoniazid in rabbits. *Indian J Pharmacol* 1998; **30**: 254–6.

Karan RS, Bhargava VK, Garg SK. Effect of *Trikatu* on the pharmacokinetic profile of indomethacin. *Indian J Pharmacol* 1999; **31**: 160–1.

Karan RS, Bhargava VK, Garg SK. Effect of *Trikatu*, an Ayurvedic prescription, on the pharmacokinetic profile of rifampicin in rabbits. *J Ethnopharmacol* 1999; **64**: 259–64.

Karan RS, Bhargava VK, Garg SK. Effect of *Trikatu*, an Ayurvedic prescription, on the pharmacokinetic profile of carbamazepine in rabbits. *Indian J Physiol Pharmacol* 1999; **43**: 133–6.

Khanuja SPS, Arya JS, Tiruppadiripuliyur RSK, Saikia D, Kaur H, Singh M. Nitrile glycoside useful as a bioenhancer of drugs and nutrients, process of its isolation from *Moringa oleifera*. 2006. United States Patent 6.

Lala LG, D'Mello PM, Naik SR. Pharmacokinetic and pharmacodynamic studies on interaction of "*trikatu*" with diclofenac sodium. *J Ethnopharmacol* 2004; **91**: 277–280.

Liu ZQ, Zhou H, Liu L, Jiang ZH, Wong YF, Xie Y, Cai X, Xu HX, Chan K. Influence of co-administrated sinomenine on pharmacokinetic fate of paeoniflorin in unrestrained conscious rats. *J Ethnopharmacol* 2005; **13**: 61–67.

Ogita A, Fujita K, Tanaka T. Enhancing effects on vacuole-targeting fungicidal activity of amphotericin B. *Front Microbiol* 2012; **3**: 100.

Qazi GN, Bedi KL, Johri RK, Tikoo MK, Tikoo AK, Sharma SC, Abdullah ST. Bioavailability/bioefficacy enhancing activity of *Cuminum cyminum* and extracts and fractions thereof. 2009. United States Patent 7514105.

Vinson JA, Al Kharrat H, Andreoli L. Effect of *Aloe vera* preparations on the human bioavailability of vitamins C and E. *Phytomedicine* 2005; **12**: 760–765.

PART B
Bioactives from Phytomedicine

Chapter 17

Phytosteroids and Related Compounds

17.1 Introduction

In higher plants, the first sterol was isolated by Hesse in 1878 from the Calabar beans (*Phytostigma venenosum*) which coined the term "phytosterine" (Cook 1958). This substance was later named stigmasterol by Windaus and Hault in 1906 from the plant genus (Cook 1958). The denomination **"phytosterol"** was proposed by Thomsin 1897 for all sterols of vegetal origin.

The word phytosteroid originate from *phyto-+steroid.* It means any steroid of plant origin. In plants, steroids play an important role as constituents of cell membranes, insect deterrents, and growth hormones (Makin et al. 1995). Phytosteroids are natural hormones precursors.

17.2 Phytosteroid Synthesis

Sterols are mostly formed from mevalonic acid. Mavalonate is changed to squalene. The latter undergoes cyclisation to produce cycloartenol in plants and lanosterol in animals. The cyclic product then gives rise to sterol. The sterol biosynthetic process is outlined below:

The chemical reactions and pathways result in the formation of phytosteroids, steroids of higher plants differing from animal steroids in having substitutions at C24 and/or a double bond at C22.

17.3 Phytosterols

Phytosterols, which encompass plant sterols and stanols, are steroid compounds similar to cholesterol, which occur in plants and vary only in carbon side chains and/or in the presence or absence of a double bond.

Sterols are a family of molecules with a specific shape and structure. Phytosterols are sterols found in plants. The sterols found in animals are called zoosterols and the best-known of these is cholesterol.

Stanol ester is a heterogeneous group of chemical compounds known to reduce the level of low-density lipoprotein (LDL) cholesterol in blood when ingested, though to a much lesser degree than prescription drugs such as statins. The starting material is phytosterols from plants.

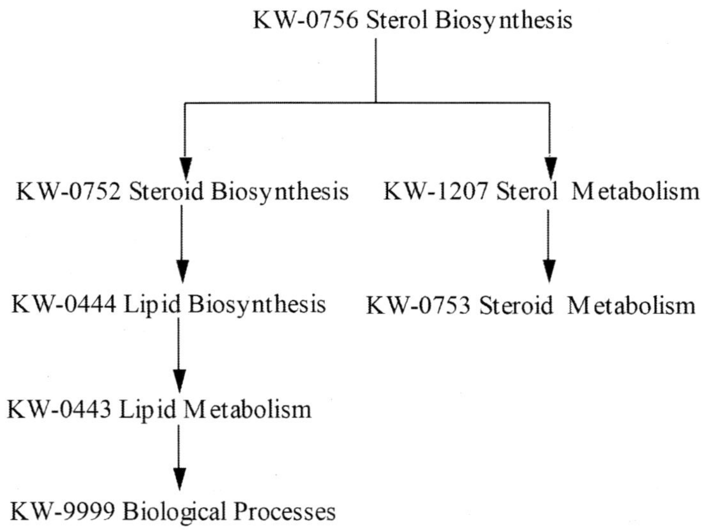

Fig. 17.1: Pathway leading to sterol biosynthesis.

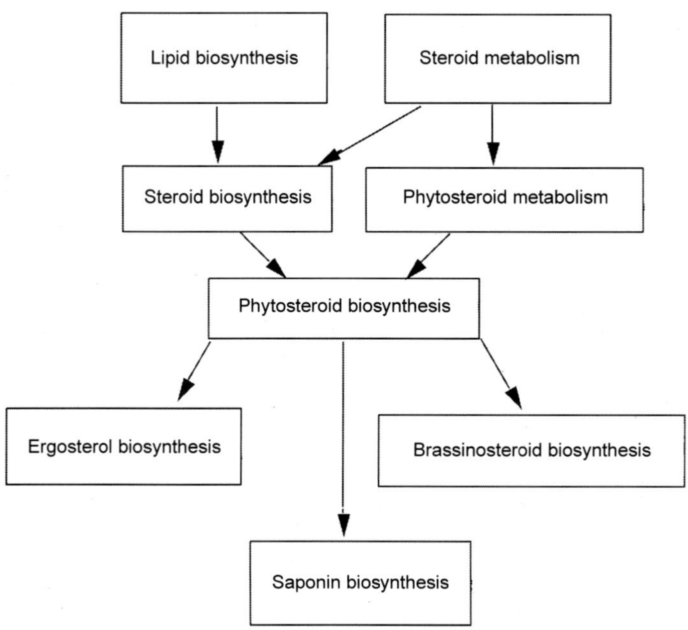

Fig. 17.2: Chemical pathways in biosynthesis of phytosterols and higher plant steroids.

17.4 Classification of Plant Sterols

Plant sterols are plant equivalents of cholesterol and have a very similar molecular structure. According to their structure, they can be divided into sterols and stanols, stanols being a saturated subgroup of sterols.

Sources of plant sterols (Tikkanen 2005)

Cereals

- Bran
- Brown rice
- Oat bran
- Rice bran
- Wheat germ
- Whole wheat

Legumes

- Dried beans
- Dried peas
- Lentils

Nuts and Seeds

- Almonds
- Peanuts
- Pecans
- Pumpkin seeds
- Sesame seeds
- Sunflower seeds
- Walnuts

Fruits and Vegetables

- Apples
- Avocados
- Blueberries
- Broccoli
- Brussels sprouts
- Cauliflower
- Dill

- Tomato
- Vegetable oils
- Wheat germ oil

Fortified Foods

- Cookies
- Energy bars
- Margarine
- Orange juice
- Yogurt drinks

17.5 Absorption of Sterols

Because of the structural differences, phytosterols are not well absorbed in animals or humans. The vast majority of the phytosterols ingested remain in the gastrointestinal tract. In healthy humans, absorption is limited to approximately 5% of the total β-sitosterol and approximately 15% of the campesterol ingested. With daily phytosterol intakes of 160 to 360 mg/day, only 0.3 to 1.7 mg/dL phytosterols are found in the blood. Plasma levels of phytosterols have been shown to as much as double with supplementation 3'. This is a small quantity compared with 240 mg/dL greater, cholesterol levels that will be present in hypercholesterolemia. Phytosterols are eliminated from the body more rapidly than cholesterol via the biliary route.

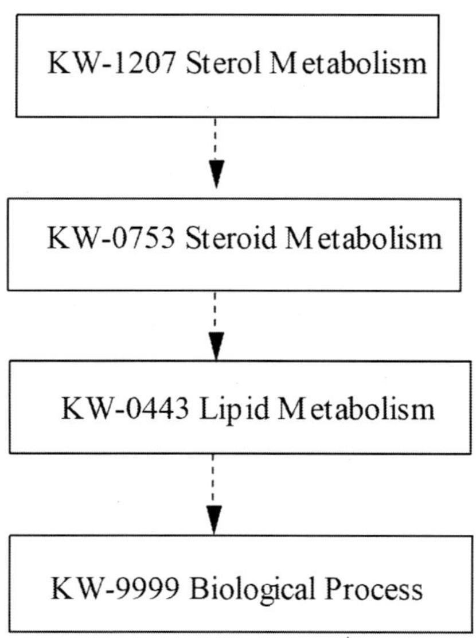

Fig. 17.3: Schematic representation of metabolism of sterols.

17.6 Cholesterol and Plant Sterols

17.6.1 The role of dietary supplementation with plant sterols and stanols in the prevention of cardiovascular disease

As per the findings of several studies, increased levels of low-density lipoprotein (LDL) cholesterol predict cardiovascular events. It was the Adult Treatment Panel II (ATP II) that introduced the concept of therapeutic lifestyle changes, and intake of plant sterols/stanols for the effective management of LDL cholesterol. Plant sterols and stanols in fat matrices effectively lower LDL cholesterol levels in hypercholesterolemic, diabetic, and healthy human volunteers.

Recent studies also show that sterols (2 g/d) lower LDL cholesterol even when incorporated in nonfat matrices. In addition, sterols may reduce biomarkers of oxidative stress and inflammation. It is postulated that plant sterols and stanol slowers the elevated levels of cholesterol probably by interfering with the uptake of both dietary as well as biliary cholesterol from the intestinal tract. Increasing evidence has accumulated encouraging the use of plant sterols and stanols in lowering LDL cholesterol levels. This therapy can be first line as well as adjunctive therapy in patients receiving statins (Devaraj and Jialal 2006).

17.6.2 Cholesterol-lowering effect of plant sterols

The chemical structure of plant sterols is similar to cholesterol except for the addition of an extra methyl or ethyl group. However, the absorption of the plant sterols is comparatively lower than cholesterol in human beings. The plant sterols reduces the absorption of cholesterol thereby reducing the serum levels of cholesterol.

Previous studies incorporating plant sterols into fat spreads have demonstrated utility in lowering elevated levels of blood cholesterol. Later on, plant sterols were added to other food matrices, including juices, nonfat beverages, milk and yogurt, cheese, meat, croissants and muffins, and cereal and chocolate bars. The beneficial physiological effects of plant sterols can be further enhanced by concomitant use of useful agants like olive oil, fish oils, fibers, soy proteins, and above all, regular exercise. It can be concluded that supplementation of the diet with plant sterols offers an effective and safe way to reduce the risk of coronary heart disease (CAD) (AbuMweis and Jones 2008).

17.6.3 Effect of plant sterols in combination with other cholesterol-lowering foods

A study assessed the effect of eliminating one out of the four dietary portfolio components. Plant sterols were selected because at 2 g/d, they have been reported to reduce low-density lipoprotein cholesterol (LDL-C) by 9% to 14%. Forty-two hyperlipidemic subjects were prescribed diets high in soy protein (22.5 g/1000 kcal), viscous fibers (10 g/1000 kcal), and almonds (23 g/1000 kcal) for 80 weeks. Subjects were instructed to take these together with plant sterols (1.0 g/1000 kcal) except between weeks 52 and 62.

> While taking the full dietary portfolio, including plant sterols, mean LDL-C reduction from baseline was 15.4% +/– 1.6% (P < .001). After sterol elimination, mean LDL-C reduction was 9.0% +/– 1.5% (P < .001). Comparable LDL-C reductions were also seen for the 18 subjects with a complete data set: on plant sterols, 16.7% +/– 3.1% (P < .001) and off plant sterols, 10.3% +/– 2.6% (P < .001), resulting in a 6.3% +/– 2.0% (P = .005) difference attributable to plant sterols.
>
> Compliance in this group of 18 was 67.0% +/– 5.9% for plant sterols and 61.9% +/– 4.8% for the other components. In combination with other cholesterol-lowering foods and against the background of a low-saturated fat diet, plant sterols contributed over one third of the LDL-C reduction seen with the dietary portfolio after one year of following dietary advice (Jenkins et al. 2008).

17.6.4 Plant sterols and stanols as cholesterol-lowering ingredients in functional foods

Plant sterols/stanols inhibit cholesterol absorption possibly by competitively inhibiting its incorporation into the mixed micelles in the small intestine although other mechanisms can not be excluded. Daily consumption of 1–2 grams of plant sterols or stanols was shown to cause 10–20% reduction in low-density lipoprotein cholesterol (LDL cholesterol). Combinations of plant sterols/stanols with certain lipid-lowering ingredients were shown to potentate their cholesterol-lowering effects and, in some cases, add triacylglycerol-lowering effects (Kamal-Eldin and Moazzami 2009).

17.6.5 Plant sterols and stanols in the treatment of dyslipidemia

There are inconsistent studies explaining effects of intake of plant sterols on serum LDL-C concentrations. The effect of plant sterols and stanols on serum LDL-C concentrations is well-characterized. There is emerging evidence regarding effect of plant sterols and stanols on triacylglycerol metabolism. This evidence has make plant sterols and stanols highly attractive for interventions in populations having metabolic syndrome. Further, there is no answer to the an ongoing debate whether increased plant sterol concentrations are associated with an increased cardiovascular disease risk or not. Two explainations have been given for this. Firstly, the potential atherogenicity of increased plant sterol concentrations might be ascribed to the formation of plant sterol oxidation products, known as oxyphytosterols. Secondly, elevated serum plant sterol concentrations should only be seen as surrogate markers for characterizing subjects with high intestinal cholesterol absorption (Baumgartner et al. 2011).

17.6.6 Plant sterols and plant stanols in the management of dyslipidaemia and prevention of cardiovascular disease

Plant sterols/stanols when taken at 2 g/day results in significant inhibition of absorption of cholesterol. They lower LDL-C levels by between 8 and 10%. The relative proportions of cholesterol versus sterol/stanol levels are similar in both plasma and tissue, with levels of sterols/stanols being 500-/10,000-fold lower than those of cholesterol, suggesting they are handled similarly to cholesterol in most cells.

> Despite possible atherogenicity of marked elevations in circulating levels of plant sterols/stanols, protective effects have been observed in some animal models of atherosclerosis. Higher plasma levels of plant sterols/stanols associated with intakes of 2 g/day in man have not been linked to adverse effects on health in long-term human studies. Significantly, at this dose, plant sterol/stanol-mediated LDL-C lowering is additive to that of statins in dyslipidaemic subjects, equivalent to doubling the dose of statin. 6–9% reported lowering of plasma triglyceride by 2 g/day in hypertriglyceridaemic patients needs further exploration (Gylling et al. 2014).

17.6.7 Adverse effect profile of sterols

17.6.7.1 Reductions in beta-carotene levels and alpha-tocopherol

Plant sterols are known to lower the plasma concentrations of carotenoids and alpha-tocopherol; however the exact of mode of action is not well established. Twenty-six normocholesterolemic men completed the double-blind, randomized, crossover study. Subjects consumed daily, for 1 week, each of the following 3 supplements: a low-fat milk-based beverage alone (control) or the same beverage supplemented with 2.2 g plant sterol equivalents provided as either free sterols or sterol esters.

> During this 1-wk supplementation period, subjects consumed a standardized diet. Both of the milks enriched with plant sterols induced a similar (60%) decrease in cholesterol absorption. Plant free sterols and plant sterol esters reduced the bioavailability of β-carotene by approximately 50% and that of α-tocopherol by approximately 20%. The reduction in β-carotene bioavailability was significantly less with plant free sterols than with plant sterol esters. At the limit of significance ($P = 0.054$) in the area under the curve, the reduction in α-tocopherol bioavailability was also less with plant free sterols than with plant sterol esters (Richelle et al. 2004).

17.6.7.2 Increased risk of cardiovascular disease

Excessive use of plant sterols has been observed to develop premature coronary artery disease in phytosterolemic patients, high risk of atherosclerotic CVDs, myocardial infarction, and even impaired endothelial functions (Choudhary and Tran 2011).

17.6.8 Sterols from algal sources

17.6.8.1 *Sterols of* Sargassum oligocystum *Montagne (Sargassaceae)*

22-dehydrocholesterol, cholesterol, fucosterol, 29-hydroperoxystigmasta-5,24(28)-dien-3β-ol, 24-hydroperoxy-24-vinylcholesterol, a mixture of 24(S)-hydroxy-24-vinylcholesterol and 24(R)-hydroxy-24-vinylcholesterol have been identified (Permeh et al. 2012).

17.6.8.2 *Sterols of* Gracilariopsispersica *Bellorin (Rhodophyceae)*

22-dehydrocholesterol, cholesterol, stigmasterol, β-sitosterol, and fucosterol have been reported (Saeidnia et al. 2012).

17.6.8.3 *Sterols of* Gracilariasalicornia *(C. Agardh) EY Dawson (Rhodophyceae)*

22-dehydrocholesterol, cholesterol, oleic acid, and stigmasterol have been identified (Nasir et al. 2011).

17.6.8.4 *Sterols of* Hypneaflagelliformis *Greville ex J. Agardh (Rhodophyceae)*

22-dehydrocholesterol, cholesterol, oleic acid, cholesterol oleate, and (22E)-cholesta-5,22-dien-3β-ol-7-one have been identified (Nasir et al. 2011).

17.6.8.5 *Sterols of* Stilophorarhizodes *(C. Agardh) J. Agardh (Phaeophyceae)*

The main sterol in *Stilophorarhizodes* is 24-methylenecholesterol, which is a biogenetic precursor of the at C-24 alkylated sterols (Zornista et al. 2003).

17.6.8.6 *Sterols of* Punctaria plantaginea *(Roth) Greville (Chordariaceae)*

The main sterol is cholesterol = 45% (Zornista et al. 2003).

17.6.8.7 *Sterols of* Punctaria latifolia *Greville (Chordariaceae)*

The main sterol is cholesterol = 76% (Zornista et al. 2003).

17.6.8.8 *Sterols of* Nizamuddiniazanardinii *(Schiffner) PC Silva (Sargassaceae)*

A hydroperoxy sterol, 24-hydroperoxy-24-vinyl cholesterol exhibiting cytotoxicity in all cell lines (IC_{50}, 3.62, 9.09, 17.96, 32.31, and 37.31 μg/mL respectively) has been isolated (Moghadam et al. 2013).

17.6.8.9 *Sterols of* Padina pavonica *(L.) Thivy (Pheophycaea)*

β-Sitosterol and campesterol have been reported (Shoubaky et al. 2014–15).

17.6.8.10 Sterols of Hormophysatriquetra *(C. Agardh) Kützing (Cystoseiraceae)*

β-Sitostanol and stigmasterol have been reported (Shoubaky et al. 2014–15).

17.6.8.11 Sterols of Chlorella pyrenoidosa *Chick (Chlorellaceae)*

Ergosterol has been reported (Klosty and Bergmann 1952).

17.6.8.12 Sterols of Chlorella ellipsoidea *Gerneck (Chlorellaceae)*

Poriferasterol, clionasterol, and 22-dihydrobrassicasterol have been reported (Patterson and Krauss 1965).

17.6.8.13 Sterols of Chlorella saccharophila *(Krüger) Migula (Chlorellaceae)*

Poriferasterol, clionasterol, and 22-dihydrobrassicasterol have been reported (Patterson and Krauss 1965).

17.6.8.14 Sterols of Chlorella vulgaris *Beyerinck (Chlorellaceae)*

Chondrilla sterol has been reported (Patterson 1967).

17.6.8.15 Sterols of Scenedesmusobliquus *(Turpin) Kützing (Scenedesmaceae)*

Chondrilla sterol has been reported (Bergmann and Feney 1950).

17.6.8.16 Sterols of Galaxaura marginata *(Ellis and Solander) JV Lamouroux (Galaxauraceae)*

Desmosterol, 24,25-epoxycholesterol, 24-hydroperoxycholesta-5,25-dien-3 beta-ol, 25-hydroperoxycholesta-5,23(E)-dien-3 beta-ol, cholesta-5,25-diene-3 beta,24-diol, and 24,25-epoxy-6 beta-hydroxycholest-4-en-3-one) were isolated. Sterols exhibited significant cytotoxicity toward several cancer cell lines (Sheu et al. 1996).

17.6.8.17 Sterols of Spirulina maxima *L. (Phormidiaceae)*

Cholesterol and β-sitosterol have been isolated and identified. The presence of sterols is related to the antimicrobial activity of the alga.

17.6.8.18 Sterols of Caulerpalentillifera *J. Agardh and C. sertularioides (SG Gmelin) Howe (Caulerpaceae)*

The sterol fraction consisted of C27–C29 steroidal alcohols with Δ5-unsaturation in the steroid core regardless of the growth conditions. The dominant (79.9%) steroid component of the sterol fraction was clionasterol (Shevchenko et al. 2009).

17.6.8.19 Sterols of Schizochytrium *sp. (Thraustochytriaceae)*

Lathosterol, ergosterol, stigmasterol, 24-ethylcholesta-5,7,22-trienol, stigmasta-7,24(24(1))-dien-3β-ol, and cholesterol have been reported. Sterol extract derived from alga *Schizochytrium* sp. possesses the same cholesterol-lowering activity as β-sitosterol (Chen et al. 2014).

17.6.8.20 Sterols of Cystoseirafoeniculacea *(Linn.) Greville (Sargassaceae)*

Fucosterol and a mixture of saringosterols have been reported (Bouzidia In Press).

17.6.8.21 Sterols of Phormidiumluridum *Gomont (Oscillatoriaceae)*

A crystalline mixture of sterols have been reported. The major component is 24-ethyl-Δ7-cholesterol (de Souza and Nes 1968).

17.6.8.22 Sterols of Endarachnebinghamiae *J. Agardh (Scytosiphonaceae)*

Sarangosterol and 23-methyl cholesta-5, 25-dien-38-ol have been reported (Ahmad et al. 1992).

17.6.8.23 Sterols of Dictyotaindica *Sonder ex Kützing (Dictyotaceae)*

Sargasterol has been reported (Ahmad et al. 1992).

17.6.8.24 Sterols of Laurenciaobtusa *(Hudson) JV Lamouroux (Rhodomelaceae)*

Cholesterol has been reported (Ahmad et al. 1992).

17.6.8.25 Sterols of Codiumiyengarii *Børgesen (Codiaceae)*

Clerosterol has been reported (Ahmad et al. 1992).

17.6.8.26 Sterols of Porphyridium cruentum *(SF Gray) Nägeli (Porphyridiaceae)*

The sterol distribution of *P. cruentum* is characterized by a predominance of cholesterol, with values as 199.0 mg 100 g-1 freeze dry weight (92.2%). The second most important sterol was stigmasterol (4.9%) followed by β-sitosterol (2.2%) (Durmaz et al. 2007).

17.6.9 Sterols from higher plants

Sterol of Tephrosia purpurea *(Linn.) Pers. (Fabaceae)*

The ethanol extract of the roots were subjected to column chromatograpy so as to yield pale yellow crystals from the petroleum ether : chloroform (1 : 1) fraction a new phytosterol ester identified by means of spectroscopy as Stigmast-5, 22-dien-3β, 21diol-3β, 21-dihexadecanoate (Sharma et al. 2008).

17.6.10 Sterols of Hypoxis rigidula Baker var. rigidula and Hypoxis hemerocallidea Fisch., CA Mey & Avé-Lall (Hypoxidaceae)

The chloroform extracts of both *H. hemerocallidea* and *H. rigidula* the presence of β-sitosterol, ergosterol and stigmasterol. HPLC analysis gave the following concentrations of phytosterols; 48.4 and 35.2 μg/ml of stigmasterol were found in *H. rigidula* and *H. hemerocallidea* respectively and 86.7 and 48.4 μg/ml of ergosterol were found in *H. rigidula* and *H. hemerocallidea* respectively (Mkhize et al. 2013).

17.6.11 Sterol of Holoptelea intergrifolia (Roxb.) Planch

17-(6-(diethylamino)decan-3-yl)-10,13-dimethyl-12,13-dihydro-10H-cyclopenta[a] phenanthren-3-ol, a new phytosterol isolated from a plant source and being reported for the first time (Sutar et al. 2014).

17.7 Phytostanols

Phytostanols are a fully-saturated subgroup of phytosterols (they contain no double bonds). They are in general produced by hydrogenation of phytosterols. Stanols often occur in dinoflagellates but are not common in other marine microalgae.

17.8 Phytosterols in Common Plant Oils

Values are mg total sterol per 100 g unrefined oil except where designated. (Data are from Spiller (1996).)

Almond: 266	Olive: 232	Sunflower: 725
Avocado: 404	Olive, refined: 176	Soybean: 327
Corn: 1,390	Peanut: 337	Soybean, refined and
Corn, refined: 952	Peanut, refined: 206	hydrogenated: 132
Flaxseed: 412	Pumpkin seed: 523	Walnut: 176
Flaxseed refined: 338	Rice bran: 3,225	Wheat germ: 1,970
Hazelnut: 120	Rice bran, refined: 1,055	Wheat germ, refined: 553

Further Readings

AbuMweis SS, Jones PJ. Cholesterol-lowering effect of plant sterols. *Curr Atheroscler Rep* 2008; **10**: 467–72.

Ahmad Viquar Uddin, Perveen Shaista, Shaiq Ali Mohammad, Uddin Shafi, Aliya Rehman, Shameel Mustafa. Sterol composition of marine algae from Karachi coast of Arabian Sea. *Pakistan J Mar Sci* 1992; **1**: 57–64.

Baumgartner S, Mensink RP, Plat J. Plant sterols and stanols in the treatment of dyslipidemia: new insights into targets and mechanisms related to cardiovascular risk. *Curr Pharm Des* 2011; **17**: 922–32.

Bergmann W, Feney RJ. Sterols of algae. I. The occurrence of chondrillasterol in *Scenzedesmiius obliquus. J Org Chem* 1950; **15**: 812–14.

Chen J, Jiao R, Jiang Y, Bi Y, Chen ZY. Algal sterols are as effective as β-sitosterol in reducing plasma cholesterol concentration. *J Agric Food Chem* 2014; **62**: 675–81.

Choudhary SP, Tran LS. Phytosterols: perspectives in human nutrition and clinical therapy. *Curr Med Chem* 2011; **18**: 4557–67.

Cook RP. Cholesterol: Chemistry, Biochemistry, and Pathology. New York: Academic Press, Inc. 1958. pp. xii+542. £5, 7s. 6d.

de Souza Noel J, Nes William R. Sterols: isolation from a Blue-Green Alga. Science 1968; **162**: 363.

Devaraj S, Jialal. The role of dietary supplementation with plant sterols and stanols in the prevention of cardiovascular disease. *Nutr Rev* 2006; **64**: 348–54.

Durmaz Y, Monteiro M, Koru E, Bandarra N. Concentration of sterols of *Porphyridium cruentum* biomass at stationary phase. *Pakistan J Biol Sci* 2007; **10**: 1144–1146.

El Shoubaky Gihan A, Salem Essam A. Terpenes and sterols composition of marine brown algae *Padina pavonica* (Dictyotales) and *Hormophysa triquetra* (Fucales). *Int J Pharmacog Phytochem Res* 2014–15; **6**: 894–900.

Gylling H, Plat J, Turley S, Ginsberg HN, Ellegård L, Jessup W, Jones PJ, Lütjohann D, Maerz W, Masana L, Silbernagel G, Staels B, Borén J, Catapano AL, De Backer G, Deanfield J, Descamps OS, Kovanen PT, Riccardi G, Tokgözoglu L, Chapman MJ. Plant sterols and plant stanols in the management of dyslipidaemia and prevention of cardiovascular disease. *Atherosclerosis* 2014; **232**: 346–60.

Jenkins DJ, Kendall CW, Nguyen TH, Marchie A, Faulkner DA, Ireland C, Josse AR, Vidgen E, Trautwein EA, Lapsley KG, Holmes C, Josse RG, Leiter LA, Connelly PW, Singer W. Effect of plant sterols in combination with other cholesterol-lowering foods. *Metabolism* 2008; **57**: 130–9.

Kamal-Eldin A, Moazzami A. Plant sterols and stanols as cholesterol-lowering ingredients in functional foods. *Recent Pat Food Nutr Agric* 2009; **1**: 1–14.

Klosty M, Bergmann W. Sterols of algae. III. The occurrence of ergosterol in *Chlorella pyrenoidosa*. *J Am Chem Soc* 1952; **74**: 1601.

Makin HLJ, Gower DB, Kirk DN. Phytosteroids. *Steroid Anal* 1995; :621–646.

Mkhize N, Mohanlall V, Odhav B. Isolation and quantification of β-sitosterol, ergosterol and stigmasterol from *Hypoxis rigidula* Baker var. *rigidula* and *Hypoxis hemerocallidea* Fisch., CA Mey & Avé-Lall (Hypoxidaceae). *Int J Sci* 2013; **2**: 118–124.

Moghadam MH, Firouzi J, Saeidnia S, Hajimehdipoor H, Jamili S, Rustaiyan A, Gohari AR. A cytotoxic hydroperoxy sterol from the brown alga, *Nizamuddinia zanardinii*. *Daru* 2013; **21**: 24.

Naima Bouzidia, Yannick Vianob, Annick Ortalo-Magnéb, Halima Seridic, Zahia Allichea, Yasmina Daghbouchea, Gérald Culiolib, Mohamed El Hattaba. Sterols from the brown alga *Cystoseira foeniculacea*: degradation of fucosterol into saringosterol epimers. *Arabian J Chem*. In press.

Nasir M, Saeidnia S, Mashinchian-Moradi A, Gohari AR. Sterols from the red algae, *Gracilaria salicornia* and *Hypnea flagelliformis*, from Persian Gulf. *Pharmacogn Mag* 2011; **7**: 97–100.

Patterson GW, Krauss RW. Sterols of Chlorella. I. The naturally occurring sterols of *Chlorella vulgaris*, *C. ellipsoidea* and *C. saccharophila*. *Plant Cell Physiol* 1965; **6**: 211–20.

Patterson GW. Sterols of Chlorella. II. The occurrence of an unusual sterol mixture in *Chlorella vulgaris*. *Plant Physiol* 1967; **42**: 1457–1459.

Permeh P, Saeidnia S, Mashinchian-Moradi A, Gohari AR. Sterols from *Sargassum oligocystum*, a brown algae from the Persian Gulf, and their bioactivity. *Nat Prod Res* 2012; **26**: 774–7.

Richelle M, Enslen M, Hager C, Groux M, Tavazzi I, Godin JP, Berger A, Métairon S, Quaile S, Piguet-Welsch C, Sagalowicz L, Green H, Fay LB. Both free and esterified plant sterols reduce cholesterol absorption and the bioavailability of beta-carotene and alpha-tocopherol in normocholesterolemic humans. *Am J Clin Nutr* 2004; **80**: 171–7.

Saeidnia S, Permeh P, Gohari AR, Mashinchian-Moradi A. *Gracilariopsis persica* from Persian Gulf contains bioactive sterols. *Iran J Pharm Res* 2012; **11**: 845–9.

Sharma SK, Vasudeva N, Rathi P, Ali M. Isolaton and identification of a new phytosterol ester from *Tephrosia purpurea* (Linn.) Pers. root. *Int J Chem Sci* 2008; **6**: 1734–1741.

Sheu JH, Huang SY, Duh CY. Cytotoxic oxygenated desmosterols of the red alga Galaxaura marginata. *J Nat Prod* 1996; **59**: 23–6.

Shevchenko NM, Burtseva Yu V, Zvyagintseva TN, Makar'eva TN, Sergeeva OS, Zakharenko AM, Isakov VV, Thi Linh Nguyen, Xuan Hoa Nguyen, Minh Ly Bui, Van Huyen Pham. Polysaccharides and sterols from green algae. *Chem Nat Comp* 2009; **45**: 1–5.

Sutar RC, Kasture SB, Kalaichelvan VK. Isolation and identification of a new phytosterol from *Holoptelea intergrifolia* (Roxb.) Planch leaves. *Int J Pharm Pharmaceut Sci* 2014; **6**: 35–357.

Tikkanen MJ. Plant sterols and stanols. *Handb Exp Pharmacol* 2005; **170**: 215–30.

Zornista GK, Dimitrova-Konaklieva SD, Stefanov KL, Simenon SP. A comparative study on the sterol composition of some brown algae from the Black Sea. *J Serb Chem Soc* 2003; **68**: 269–275.

Chapter 18

Botany of Phytosteroids Containing Medicinal Plants

18.1 *Aletris farinosa* Linn. (Nartheciaceae)

Habitat: Eastern United States.

Botany: *A. farinosa* is a perennial herb spreading by means of underground rhizomes and forming rosettes of leaves. Leaves are narrow, up to 20 cm long, bright yellowish-green. Flowering stalks can be as much as 100 cm tall. Flowers are white, up to 10 mm long. Fruit is a dry capsule tapering at the tip.

18.2 *Costus igneus* N.E. Br (Costaceae)

Habitat: Native to South and Central America.

Botany: *C. igneus* is a perennial, upright, spreading plant reaching about two feet tall, with the tallest stems falling over and lying on the ground. Leaves are simple, alternate, entire, oblong, evergreen, 4–8 inches in length with parallel venation. The large, smooth, dark green leaves of this tropical evergreen have light purple undersides and are spirally arranged around stems, forming attractive, arching clumps arising from underground root stocks. Beautiful, 1.5-inch diameter, orange flowers are produced in the warm months, appearing on cone-like heads at the tips of branches. Fruits are inconspicuous, not showy, less than 0.5 inch, and green-colored.

18.3 *Costus pictus* D. Don (Costaceae)

Habitat: Native to Mexico.

Botany: *C. pictus* has long narrow leaves with a characteristic wavy edges. The bases of the sheaths are mottled with markings that have earned the plant the synonym of Costus hieroglyphica. The inflorescences form both at the end of a leafy stem, and less often radically on a short nearly leafless stem. *C. pictus* can be recognized by its yellow flowers with red spots and stripes.

18.4 *Costus speciosus* Smith (Costaceae)

Habitat: Throughout tropical Africa, Asia, and the Americas.

Botany: *C. speciosus* is a succulent, erect, perennial, ornamental, herbaceous plant, root stock tuberous stem, sub-woody at the base, thick creeping rhizomes (120–300 cm height) growing up to 2–2.7 m height with long lanceolate leaves and white fragrant flowers in terminal clusters. It is tall and dramatic landscape plant with large dark green, subsessile, elliptic or obovate leaves arranged on the stalk in spiral form.

18.5 *Dioscorea alata* L. (Dioscoreaceae)

Habitat: Asian tropics.

Botany: Vigorously twining herbaceous vine, from massive underground tuber. Stems to 10 m (30 ft) or more in length, freely branching above; internodes square in cross section, with corners compressed into "wings," these often red-purple tinged. Aerial tubers (bulbils) formed in leaf axils, elongate, to 10 cm (4 in) x 3 cm (1.2 in), with rough, bumpy surfaces. Leaves long petioled, opposite (often with only one leaf persistent); blades to 20 cm (8 in) or more long, narrowly heart shaped, with basal lobes often angular. Flowers small, occasional, male and female arising from leaf axils on separate plants, male flowers in panicles to 30 cm (1 ft) long, female flowers in smaller spikes. Fruit a 3-parted capsule; seeds winged.

18.6 *Dioscorea colletti* Hook. f. (Dioscoreaceae)

Habitat: Mountain slopes; 200–3200 m. S Anhui, Fujian, N Guangdong, NE Guangxi, Guizhou, S Henan, Hubei, Hunan, Jiangxi, W Sichuan, N Taiwan, Yunnan, Zhejiang, India, Laos, Myanmar, Thailand, Vietnam.

Botany: Rhizome horizontal, ginger-shaped, variable in length, ca. 2 cm thick; transverse section yellow; roots slender, fibrous. Stem twining to left, glabrous, sometimes densely and short, yellow, and hairy. Leaves alternate, simple; petiole 4–7 cm; leaf blade drying blackish, triangular to ovate or ovate-lanceolate, 5–19(–22) × 3–13(–16) cm, usually membranous, abaxially sometimes hispidulous especially along veins, base cordate to subtruncate, margin subentire or slightly undulate, sometimes transparent, apex acuminate. Male spikes solitary or 2 or 3 together. Male flowers: solitary or in cymules of 2 or 3, sessile; bracts ovate-lanceolate; bracteoles ovate; perianth yellow, usually drying black, saucer-shaped; stamens 3, inserted in perianth tube, filaments short, anthers ovoid, connectives elongate at anthesis, fork-shaped; staminodes filiform. Female spike solitary, to 5 cm. Female flowers: staminodes present; ovary terete. Capsule reflexed, brown, shiny, broadly obovoid or ellipsoid, usually somewhat angular, 1.6–2.1 cm, apex truncate; wings 0.8–1.1 cm wide. Seeds inserted near middle of capsule, winged all round. Fl. May–Aug, fr. Jun–Oct.

18.7 *Dioscorea composita* L. f. (Dioscoreaceae)

Habitat: Native to Mexico, live in warm climate from sea level to 1500 m.

Botany: Climbing plant that comes from a similar result to a white or pink colored sweet potato. The leaves are heart-shaped, are green and stiff and having longitudinal ribs (7 to 9). The flowers are green and small, male are grouped in number of 2–3, at the junction of the stem and leaves, female are hanging clusters. The fruits are elongated and contain two flat, winged seeds.

18.8 *Dioscorea deltoidea* Wall. (Dioscoreaceae)

Habitat: Himalayas, from Kashmir to Assam, Indo-China and W. China, at altitudes of 450–3100 m.

Botany: *D. deltoidea* is a hairless vine, twining clockwise. Tubers are ligneous, irregular. Alternately arranged leaves are simple, 5–11.5 cm long, 4–10.5 cm broad, ovate or triangular-ovate, often heart-shaped, the basal lobes rounded or sometimes dilated outwards, 7–9-nerved, long-pointed, hairless above, velvety on the nerves beneath. Leaf-stalks are 5–10 cm long, slender. Male flower spikes occur solitary in leaf axils, simple or sometimes branched, slender, lax, 7.5–25 cm long. Flowers are in small distant clusters; stamens 6, antheriferous. Female spikes are solitary, slender, up to 15 cm long, few-flowered. Capsule is 2 cm long, 3 cm broad, obovate or obcordate.

18.9 *Dioscorea floribunda* M. Martens & Galeotti (Dioscoreaceae)

Habitat: Himalayas, from Kashmir to Assam, Indo-China and W. China, at altitudes of 450–3100 m.

Botany: A perennial, climbing plant producing annual stems from a tuberous rootstock.

18.10 *Dioscorea gracillima* Miq. (Dioscoreaceae)

Habitat: Japan.

Botany: Twining herbaceous vine. Rhizome occur, bamboo-like, irregular shape, surface filamentous fibrous roots. Stems L, hairless. Leaves alternate, sometimes at the stem base 3–4 whorled, ovate leaves heart-shaped, apex acuminate, base heart-shaped, wide heart-shaped or subtruncate, entire or wavy, sometimes obvious erose edge. Male inflorescences spicate, solitary, and axillary, usually irregularly branched; male flowers sessile, solitary, rarely 2–3 flowers clustered, was born in the base of inflorescence; bracts ovate, film quality, bractlets relatively short and narrow bracts; perianth dish, top 6 crack, lobes oblong, when the bloom spreading; fertile stamens 3, anther width of 1/2, infertility stamens 3, clavate. Both alternate, was born in the edge of the torus. Female inflorescences with male inflorescences similar; there are six female staminodes. Capsule three prism, top truncate, each furrow winged, long oval, large and small, usually 1.8–2.8 cm long, 1–1.3 cm wide; seeds 2 per room, was born in the central axis, surrounded by film-like wings. Flowering from May to August, fruiting from June to October.

18.11 *Dioscorea mexicana var. sessiliflora* (Uline) Matuda (Dioscoreaceae)

Habitat: The state of San Luis Potosí in north eastern Mexico south to Panama.

Botany: *D. mexicana* is having either a partly to completely above-ground dome-shaped caudex with a thick, woody outer layer up to 3 feet (90 cm) in diameter and 8–10 inches (20 to 25 cm) in height. The caudex of *D. mexicana* is divided into regular polygonal plates that become protuberant with age, and separated by deep fissures. The vigorous annual vines which may reach 30 feet (9 m) long before dying back in winter, that grow up from the caudex, bear heart-shaped leaves.

18.12 *Dioscorea nipponica* Makino (Dioscoreaceae)

Habitat: Mixed forests, scrub forests, warm-temperature transitional areas; 100–1800 m.

Botany: Rhizome horizontal, many branched, cylindric, more than 1.5 cm thick; cork layer persistent or readily detached. Stem twining to left, drying green or reddish brown, to 5 m, glabrescent. Leaves alternate, simple; petiole 10–20 cm; leaf blade shiny, drying yellowish green, broadly cordate to palmately unequally 3–7-lobed, very variable in shape, 7–15 × 4–13 cm, glabrous or sparsely minutely setose especially along veins, basal veins 7 or 9, outermost ones often forked, base cordate, margin undulate to prominently bluntly toothed or lobed, apex acuminate. Male spike solitary, to 17 cm, rarely with occasional cymules expanded to form lateral branches. Male flowers: usually in cymules or umbellules of 2–4(–7), solitary distally on inflorescence, sessile or shortly pedicellate; bracts lanceolate, slightly shorter than perianth; bracteoles ± obsolete; perianth saucer-shaped, lobes obtuse at apex; stamens 6, inserted at middle of perianth lobes, anthers introrse. Female flowers: staminodes filiform; stigma 3-lobed. Capsule reflexed at maturity, light brown, purplish speckled, ellipsoid-oblanceolate, 1.5–2 cm, base rounded, apex shallowly emarginate; wings 0.7–0.8 cm wide. Seeds inserted near base of capsule, sometimes only one fertile, winged all round but wing much wider toward capsule apex. Fl. Jun–Aug, fr. Aug–Oct.

18.13 *Dioscorea opposita* Thunb. (Dioscoreaceae)

Habitat: Native to Myanmar (Burma) and to the Indian Subcontinent (India, Sri Lanka, Bangladesh).

Botany: Herbaceous, high climbing vines to 65 feet (20 m) long, infestations covering shrubs and trees. Twining and sprawling stems with long-petioled heart-shaped leaves. Spreading by dangling potato-like tubers (bulbils) at leaf axils and underground tubers. Monocots.

18.14 *Dioscorea septemloba* Thunb. (Dioscoreaceae)

Habitat: Southern China.

Botany: *D. septemloba* is a perennial twining vine.

18.15 *Dioscorea villosa* L. (Dioscoreaceae)

Habitat: Native to eastern North America.

Botany: This is an herbaceous, slender vine, found throughout the United States, but more common in the central and southern portions. The stem is a smooth green twiner, about the size of a goose-quill, twining from the right to the left, over fences, bushes, etc. The leaves are symmetrical and heart-shaped, gradually tapering to a sharp, acuminate point, and are borne on leaf stalks from 2 to 4 inches long. The lower leaves are in whorls of 4 or 5, with intervals of from 6 inches to a foot between, while those on the upper part of the vine are irregularly alternate. The margins of the leaves are entire and wavy in the larger leaves. The veins are generally 9, quite prominent, and gradually diverge from the top of the leaf-stalk. The under side of the leaves is clothed with a thick pubescence. The flowers appear in June or July, are dioecious, very small, and greenish-yellow. The male flowers are in compound loose spikes, with from 3 to 5 slender branches; the perianth is 6-parted, sessile, flattened, and has, near the base, 6 minute stamens. The female flowers are placed at intervals of ¼ of an inch or ½ an inch apart, in simple, drooping, axillary spikes, consisting each, of from 4 to 8 flowers. The ovary is sessile, slender, about ¼ of an inch in length, bearing at the summit a 6-parted, small perianth, and 3 short styles. The female flowers are succeeded by dry, brown fruit, which remains hanging among the limbs of shrubs in winter for some time after the herbaceous stems of the plant have perished. They are sharply 3-angled, and have 3 cells, each cell bearing 2 (or often, by abortion, 1) flat, membranous-winged seed.

18.16 *Dioscorea zingiberensis* C.H. Wright (Dioscoreaceae)

Habitat: Cultivated in China.

Botany: Plants sometimes monoecious. Rhizome horizontal, sometimes irregularly branched, subcylindric, 1–1.5 cm thick; cork dull brown, rough; transverse section yellow. Stem twining to left, glabrous, smooth. Leaves alternate, simple; petiole 2.5–6 cm; leaf blade adaxially green, often irregularly spotted, drying dark grayish brown, narrowly peltate, triangular-ovate, usually ± 3-lobed by enlargement of basal lobes, 4.5–10 × 4–8.5 cm, papery, glabrous, base cordate with broadly rounded sinus, apex rounded and cuspidate to acuminate; lateral lobes reflexed, rounded. Male spikes solitary or 2 or 3 together, 5–10 cm, very slender, often borne along specialized, panicle like, lateral shoots with reduced leaves. Male flowers: solitary or in cymules of 2 or 3, sessile; bracts 3 or 4, brown, membranous; perianth purplish red, drying black, lobes spreading at anthesis, 1.2–1.5 × 0.8–1 mm; stamens 6, inserted at margin of receptacle, filaments extremely short. Female spike to 8 cm. Female flowers: staminodes filiform. Capsule reflexed, long pedicellate, drying blue black, obovoid, 1.4–2 cm, about as long as wide, pruinose, base ± truncate, apex emarginate; wings 0.8–1.2 cm wide. Seeds inserted near middle of capsule, winged all round. Fl. May–Aug, fr. Sep–Oct.

18.17 *Helicteres isora* L. (Sterculiaceae)

Habitat: Asia including India, South china, Malay Peninsula, Java, and Saudi Arabia. Also, found in Australia.

Botany: Large shrubs or small trees; bark pale greyish, finely wrinkled; young shoots stellate-tomentose. Leaves 5–12 x 3–8 cm, obovate to suborbicular, base cordate, margin irregularly crenate-serrate, apex acute or acuminate, 3–5-nerved at the base, scabrous above and stellately tomentose below; petioles to 1.2 cm long. Flowers axillary, solitary or in few-flowered cymes; bracts 2–3 mm long, linear, 2 brown glands present in the axil of bracts; pedicel to 6 mm long. Calyx slightly yellow, persistent, tubular, splitting in to 5 irregular lobes; tube 1.5–2 cm long, densely stellate hairy without. Petals 5, unequal, 2–2.5 cm long, obovate, clawed, crimson, fading to pale blue. Staminal column 3–3.5 cm long, cylindric; stamens 10; staminodes 5. Ovary 2–2.5 mm long, placed at tip of gynophore, 5-lobed, 5-celled; ovules many; style 5; stigma subulate. Follicles 5, 4–6 cm long, spirally twisted, stellate-tomentose, beaked. Seeds 2–3 mm long, angular, black, wrinkled.

18.18 *Paris polyphylla* Sm. (Melanthiaceae)

Habitat: The Himalayas on the border of China and India.

Botany: *P. polyphylla* has a small purple flower nestled in the middle of a symmetrical whorl of leaves. Long yellowish-green petals sprout from the flower center like cat whiskers, and in the fall, the flowers are replaced by scarlet berries.

18.19 *Smilax china* L. (Smilacaceae)

Habitat: Native of China and Japan.

Botany: *S. china* has a hard, large, knotty, uneven rhizome, blackish externally, pale coloured or whitish internally. Stem without support, about 3 feet high, but growing much taller if it has a bush to cling to. Leaves thin, membranous, round, five-nerved acute or obtuse at each end, mucronate at points. Stipules distinct obtuse; umbels greenish yellow, small ten-flowered; fruit red, size of bird cherry.

18.20 *Solanum dulcamara* L. (Solanaceae)

Habitat: Native to most of Europe, North Africa and eastern Asia.

Botany: *S. dulcamara* is a semi-woody herbaceous perennial vine, which scrambles over other plants, capable of reaching a height of 4 m where suitable support is available, but more often 1–2 meters high. The leaves are 4–12 cm long, roughly arrowhead-shaped, and often lobed at the base. The flowers are in loose clusters of 3–20, 1–1.5 cm across, star-shaped, with five purple petals and yellow stamens and style pointing forward. The fruit is an ovoid red berry about 1 cm long, soft and juicy, with the aspect and odor of a tiny tomato.

18.21 *Solanum khasianum* C.B. Clarke (Solanaceae)

Habitat: Khasia Mountains in Assam (India).

Botany: *S. khasianum* is a stout, branched, woody shrub attaining a height of 0.75 to 1.5 m. The stem has spines, the leaves are ovate to lobed with spines on both the surfaces, the flowers are hermophrodite, borne on axillary clusters, white; the berries are yellowish when ripe or greenish; the seeds are small, brown in color, and abundant, embedded in a sticky mucilage.

18.22 *Solanum laciniatum* Ait (Solanaceae)

Habitat: In temperate regions of New South Wales, the Australian Capital Territory, Victoria, South Australia, Tasmania, and New Zealand.

Botany: Fleshy shrub to 4 m tall bearing dark green thin wide leaves that are divided into 1–3 large sharp lobes and with large purplish ruffled flowers that have a projecting yellow centre. Leaves 10–80 cm long by 4–6 cm wide. Flowers dished, up to 50 cm wide. Fruit yellow or orange, 23–30 mm long.

18.23 *Solanum nigrum* L. (Solanaceae)

Habitat: Throughout India.

Botany: *S. nigrum* is an annual weed that grows up to 60 cm tall, is branched and usually erect, growing wild in wastelands and crop fields. Alternate leaves are ovate deep green with an indented margin and acuminate at the tip. Flowers are white with yellow colored centre. The berries are green at the early stage and turn to orange or black when ripened.

18.24 *Solanum incanum* L. (Solanaceae)

Habitat: Native to Sub-Saharan Africa and the Middle East, eastwards to India.

Botany: Herb or shrub up to 1.8 m height with spines on the stem, leaves, stalks and calyces, and with velvet hairs on the leaves. Leaves are alternate, flowers often borne in the leaf axilles, sometimes solitary or in few-flowered clusters. The calyx is united, corolla regular, bell- or wheel shaped. Five stamens are inserted on its throat. Fruits are yellow at the beginning, later on black.

18.25 *Solanum trilobatum* L. (Solanaceae)

Habitat: Throughout India, growing wild.

Botany: *S. trilobatum* is a prickly diffuse, bright green perennial herb, woody at the base, 2–3 m height, found throughout India, mostly in dry places as a weed along roadsides and waste lands. The plant has many branched spiny scandent shrubs. Leaves are deltoid or triangular, irregularly lobed. Flowers are purplish-blue, in cymes. Berry are globose, red or scarlet.

18.26 *Solanum torvum* Sw. (Solanaceae)

Habitat: Throughout the world's tropical regions.

Botany: *S. torvum* is a broadleaved, evergreen, shrub or small tree, growing up to 16 ft tall. The stems are armed with stout, straight or lightly curved prickles. The alternate leaves are elliptical in shape, have prickles along the mid vein, and range from unlobed to strongly lobed. The small, white flowers occur in large clusters. Fruit are small yellow berries with 210–220 seeds. Seeds ovate to broadly ovate, broadly elliptic, or nearly circular, infrequently C-shaped in outline.

18.27 *Solanum xanthocarpum* L. (Solanaceae)

Habitat: Throughout India.

Botany: *S. xanthocarpum* is a very spiny diffused herb, with a height of up to 1.2 meters. The young branches are densely covered with minute star-shaped hair, while the mature branches are zigzag, covered with yellow, sharp shining prickles, and spread close to the ground. The midribs and other nerves of the leaves have sharp yellow prickles and grow up to 10 cm in length. The purple flowers, that are 2 cm long with five petals, can be seen in small bunches, sometimes opposite to the leaves.

18.28 *Trillidium govanianum* (D. Don.) Kunth. (Melanthiaceae)

Habitat: Scattered localities in the Himalayas, Bhutan, Nepal, and China, at altitudes of 2700–4000 m.

Botany: *T. govanianum* is a perennial herb. Stocky 15 cm purple-red stems carry 3 green leaves just below a single, small, starry flower of deep red and green color. This species is unusual in that the sepals and petals resemble each other, giving a six-petalled appearance. These contrast well with the prominent yellow anthers and red style, that sits atop the flower. The flower is followed by a grape-red fruit, sitting at the center of the three leaves.

Chapter 19

Pharmacology of β-sitosterol and other Sterols

19.1 Introduction

β-sitosterol (Fig. 19.1) is present in all plant lipids and is used for steroid synthesis. Chemical structure of β-sitosterol is similar to cholesterol. Sitosterols are white, waxy powders with a characteristic odor. Sitosterols are hydrophobic and soluble in alcohols.

19.2 Other Name

22, 23-Dihydrostigmasterol, Stigmast-5-en-3-ol, β-Sitosterin.

19.3 Pharmacology

19.3.1 Analgesic and anti-inflammatory

Various extracts of leaves of *Nyctanthes arbortristis* Linn. (Oleaceae) were screened for analgesic activity by hot plate test and acetic acid-induced writhings and anti-

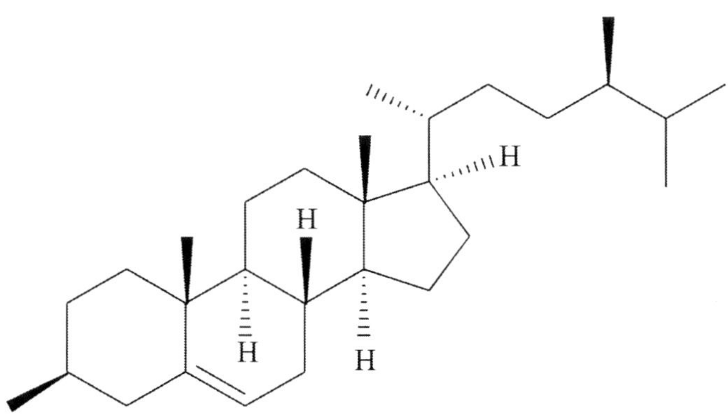

Fig. 19.1: Structure of β-sitosterol.

inflammatory activity by carrageenan-induced hind paw edema method at the dose of 50 mg/kg, i.p. Petroleum ether extract was found to be most active and hence subjected to activity-guided fractionation. Results showed that β-sitosterol (5, 10, and 20 mg/kg, i.p.) was responsible for the significant and dose-dependent activity comparable with the standard extract. β-sitosterol from *N. arbortristis* leaves might be responsible for analgesic and anti-inflammatory activity.

19.3.2 Antifertility

The effects of β-sitosterol on fertility, epididymal sperm counts and testicular and accessory reproductive organ weights were evaluated in male albino rats. The effects were studied at two dosages (0.5 and 5 mg/kg per day rat subcutaneously) for 16, 32, and 48 days. The antifertility effect of β-sitosterol was pronounced only at the high dose level, but there was a significant decrease in testicular weight and sperm concentrations after long-term treatment with the low dose of β-sitosterol.

The weights of all accessory sex tissues except caput epididymis increased following low dose sitosterol treatment. High dose treatment reduced the sperm concentrations as well as the weights of testis and accessory sex tissues in a time-dependent manner. Withdrawal of treatment for 30 days restored only the weights of accessory sex tissues to near normal conditions (Malini and Vanithakumari 1991).

19.3.3 Bioactivity

β-sitosterol and β-sitosteryl-β-D-glucoside were isolated as analgesic constituents from the leaves of *Mentha cordifolia* Opiz. The acetic acid-induced writhing test showed that β-sitosterol and β-sitosteryl-β-D-glucoside decreased the number of squirms induced by acetic acid by 70.0% and 73.0%, respectively, at a dose of 100 mg/kg mouse.

Statistical analysis using the Kruskall Wallis one-way analysis of variance by ranks showed that these isolates approximate the analgesic activity of mefenamic acid at a 0.001 level of significance. The hot plate method confirmed their analgesic activities, as β-sitosterol and β-sitosteryl-β-D-glucoside exhibited a 300% and 157% increase in pain tolerance, respectively, while mefenamic acid, a known analgesic, showed a 171% increase. Neither isolate exhibited anti-inflammatory activity using the carrageenan-induced mouse paw oedema assay.

β-sitosterol also exhibited anthelminthic and antimutagenic activities. *In vitro* tests using live *Ascarissuum* as test animals showed that the behaviour of worms treated with beta-sitosterol approximated that of the positive controls, Combantrin and Antiox. An *in vivo* micronucleus test showed that β-sitosterol inhibited the mutagenicity of tetracycline by 65.3% at a dose of 0.5 mg/kg mouse. At the same dose, it did not exhibit chromosome-breaking activity (Villaseñor et al. 2002).

19.3.4 Cardio protective

A study examined the effect of β-sitosterol (0.1–200 microM) on (i) the expression of vascular adhesion molecule 1 and intracellular adhesion molecule 1 by cell ELISA and (ii) the attachment of monocytes (U937 cells) in tumor necrosis factor-alpha (TNF-alpha)-stimulated human aortic endothelial cells (HAECs) by adhesion assay.

The effect on nuclear factor-kB phosphorylation was also examined via a cell-based ELISA kit. Results showed that β-sitosterol inhibits significantly vascular adhesion molecule 1 and intracellular adhesion molecule 1 expression in TNF-alpha-stimulated HAEC as well as the binding of U937 cells to TNF-alpha-stimulated HAEC and attenuates the phosphorylation of nuclear factor-kB p65 (Loizou et al. 2010).

19.3.5 Chemopreventive

19.3.5.1 Inhibition of HT-29 human colon cancer cell growth

Tumor cells were grown in DMEM containing 10% FBS and supplemented with sterols (cholesterol or β-sitosterol) at final concentrations up to 16 microM. The sterols were supplied to the media in the form of sterol cyclodextrin complexes. The study indicated that 8 and 16 microM β-sitosterol were effective at cell growth inhibition as compared to cholesterol or to the control (no sterol supplementation).

After supplementation with 16 microM β-sitosterol for 9 days, cell growth was only one-third that of cells supplemented with equimolar concentration of cholesterol. No effect was observed on total membrane phospholipid concentration. At 16 microM β-sitosterol supplementation, membrane cholesterol was reduced by 26%. Cholesterol supplementation resulted in a significant increase in the cholesterol/phospholipid ratio compared to either beta-sitosterol supplemented cells or controls.

There was a 50% reduction in membrane sphingomyelin of cells grown in 16 microM β-sitosterol. Additional changes were observed in the fatty acid composition of minor phospholipids of beta-sitosterol supplemented cells, such as sphingomyelin, phosphatidylserine, and phosphatidylinositol. Only in the case of phosphatidylinositol was there an effect of these fatty acid changes on the unsaturation index, β-sitosterol incorporation resulted in an increase in the U.I. (Awad et al. 1996).

19.3.5.2 Effect of β-sitosterol on the expression of a multifunctional growth factor (TGF-beta1) and the activity of PKC-alpha membrane in stromal cells of the human prostate in vitro

β-sitosterol was able to induce the expression and secretion of TGF-beta1 significantly between 1.26- and 1.86-fold compared to a cholesterol and the nonsupplemented control in 6 of 8 individual cultures. The total amount of secreted TGF-beta1 varied in cells from different patients. Based on its presence in both membrane fraction and cytosol, PKC-alpha appeared to be constitutively expressed in stromal cells. In the absence of β-sitosterol PKC-alpha was predominantly found in its membrane-associated active form. Following a culture with β-sitosterol, a translocation of PKC-alpha from the membrane to the cytosol was observed. This effect was specific for β-sitosterol as compared to cholesterol (Kassen et al. 2000).

19.3.5.3 Chemopreventive potential in experimental colon cancer model

The chemopreventive potential of β-sitosterol in colon carcinogenesis was assessed by injecting 1,2-dimethylhydrazine (DMH, 20 mg/kg b.w.) into male Wistar rats and supplementing this with β-sitosterol throughout the experimental period of 16 weeks at 5, 10, and 20 mg/kg b.w. β-sitosterol induced significant dose-dependent growth inhibition of COLO 320 DM cells (IC50 266.2 μM), induced apoptosis by scavenging reactive oxygen species, and suppressed the expression of β-catenin and PCNA antigens in human colon cancer cells. β-sitosterol supplementation reduced the number of aberrant crypt and crypt multiplicity in DMH-initiated rats in a dose-dependent manner with no toxic effects (Baskar et al. 2010).

19.3.5.4 Effect on prostate cancer cell lines

The aim of this *in-vitro* study was to evaluate the inhibitory effect of different cocoa polyphenols extracts, alone or combined with β-sitosterol, on two human prostate cancer cell lines (nonmetastatic 22Rv1 cells and metastatic DU145 cells) and a normal human prostate cell line (RWEP-1). A synergy between β-sitosterol and cocoa polyphenols extract was also researched.

> Cells were treated independently with five products from 1 to 72 h: (1) synthetic β-sitosterol, (2) a cocoa polyphenols extract supplemented with beta-sitosterol, (3) three different cocoa polyphenols extracts naturally containing β-sitosterol. In the experiment, β-sitosterol was tested from 10(–6) to 10(–3)%; cocoa polyphenols extract supplementation was with 0.72% β-sitosterol; finally cocoa polyphenols extracts were added to the cells at very low concentrations ranging from 0.001 to 0.2%.
>
> The growth and viability of cells were measured using colorimetric assay at 1, 3, 6, 24, 48, and 72 h of treatment. IC50 and IC100 corresponding to the concentration leading to a decrease of 50% and 100% of cell growth were determined. At the highest tested concentration, cocoa polyphenols extracts induced a complete inhibition of growth of metastatic and nonmetastatic cancer cell lines. In addition, cocoa polyphenols extracts were more active against local cancer cells than against metastatic cells.
>
> Moreover, at the highest tested concentration, cocoa polyphenols extracts are not effective on a normal prostate cell lines. β-sitosterol induced low growth inhibition of both cancer cell lines. Cocoa polyphenols extracts, however, were significantly more active and showed a strong and fast inhibition of cell growth than β-sitosterol alone. No synergy or addition was observed when β-sitosterol was tested together with the cocoa polyphenols extract (Jourdain et al. 2006).

19.3.6 Hypocholesterolemic

19.3.6.1 Effects of β-sitosterol on the concentrations of serum and liver cholesterol and serum apolipoproteins in rats fed butter fat

Male rats were fed on semipurified cholesterol-free diets containing butter fat with or without supplementary β-sitosterol. The expected rise of serum cholesterol caused by butter fat, as compared with safflower oil, was not able to be demonstrated, and hence the hypocholesterolemic effect of β-sitosterol as well.

However, the plant sterol effectively lowered the liver cholesterol level. Similar responses were also observed in mice. The distribution of cholesterol in serum lipoproteins remained unchanged among different dietary regimens. Butter fat increased the concentration of serum apoA-I in relation to safflower oil. There was possibly a trend toward higher serum apoA-I with supplementation of β-sitosterol in a butter-fat diet.

The effect of β-sitosterol on serum apoB was rather variable. The observation strongly suggests that alteration in cholesterol metabolism in these rodents may not satisfactorily be estimated by the serum cholesterol parameter alone when diets free of cholesterol are fed. The concentration of hepatic cholesterol and serum apolipoproteins seems a more apposite measure for this purpose (Sugano et al. 1982).

19.3.6.2 Effect of micellar β-sitosterol on cholesterol metabolism in CaCo-2 cells

CaCo-2 cells were used to address the effect of the plant sterol, beta-sitosterol, on cholesterol trafficking, cholesterol metabolism, and apoB secretion. Compared to cells incubated with micelles (5 mMtaurocholate and 250 microM oleic acid) containing cholesterol, which caused an increase in the influx of plasma membrane cholesterol to the endoplasmic reticulum and increased the secretion of cholesteryl esters derived from the plasma membrane, β-sitosterol did not alter cholesterol trafficking or cholesteryl ester secretion.

Including β-sitosterol in the micelle together with cholesterol attenuated the influx of plasma membrane cholesterol and prevented the secretion of cholesteryl esters derived from the plasma membrane. Stigmasterol and campesterol had effects similar to beta-sitosterol, although campesterol did not promote a modest influx of plasma membrane cholesterol.

Including β-sitosterol in the micelle with cholesterol decreased the uptake of cholesterol. Compared to cholesterol, 60% less beta-sitosterol was taken up by CaCo-2 cells. Cholesterol synthesis and HMG-CoA reductase activities were decreased in cells incubated with β-sitosterol. This was associated with a decrease in reductase mass and mRNA levels. Cholesteryl ester synthesis and ACAT activities were unaltered by β-sitosterol. Both stigmasterol and campesterol decreased reductase activity, but only campesterol increased ACAT activity. β-sitosterol did not affect the secretion of apoB mass (Field et al. 1997).

19.3.7 Metabolism

The metabolism of β-sitosterol was compared to that of cholesterol in 12 patients. Sterol balance methods were supplemented by radio sterol studies, with the following results.

(a) Plasma concentrations of β-sitosterol ranged from 0.30 to 1.02 mg/100 ml plasma in patients on intakes of β-sitosterol typical of the American diet. Plasma levels were raised little when intakes were increased greatly, and on fixed intakes they were constant from week to week. On diets devoid of plant sterols, the plasma and feces rapidly became free of beta-sitosterol.

(b) The percentage of esterified β-sitosterol in the plasma was the same as for cholesterol. However, the rate of esterification of beta-sitosterol was slower than that for cholesterol.

(c) Specific activity-time curves after simultaneous pulse labeling with β-sitosterol-(3)H and cholesterol-(14)C conformed to two-pool models. The two exponential half-lives of β-sitosterol were much shorter than for cholesterol, and pool sizes were much smaller.
Values of turnover for β-sitosterol obtained by the sterol balance method agreed closely with those derived by use of the two-pool model. Absorption of β-sitosterol was 5% (or less) of daily intake, while cholesterol absorption ranged from 45 to 54% of intake.

(d) About 20% of the absorbed β-sitosterol was converted to cholic (Fig. 19.2) and chenodeoxycholic acids (Fig. 19.3). The remainder was excreted in bile as free sterol; this excretion was more rapid than that of cholesterol.

(e) The employment of β-sitosterol as an internal standard to correct for losses of cholesterol in sterol balance studies is further validated by the results presented here (Salen et al. 1970).

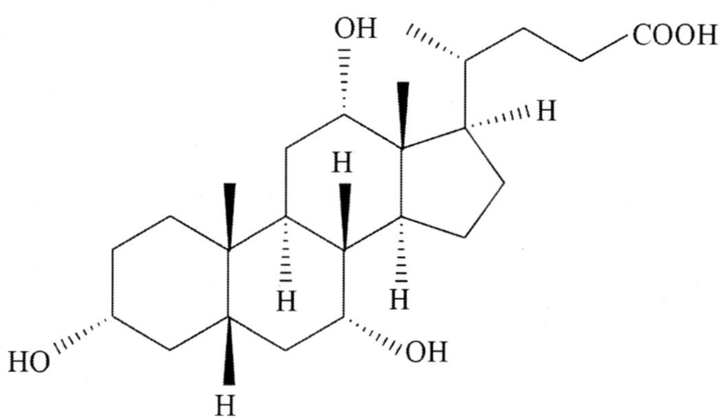

Fig. 19.2: Structure of cholic acid.

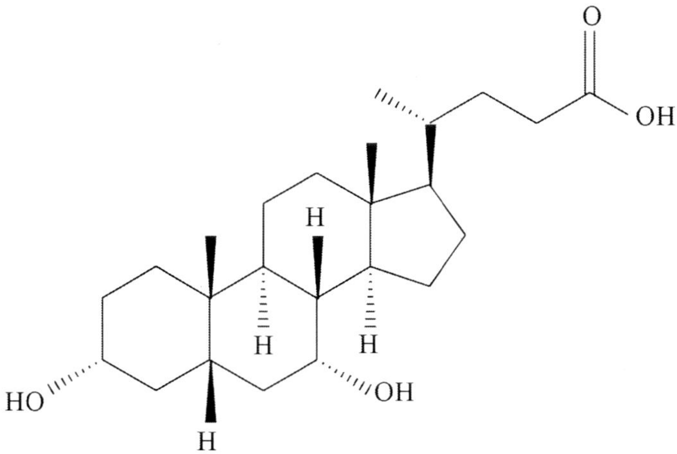

Fig. 19.3: Structure of chenodeoxycholic acid.

19.3.8 Toxicity

19.3.8.1 Chronic toxicity

Chronic administration of β-sitosterol subcutaneously to rats for 60 days was well tolerated and there was no clear cut evidence of any gross or microscopic lesions either in the liver or kidney. Liver and kidney function tests were assessed by determining the blood/serum parameters like haemoglobin, blood glucose, serum protein, serum bilirubin, serum cholesterol, serum GPT, and serum GOT.

All the parameters were in the normal range except serum protein and serum cholesterol. Serum cholesterol was the only variable which depleted markedly in both sexes in a dose-dependent manner suggesting intrinsic hypocholesterolemic effect of the sterol (Malini and Vanithakumari 1990).

19.3.8.2 Genotoxicity

The present study evaluated the genotoxic and cytotoxic potential of sterols by determining the capacity of the compounds to induce sister chromatid exchanges (SCE), or to alter cellular proliferation kinetics (CPK) and the mitotic index (MI) in mouse bone marrow cells. Besides, it also determined their capacity to increase the rate of micronucleated polychromatic erythrocytes (MNPE) in peripheral mouse blood, and the relationship polychromatic erythrocytes/normochromatic erythrocytes (PE/NE) as an index of cytotoxicity.

For the first assay, four doses of each compound were tested: 200, 400, 600, and 1000 mg/kg in case of BS, and 100, 200, 300, and 600 mg/kg for PT. The results in regard to both agents showed no SCE increase induced by any of the tested doses, as well as no alteration in the CPK, or in the MI. With respect to the second assay, the results obtained with the two agents were also negative for both the MNPE and the PE/NE index along the daily evaluation made for four days. In the present study, the highest tested dose corresponded to 80% of the LD50 obtained for BS and to 78% in the case of PT (Paniagua-Pérez et al. 2005).

19.3.9 Clinical study

19.3.9.1 Arteriosclerotic heart disease

In this study, 15 young men with previous myocardial infarction were given 12 to 18 Gm/day of β-sitosterol with resultant sustained reductions of serum cholesterol and beta lipoprotein lipid. Lipoprotein lipid fractionation was performed by paper electrophoresis.

> These changes occurred irrespective of initial serum cholesterol or content of diet. Control observations of the effects of diet, weight maintenance, and the inclusion of placebos brought about increased confidence that the changes were due to the administered sitosterol (Farquhar et al. 1956).

19.3.9.2 Benign prostatic hyperplasia

A. In a randomised, double-blind, placebo-controlled multicentre study, 200 patients with symptomatic benign prostatic hyperplasia were treated with either 20 mg β-sitosterol three times per day or placebo.

   Primary end-point was a difference of modified Boyarsky score between treatment groups after six months; secondary end-points were changes in International Prostate Symptom Score (IPSS), urine flow, and prostate volume. Modified Boyarsky score decreased significantly with a mean of –6.7 (SD 4.0) points in the β-sitosterol-treated group versus –2.1 (3.2) points in the placebo group $p < 0.01$.

   There was a decrease in IPSS (–7.4 [3.8] points in the β-sitosterol-treated group vs –2.1 [3.8] points in the placebo group) and changes in urine flow parameters: beta-sitosterol treatment resulted in increasing peak flow (15.2 [5.7] mL/s from 9.9 [2.5] mL/s), and decrease of mean residual urinary volume (30.4 [39.9] mL from 65.8 [20.8] mL). These parameters did not change in the placebo group ($p < 0.01$). No relevant reduction of prostatic volume was observed in either group (Berges et al. 1995).

B. A randomized, double-blind, and placebo-controlled clinical trial was conducted to assess the efficacy and safety of 130 mg free β-sitosterol daily, using the international prostate symptom score (IPSS) as the primary outcome variable. In total, 177 patients with BPH were recruited for 6 months of treatment in 13 study centres.

   In addition to the relative difference in the IPSS, changes in quality of life, peak urinary flow rate (Q max) and post-void residual urinary volume (PVR) were recorded. The drug used in the trial consisted of a chemically defined extract of phytosterols, derived for example from species of *Pinus*, *Picea* or *Hypoxis*, with β-sitosterol as the main component.

There were significant ($P < 0.01$) improvements over placebo in those treated with beta-sitosterol; the mean difference in the IPSS between placebo and beta-sitosterol, adjusted for the initial values, was 5.4 and in the quality-of-life index was 0.9. There were also significant improvements in the secondary outcome variables, with an increase in Q max (4.5 mL/s) and decrease in PVR (33.5 mL) in favour of beta-sitosterol when adjusted for the changes after placebo (Klippel et al. 1997).

C. A study determined the long-term effects of phytotherapy with β-sitosterol for symptomatic benign prostatic hyperplasia. At 18 months after enrolment in a 6-month multicentre double-blind placebo-controlled clinical trial with β-sitosterol (reported previously), patients were re-evaluated using the modified Boyarsky score, the International Prostate Symptom Score and quality-of-life index, the maximum urinary flow rate (Q max), and post void residual urine volume (PVR). In this open extension of the original trial (after six months of treatment or placebo), patients were free to chose their further treatment for BPH.

In all, 117 patients (59%) were eligible for analysis during the follow-up. Of the former β-sitosterol group, 38 patients who continued β-sitosterol treatment had stable values for all outcome variables between the end of the double-blind study and after 18 months of follow-up.

The 41 patients choosing no further therapy had slightly worse symptom scores and PVR, but no changes in Q max. Of the former placebo group, 27 patients who started β-sitosterol after the double-blind trial improved to the same extent as the treated group for all outcome variables. The 18 patients choosing no further therapy showed no signs of improvement (Berges et al. 2000).

19.3.10 Sitostanol (Stigmastanol)

19.3.10.1 Introduction

Sitostanol (Stigmastanol) (Fig. 19.4) is a phytosterol found in a variety of plant sources. Sitostanol (Fig. 19.4) is the product of the reduction of β-sitosterol (Crompton et al. 1985) and the hydrogenation of stigmasterol (Paxena 2007). Sitostanol is made from vegetable oils or the oil from pine tree wood pulp, and is then combined with canola oil.

19.3.10.2 Other names

(3β)-Stigmastan-3-ol; (3β,5α)-Stigmastan-3-ɔl; β-Sitostanol; Dihydrositosterin; Dihydrositosterol; Dihydro-β-sitosterol; Fucostanol; Spinastanol; 24α-Ethylcholestanol.

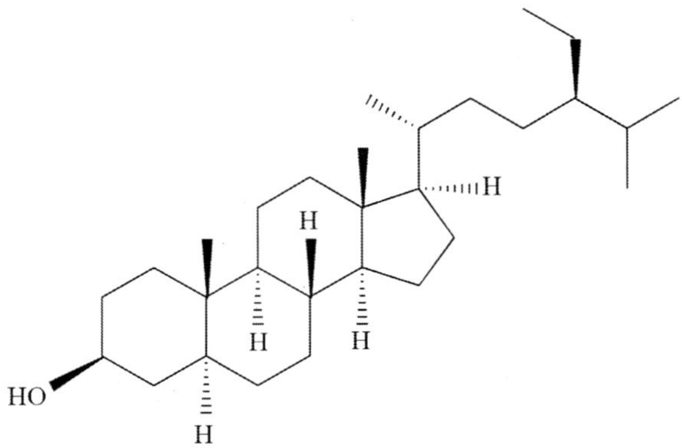

Fig. 19.4: Structure of sitostanol.

19.3.10.3 Pharmacology

Similar to sterol esters and stanol esters, stigmastanol inhibits the absorption of cholesterol from the diet. Animal studies suggest that it also inhibits biosynthesis of cholesterol in the liver (Heinemann et al. 1988; Batta et al. 2006). Consumption of the sitostanol-containing mixture (1% dietary levels) caused a compensatory increase in cholesterol synthesis as indicated by elevated ($P < 0.05$) lathosterol/cholesterol ratios in plasma and hepatic cholesterol fractional synthesis rate (FSR) ($P < 0.02$). Both sitostanol and sitostanol-free mixtures at 0.5% or 1% dietary intake levels increased plasma campesterol and beta-sitosterol levels, while plasma sitostanol levels were negligible (Ling and Jones 1995).

19.3.10.4 Sitostanol and FDA

The U.S. Food and Drug Administration (FDA) allows manufacturers of products that contain sitostanol or related plant chemicals (stanol esters) to claim that the product lowers the risk of getting coronary heart disease (CHD). The FDA reasons that sitostanol and other plant stanol esters along with a diet low in saturated fat and cholesterol might reduce the risk of CHD by lowering blood cholesterol levels.

19.3.10.5 Clinical study

The effects of two different plant sterols on intestinal cholesterol absorption were compared in normal volunteers by an intestinal perfusion study during a control period followed by high dose infusion of sitosterol or sitostanol (3.6 mumol/min), to which subjects were allocated in a randomized manner.

> Cholesterol absorption during the control period was similar in the two groups, averaging 0.88 +/– 0.48 mumol/min (32 +/– 11%) for group I (sitosterol) and 0.68 +/– 0.33 mumol/min (29 +/– 9%) for group II (sitostanol). The infusion of a high dose of sitosterol resulted in a significant reduction of cholesterol absorption to 0.47 mumol/min (16%). Following the same dose of sitostanol,

cholesterol absorption diminished significantly to 0.15 +/– 0.11 mumol/min (5.1 +/– 2.9%).

Overall cholesterol absorption declined during sitosterol infusion by almost 50%, whereas sitostanol infusion caused a reduction of cholesterol absorption by almost 85%. These findings of a more effective inhibition of cholesterol absorption by sitostanol might confirm the observation recorded by others that an increase in hydrophobicity of a plant sterol results in a higher affinity but lower capacity to mixed micells. This may cause an effective displacement of cholesterol from micellar binding and therefore diminished cholesterol absorption (Heinemann et al. 1991).

19.3.11 Sterolins

Sterolins are glucosides, which are molecular structures joined to the sterol. Sterolin is easily destroyed, and without it, the sterol does not have the same immune-enhancing benefits. In nature, plants never contain sterols only. The sterols are always associated with their glucoside sterolin.

19.3.12 *β-sitosterol glucoside* (Fig. 19.5)

The phytosterols, beta-sitosterol (BSS), and its glucoside (BSSG) enhance the *in vitro* proliferative response of T-cells stimulated by sub-optimal concentrations of phytohaemagglutinin (PHA) several fold at extremely low concentrations (femtogram level). A 100:1 (mass:mass) ratio of BSS:BSSG (termed essential sterolin formulation, ESF) showed higher stimulation than the individual sterols at the same concentration (Bouic et al. 1996).

β-sitosterol, β-sitosterol glucoside, and a mixture of β-sitosterol and β-sitosterol glucoside have modulating effect on the growth of estrogen-responsive breast cancer vells *in vitro* and in ovariectomizedathymic mice (Ju et al. 2004).

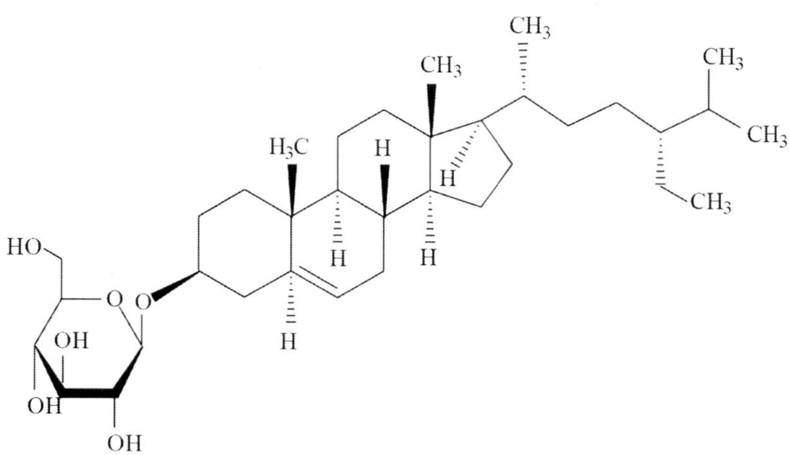

Fig. 19.5: Structure of β-sitosterol glucoside.

19.3.13 Campesterol

19.3.13.1 Introduction

Campesterol (Fig. 19.6) is a phytosterol whose chemical structure is similar to that of cholesterol.

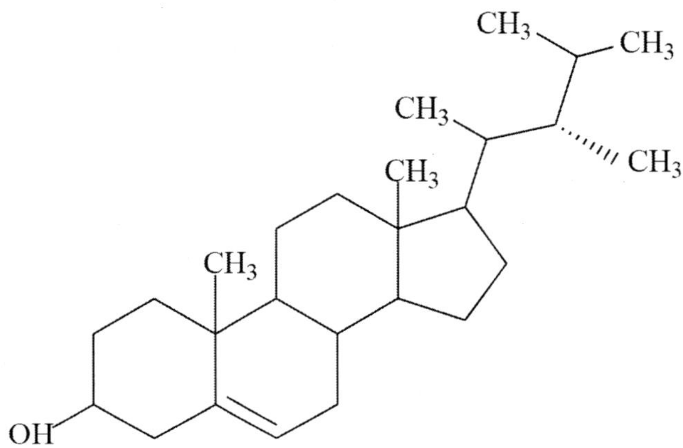

Fig. 19.6: Structure of campesterol.

19.3.13.2 Sources

Many vegetables, fruits, nuts, and seeds contain campesterol, but in low concentrations. Banana, pomegranate, pepper, coffee, grapefruit, cucumber, onion, oat, potato, and lemon grass (citronella) are few examples of common sources containing campesterol at ~1–7 mg/100 g of the edible portion. In contrast canola and corn oil contain as much as 16–100 mg/100 g.

19.3.13.3 Pharmacology

19.3.13.3.1 Antiangiogenic

A study investigated the effect of campesterol on basic fibroblast growth factor (bFGF)-induced angiogenesis *in vitro* in human umbilical vein endothelial cells (HUVECs) and an *in vivo* chorioallantoic membrane (CAM) model.

> Campesterol isolated from an ethylacetate fraction of *Chrysanthemum coronarium* L. (Asteraceae) showed a weak cytotoxicity in non-proliferating HUVECs. Within the non-cytotoxic concentration range, campesterol significantly inhibited the bFGF-induced proliferation and tube formation of HUVECs in a concentration-dependent manner, while it did not affect the motility of HUVECs. Furthermore, campesterol effectively disrupted the bFGF-induced neovascularization in chick chorioallantoic membrane (CAM) *in vivo* (Choi et al. 2007).

19.3.13.3.2 Antioxidant

In the present study, the antioxidant effects of phytosterol and its components, β-sitosterol, stigmasterol, and campesterol, against lipid peroxidation were examined by making a comparison with 2, 2, 5, 7, 8-pentamethyl-6-chromanol (PMC).

> It was found the sterols exerted antioxidant effects on the oxidation of methyl linoleate in solution and its effect decreased in the order of: PMC >> phytosterol approximately campesterol approximately beta-sitosterol > stigmasterol. Phytosterol also suppressed the oxidation and consumption of alpha-tocopherol in beta-linoleoyl-gamma-palmitoylphosphatidylcholine (PLPC) liposomal membranes, the effects being more significant than dimyristoyl PC of the same concentration. Stigmasterol accelerated the oxidation of both methyl linoleate in solution and PLPC liposomal membranes in aqueous dispersions, which was ascribed to the oxidation of allylichydrogens at the 21- and 24-positions (Yoshida and Niki 2003).

19.3.13.3.3 Adverse effect

There is no data to suggest that modestly elevated levels of campesterol have a negative cardiac impact (Calpe-Berdiel et al. 2009).

19.3.13.3.4 Effect of dalcetrapib on non-cholesterol sterol markers of cholesterol homeostasis

A recent trial with dalcetrapib (Fig. 19.7), a cholesteryl esterase transport protein (CETP) inhibitor. Cholesteryl esterase transport protein inhibitor is a new class of cholesterol lowering medications currently in development. It showed that this agent may have the potential to increase levels of campesterol through increasing intestinal absorption. Dalcetrapib specifically increased markers of cholesterol absorption, most likely reflecting nascent HDL lipidation by intestinal ABCA1, without affecting markers of synthesis (Niesor et al. 2011).

19.3.14 Stigmasterol

19.3.14.1 Introduction

Stigmasterol (Fig. 19.8) is an unsaturated plant sterol, also known as Wulzen anti-stiffness factor.

19.3.14.2 Sources

Stigmasterol is found in various vegetables, legumes, nuts, seeds, and unpasteurized milk. Pasteurization will inactivate stigmasterol. Edible oils contains higher amount than vegetables.

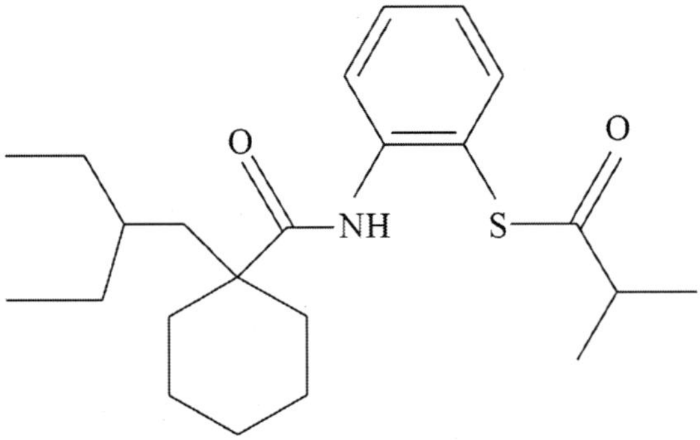

Fig. 19.7: Structure of dalcetrapib.

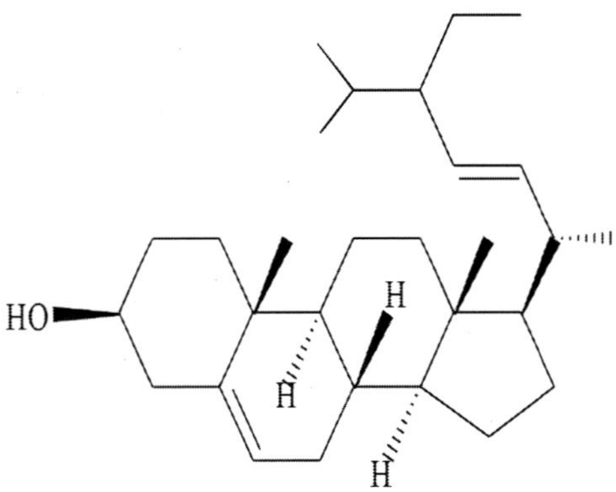

Fig. 19.8: Structure of stigmasterol.

19.3.14.3 Pharmacology

19.3.14.3.1 Anti-osteoarthritic

A study shows that stigmasterol inhibits several pro-inflammatory and matrix degradation mediators typically involved in OA-induced cartilage degradation, at least in part through the inhibition of the NF-κB pathway. These promising results justify further *ex vivo* and *in vivo* investigations with stigmasterol (Gabay et al. 2010).

19.3.14.3.2 Thyroid inhibitory, antiperoxidative, and hypoglycaemic

Administration of stigmasterol, isolated from the bark of *Buteamonosperma* at 2.6 mg/kg/d for 20 days reduced serum triiodothyronine (T(3)), thyroxin (T(4)), and glucose concentrations as well as the activity of hepatic glucose-6-phophatase (G-6-Pase) with a concomitant increase in insulin indicating its thyroid inhibiting and hypoglycemic properties. A decrease in the hepatic lipid peroxidation (LPO) and an increase in the activities of catalase (CAT), superoxide dismutase (SOD), and glutathione (GSH) suggested its antioxidative potential. The highest concentration tested (5.2 mg/kg) evoked pro-oxidative activity (Panda et al. 2009).

19.3.14.3.3 An antagonist of the bile acid nuclear receptor FXR

The study demonstrated that stigmasterol is a potent *in vitro* antagonist of the NR for bile acids FXR. In HepG2 cells, stigmasterol acetate (StigAc), (Fig. 19.9) a water-soluble stigmasterol derivative, suppressed ligand-activated expression of FXR target genes involved in adaptation to cholestasis (i.e., BSEP, FGF-19, OST alpha/beta). Furthermore, StigAc antagonized BA-activated, FXR target genes SHP and BSEP in FXR +/+, but not in FXR –/– mouse hepatocytes. Both Stig and StigAc inhibited BA-activated, FXR-dependent reporter gene expression in transfected HepG2 cells, whereas the most prevalent phytosterol in lipids, beta-sitosterol, had no inhibitory effect (Carter et al. 2007).

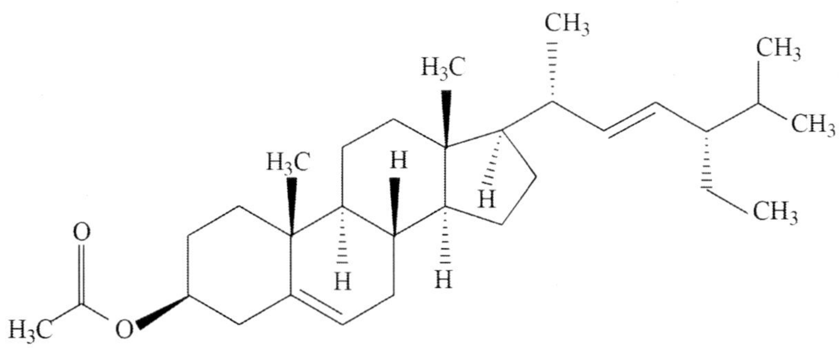

Fig. 19.9: Structure of stigmasterol acetate (StigAc).

19.3.14.3.4 Hypolipidemic

Stigmasterol, when fed, lowers plasma cholesterol levels, inhibits intestinal cholesterol and plant sterol absorption, and suppresses hepatic cholesterol and classic bile acid synthesis in Wistar as well as WKY rats.

19.3.14.4 Toxicity

The Panel on Food Additives and Nutrient Sources added to Food (ANS) concluded that stigmasterol-rich plant sterols are not of concern with respect to genotoxicity. Toxicity studies on phytosterols and phytosterol esters were limited to 90-day subchronic toxicity studies and a 2-generation reproductive toxicity study in rats. No chronic toxicity, carcinogenicity, or developmental toxicity studies conducted with phytosterols, phytostanols, and their esters were identified. For adults, mean and 95th percentile exposures to stigmasterol-rich plant sterols from the proposed uses and use levels were 0.01–0.2 mg/kg bw/day and 0.4–7.4 mg/kg bw/day, respectively.

Using the lowest NOAEL values of 1.54 g phytosterols/kg bw/day (335 mg stigmasterol/kg bw/day) the calculated Margin of Safety (MOS) values amount to 7700–154000 at the mean and 208–3850 at the 95th percentile for the phytosterols and to 1675–33500 at the mean and 45–838 at the 95th percentile for stigmasterol. The panel considered these MOS values adequate and concluded that the proposed use and use levels of stigmasterol-rich plant sterols in ready-to-freeze alcoholic cocktails as a stabiliser would not be of safety concern (Anonymous 2012).

19.3.15 Brassicasterol

Brassicasterol (Fig. 19.10) is a sterol found in certain plants and other foods such as seafood. Brassicaterol is found in high concentrations in brassica, also known as rapeseed oil. There are no human trials examining the health benefit of brassicasterol when given as a supplement by itself, but it would make sense that brassicaterol would have similar physiological effects as other sterols such as sitosterol and stigmasterol.

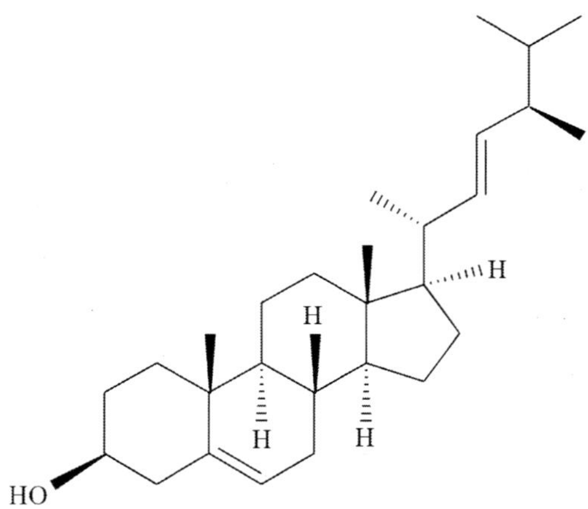

Fig. 19.10: Structure of brassicasterol.

19.3.16 Ergosterol

Ergosterol (Fig. 19.11) is found in cell membranes of fungi and protozoa, serving many of the same functions that cholesterol serves in animal cells. Because many fungi and protozoa cannot survive without ergosterol, the enzymes that create it have become important targets for drug discovery. Ergosterol is a provitamin form of vitamin D2; exposure to ultraviolet (UV) light causes a chemical reaction that produces vitamin D2.

19.3.17 Fucosterol

Fucosterol (Fig. 19.12) from the marine algae *Pelvetiasiliquosa*, exhibited a significant decrease in serum transaminase activities elevated by hepatic damage induced by CCl4-intoxication in rats (Lee et al. 2003). An investigation of a

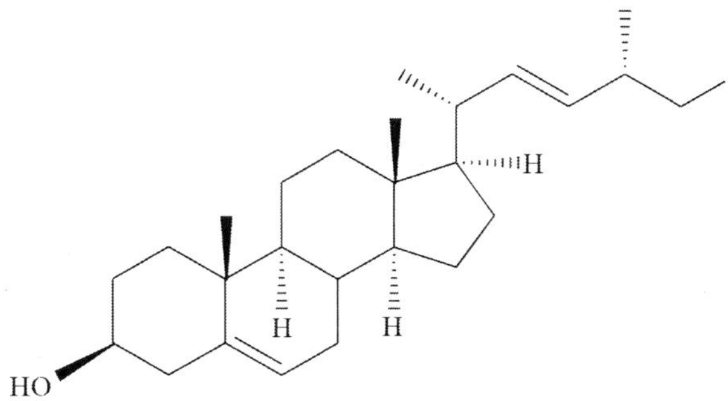

Fig. 19.11: Structure of ergosterol.

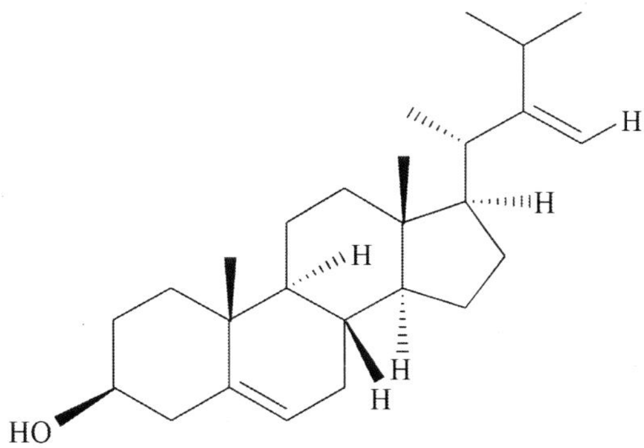

Fig. 19.12: Structure of fucosterol.

component from the hexane fraction of *Sargassum angustifolium* yielded a steroidal metabolite, fucosterol, with cytotoxicity in T47D and HT29 (IC_{50} 27.94 ± 9.3 and 70.41 ± 7.5 μg/ml) (Khanavi et al. 2012). Anti-inflammatory of fucosterol isolated from *Undaria pinnatifida* are associated with the suppression of the NF-κB and p38 MAPK pathways (Yoo et al. 2012). Fucosterol isolated from *Eiseniabicyclis* and *Eckloniastolonifera* with anti-diabetic activity (Jung et al. 2013).

19.3.18 *Isofucosterol* (Fig. 19.13)

Three compounds (chlorophyll a, isofucosterol, and saringosterol) were isolated from chloroform fraction of *Sargassum thunbergii* extract. The three compounds had two- to four fold lower lipase inhibitory activity than that of the CHCl3:MeOH (C:M) (100:1) fraction (fraction I, 83.78% at 1 mg/mL) (Kim et al. 2014).

19.2.19 *Saringosterol*

Saringosterol (Fig. 19.14), a derivative of fucosterol, was discovered in *Lessonia nigrescens* Bory de Saint-Vincent (Lessoniaceae), and *Sargassum ringgoldianum* Harvey (Sargassaceae). Saringosterol isolated from *L. nigrescens* has activity against

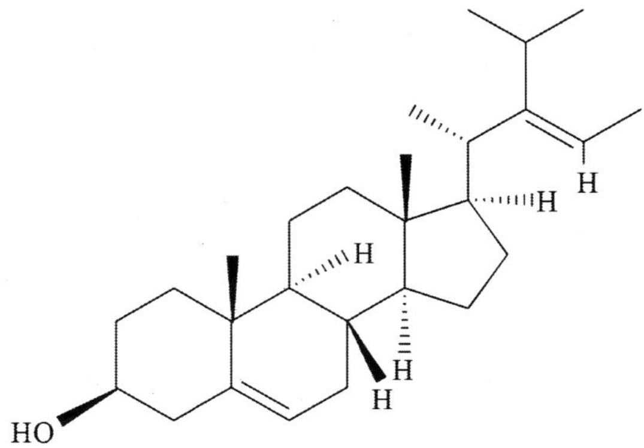

Fig. 19.13: Structure of isofucosterol.

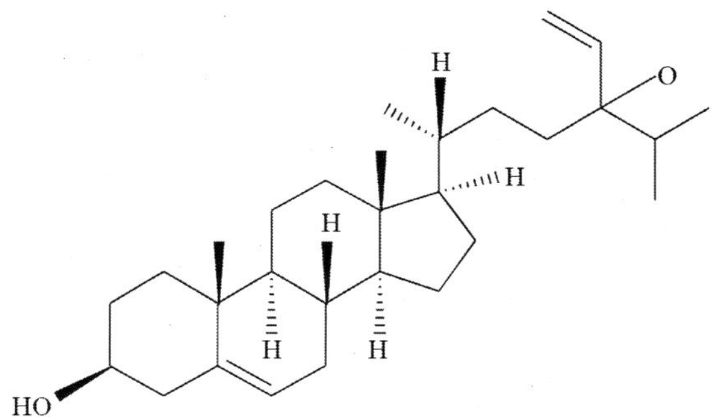

Fig. 19.14: Structure of saringosterol.

Mycobacterium tuberculosis (Wächter et al. 2001). It has high lipase inhibitory activity (Kim et al. 2014).

19.2.20 Avenasterol

Avenasterol is a natural, non-cholesterol sterol (Fig. 19.15). It is found in *Alhagicamelorum* Fisch. Ex DC. (Fabaceae), *Galiumaparine* Linn. (Rubiaceae), *Lagenariasiceraria* (Molina) Standley (Cucurbitaceae), and *Nigella sativa* Linn. (Ranunculaceae), and *Sclerocaryabirrea* Marula (Anacardiaceae).

19.2.21 Saoussazine (28-hydroxy-24-methyl-β-sitosterol)

A new sterol (Fig. 19.16) has been reported from the brown alga *Cytoseira compressa* (Cystoseiraceae).

19.2.22 Antrosterol

Antrosterol (Fig. 19.17) is an isomeric form of ergosterol. The hepatoprotective potential of antrosterol from *Antrodiacamphorata* (M. Zang and C.H. Su) Sheng H.

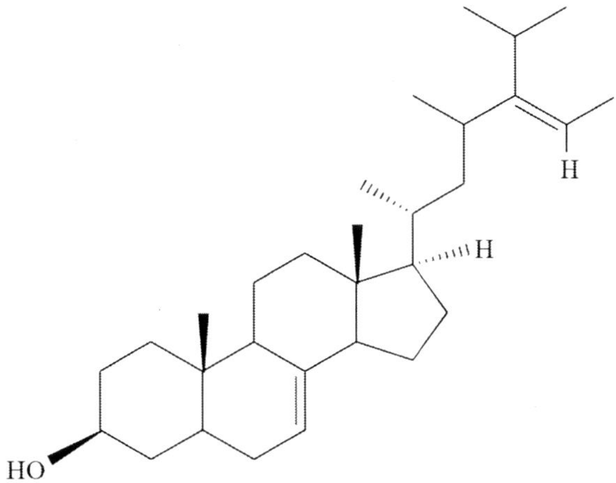

Fig. 19.15: Structure of avenasterol.

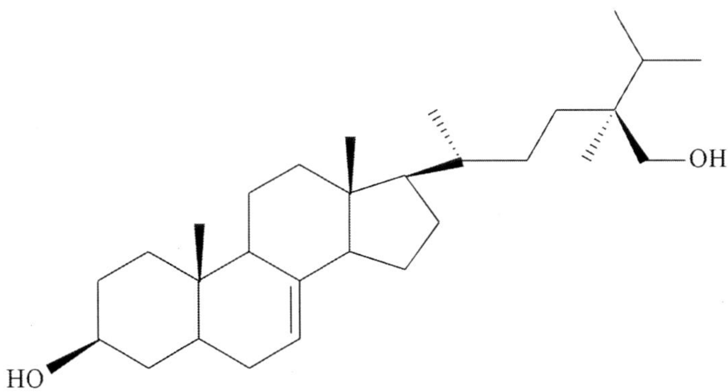

Fig. 19.16: Structure of saoussazine.

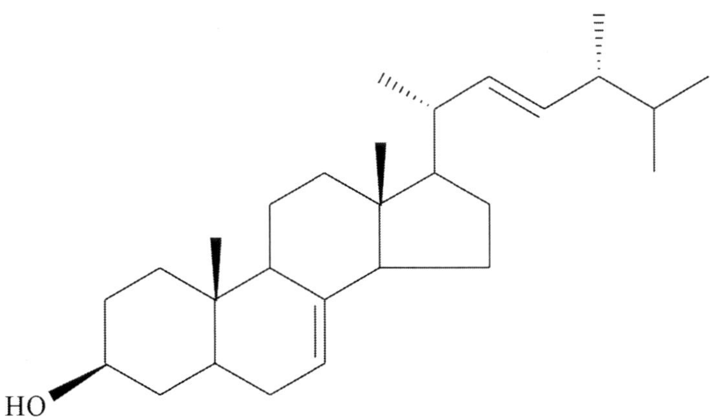

Fig. 19.17: Structure of antrosterol.

Wu, Ryvarden and T.T. Chang (incertaesedis) against carbon tetrachloride-induced liver damage was evaluated in preventive models in mice. Antrosterol was supposed to demonstrate hepatoprotective effect in mice through their anti-inflammation capacity (Huanga et al. 2012).

Further Readings

Anonymous. Scientific opinion on the safety of stigmasterol-rich plant sterols as food additive. *EFSA Journal* 2012; **10**: 2659.

Awad AB, Chen YC, Fink CS, Hennessey T. beta-Sitosterol inhibits HT-29 human colon cancer cell growth and alters membrane lipids. *Anticancer Res* 1996; **16**: 2797–804.

Baskar AA, Ignacimuthu S, Paulraj GM, Numair KS. Chemopreventive potential of β-Sitosterol in experimental colon cancer model—an *in vitro* and *in vivo* study. *BMC Complementary and Alternative Medicine* 2010; **10**: 24.

Batta AK, Xu G, Honda A, Miyazaki T, Gerald S. Stigmasterol reduces plasma cholesterol levels and inhibits hepatic synthesis and intestinal absorption in the rat. *Metabol Clin Exp* 2006; **55**: 292–299.

Berges RR, Windeler J, Trampisch HJ, Senge T. Randomised, placebo-controlled, double-blind clinical trial of beta-sitosterol in patients with benign prostatic hyperplasia. Beta-sitosterol Study Group. *Lancet* 1995; **345**: 1529–32.

Berges RR, Kassen A, Senge T. Treatment of symptomatic benign prostatic hyperplasia with beta-sitosterol: an 18-month follow-up. *BJU Int* 2000; **85**: 842–6.

Bouic PJ, Etsebeth S, Liebenberg RW, Albrecht CF, Pegel K, Van Jaarsveld PP. beta-Sitosterol and beta-sitosterol glucoside stimulate human peripheral blood lymphocyte proliferation: implications for their use as an immunomodulatory vitamin combination. *Int J Immunopharmacol* 1996; **18**: 693–700.

Calpe-Berdiel L, Méndez-González J, Blanco-Vaca F, Carles Escolà-Gil J. Increased plasma levels of plant sterols and atherosclerosis: A controversial issue. *Curr Atheroscler Rep* 2009; **11**: 391–8.

Carter BA, Taylor OA, Prendergast DR, Zimmerman TL, Von Furstenberg R, Moore DD, Karpen SJ. Stigmasterol, a soy lipid-derived phytosterol, is an antagonist of the bile acid nuclear receptor FXR. *Pediatr Res* 2007; **62**: 301–6.

Choi JM, Lee EO, Lee HJ, Kim KH, Ahn KS, Shim BS, Kim NI, Song MC, Baek NI, Kim SH. Identification of campesterol from *Chrysanthemum coronarium* L. and its antiangiogenic activities. *Phytother Res* 2007; **21**: 954–9.

Crompton T, William D, Nickol BB. Biology of the Acanthocephala. Cambridge University Press, 1985; p. 185.

Farquhar JW, Smith RE, Dempsey ME. The effect of beta-sitosterol on the serum lipids of young men with arteriosclerotic heart disease. *Circulation* 1956; **14**: 77–82.

Field FJ, Born E, Mathur SN. Effect of micellar beta-sitosterol on cholesterol metabolism in CaCo-2 cells. *J Lipid Res* 1997; **38**: 348–60.

Gabay O, Sanchez C, Salvat C, Chevy F, Breton M, Nourissat G, Wolf C, Jacques C, Berenbaum F. Stigmasterol: a phytosterol with potential anti-osteoarthritic properties. *Osteoarthritis Cartilage* 2010; **18**: 106–16.

Heinemann T, Pietruck B, Kullak-Ublick G, von Bergmann K. Comparison of sitosterol and sitostanol on inhibition of intestinal cholesterol absorption. *Supplements* 1988; **26**: 117–122.

Heinemann T, Kullak-Ublick GA, Pietruck B, von Bergmann K. Mechanisms of action of plant sterols on inhibition of cholesterol absorption. Comparison of sitosterol and sitostanol. *Eur J Clin Pharmacol* 1991; **40**: 59–63.

Huanga G-J, Dengb J-S, Huanga S-S, Shaoc Y-Y, Chenc C-C, Kuo Y-H. Protective effect of antrosterol from *Antrodia camphorata* submerged whole broth against carbon tetrachloride-induced acute liver injury in mice. *Food Chem* 2012; **132**: 709–716.

Jourdain C, Tenca G, Deguercy A, Troplin P, Poelman D. *In-vitro* effects of polyphenols from cocoa and beta-sitosterol on the growth of human prostate cancer and normal cells. *Eur J Cancer Prev* 2006; **15**: 353–61.

Ju Young H, Clausen Laura M, Allred Kimberly F, Almada Anthony L, Helferich William G. β-Sitosterol, β-Sitosterol Glucoside, and a mixture of β-Sitosterol and β-Sitosterol glucoside modulate the growth of estrogen-responsive breast cancer cells *in vitro* and in ovariectomized athymic mice. *J Nutr* 2004; **134**: 1145–1151.

Jung HA, Islam MN, Lee CM, Oh SH, Lee S, Jung JH, Choi JS. Kinetics and molecular docking studies of an anti-diabetic complication inhibitor fucosterol from edible brown algae *Eisenia bicyclis* and *Ecklonia stolonifera*. *Chem Biol Interact* 2013; **206**: 55–62.

Kassen A, Berges R, Senge T. Effect of beta-sitosterol on transforming growth factor-beta-1 expression and translocation protein kinase C alpha in human prostate stromal cells *in vitro*. *Eur Urol* 2000; **37**: 735–41.

Khanavi M, Gheidarloo R, Sadati N, Ardekani MR, Nabavi SM, Tavajohi S, Ostad SN. Cytotoxicity of fucosterol containing fraction of marine algae against breast and colon carcinoma cell line. *Pharmacogn Mag* 2012; **8**: 60–4.

Kim KB, Kim MJ, Ahn DH. Lipase inhibitory activity of chlorophyll a, isofucosterol and saringosterol isolated from chloroform fraction of *Sargassum thunbergii*. *Nat Prod Res* 2014; **28**: 1310–2.

Klippel KF, Hiltl DM, Schipp B. A multicentric, placebo-controlled, double-blind clinical trial of beta-sitosterol (phytosterol) for the treatment of benign prostatic hyperplasia. German BPH-Phyto Study group. *Br J Urol* 1997; **80**: 427–32.

Lee S, Lee YS, Jung SH, Kang SS, Shin KH. Anti-oxidant activities of fucosterol from the marine algae *Pelvetia siliquosa*. *Arch Pharm Res* 2003; **26**: 719–22.

Ling WH, Jones PJ. Enhanced efficacy of sitostanol-containing versus sitostanol-free phytosterol mixtures in altering lipoprotein cholesterol levels and synthesis in rats. *Atherosclerosis* 1995; **118**: 319–31.

Loizou S, Lekakis I, Chrousos GP, Moutsatsou P. Beta-sitosterol exhibits anti-inflammatory activity in human aortic endothelial cells. *Mol Nutr Food Res* 2010; **54**: 551–8.

Malini T, Vanithakumari G. Rat toxicity studies with beta-sitosterol. *J Ethnopharmacol* 1990; **28**: 221–34.

Malini T, Vanithakumari G. Antifertility effects of beta-sitosterol in male albino rats. *J Ethnopharmacol* 1991; **35**: 149–153.

Niesor EJ, Chaput E, Staempfli A, Blum D, Derks M, Kallend D. Effect of dalcetrapib, a CETP modulator, on non-cholesterol sterol markers of cholesterol homeostasis in healthy subjects. *Atherosclerosis* 2011; **219**: 761–7.

Nirmal SA, Pal SC, Mandal SC, Patil AN. Analgesic and anti-inflammatory activity of β-sitosterol isolated from *Nyctanthes arbortristis* leaves. *Inflammopharmacology* 2012; **20**: 219–24.

Panda S, Jafri M, Kar A, Meheta BK. Thyroid inhibitory, antiperoxidative and hypoglycemic effects of stigmasterol isolated from *Butea monosperma*. *Fitoterapia* 2009; **80**: 123–6.

Paniagua-Pérez R, Madrigal-Bujaidar E, Reyes-Cadena S, Molina-Jasso D, Pérez Gallaga J, Silva-Miranda A, Velazco O, Hernández N, Chamorro G. Genotoxic and cytotoxic studies of beta-sitosterol and pteropodine in mouse. *J Biomed Biotechnol* 2005; **3**: 242–247.

Paxena PB. Chemistry of Alkaloids. U.K. Discovery Publishing House, 2007; p. 231.

Salen G, Ahrens EH Jr, Grundy SM. Metabolism of beta-sitosterol in man. *J Clin Invest* 1970; **49:** 952–67.

Sugano M, Ikeda I, Imaizumi K, Watanabe M, Andoh M. Effects of beta-sitosterol on the concentrations of serum and liver cholesterol and serum apolipoproteins in rats fed butter fat. *J Nutr Sci Vitaminol* (Tokyo) 1982; **28**: 117–26.

Villaseñor IM, Angelada J, Canlas AP, Echegoyen D. Bioactivity studies on beta-sitosterol and its glucoside. *Phytother Res* 2002; **16:** 417–21.

Wächter GA, Franzblau SG, Montenegro G, Hoffmann JJ, Maiese WM, Timmermann BN. Inhibition of Mycobacterium tuberculosis growth by saringosterol from Lessonia nigrescens. *J Nat Prod* 2001; **64**: 1463–4.

Yoo MS, Shin JS, Choi HE, Cho YW, Bang MH, Baek NI, Lee KT. Fucosterol isolated from *Undaria pinnatifida* inhibits lipopolysaccharide-induced production of nitric oxide and pro-inflammatory cytokines via the inactivation of nuclear factor-κB and p38 mitogen-activated protein kinase in RAW264.7 macrophages. *Food Chem* 2012; **135**: 967–975.

Yoshida Y, Niki E. Antioxidant effects of phytosterol and its components. *J Nutr Sci Vitaminol* (Tokyo) 2003; **49**: 277–80.

Chapter 20

Pharmacology of Disogenin and Related Compounds

20.1 Introduction

Disogenin (Fig. 20.1) is a steroid sapogenin, extracted from the tubers of genus Dioscorea and other plants. The sugar-free (aglycone), disogenin is used for the commercial synthesis of cortisone, pregnenolone, progesterone, and other steroid products. (25R)-5-Spirosten-3β-ol, 3β-Hydroxy-5-spirostene is synonym for disogenin.

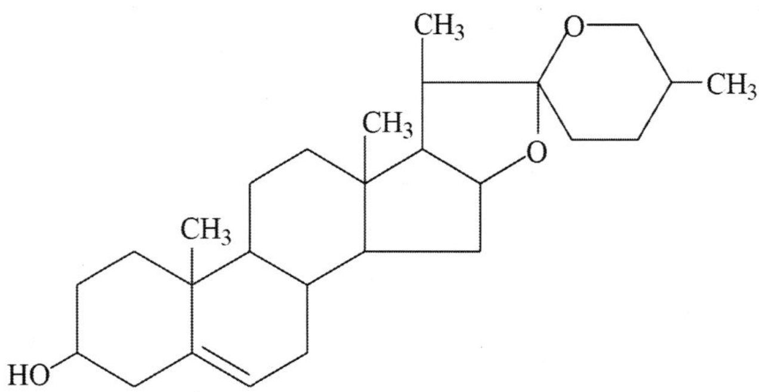

Fig. 20.1: Structure of diosgenin.

20.2 Pharmacology

20.2.1 Anticancer

20.2.1.1 Inhibition of cell growth in the HT-29 human colon cancer cell line

Dietary disogenin at 0.1% and 0.05% inhibited total colonic ACF and multicrypt foci formation in a dose-dependent manner. Results from the *in vitro* experiments indicated that disogenin inhibits cell growth and induces apoptosis in the HT-29 human colon cancer cell line in a dose-dependent manner. Furthermore, disogenin induced apoptosis in HT-29 cells at least in part by inhibition of bcl-2 and by induction of caspase-3 protein expression (Raju et al. 2004).

20.2.1.2 Inhibition of laryngocarcinoma HEp-2 and melanoma M4Beu cells

Disogenin had potential antiproliferative effect on different types of cancer cells. Disogenin-induced apoptosis is caspase-3 dependent with a fall of mitochondrial membrane potential, nuclear localization of AIF, and poly (ADP-ribose) polymerase cleavage. Disogenin treatment also induces p53 activation and cell cycle arrest in the different cell lines studied (Corbiere et al. 2004).

20.2.1.3 Induction of apoptosis in HCT-116 human colon carcinoma cells

The IC(50) cytotoxic dose of disogenin in HCT-116 was approximately 35 microM after 24 h, while concentrations of approximately 32 microM or greater decreased the percent viable cells by 50%. Higher doses of disogenin (30–40 microM) effectively inhibited recovery of cells for up to 24 h post-treatments (Raju and Bird 2007).

20.2.1.4 Inhibition of N-Methyl-N-nitrosourea-induced breast carcinoma

Breast cancer was induced in female Sprague Dawley rats by an intraperitoneal administration of a single dose of NMU (a concentration of 50 mg/kg body weight) diluted in 0.9% saline, and the rats were treated with oral disogenin, 20 mg/kg body weight, for 45 days. The disogenin treatment remarkably downregulated the peroxidation reaction and marker enzymes and extraordinarily enhanced the indigenous antioxidant defense system (Jagadeesan et al. 2012).

20.2.1.5 Squamous cell carcinoma and sarcoma 180-induced tumors

The study investigated the antineoplastic activity of thymoquinone (Fig. 20.2) and disogenin against squamous cell carcinoma *in vitro* and sarcoma 180-induced tumors *in vivo*. Thymoquinone and disogenin inhibited cell proliferation and induced cytotoxicity in A431 and Hep2 cells. Both inhibited Akt and JNK phosphorylations, thus inhibiting cell proliferation while inducing apoptosis (Das et al. 2012).

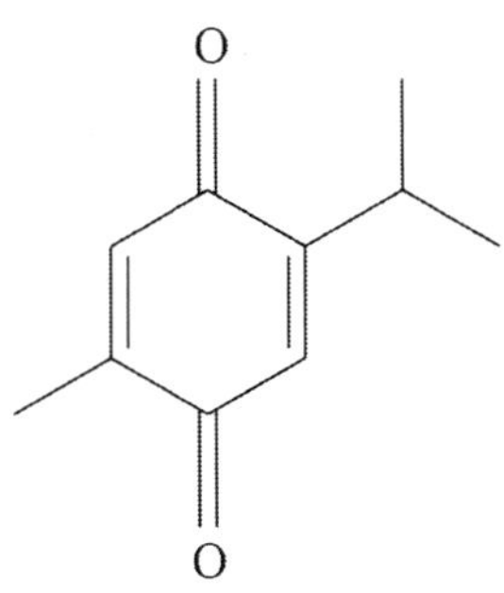

Fig. 20.2: Structure of thymoquinone.

20.2.1.6 Down-regulation of hTERT expression

A investigatory study was aimed at the inhibitory effect of disogenin on human telomerase reverse transcriptase gene (hTERT) expression which is critical for telomerase activity. MTT-assays and qRT-PCR analysis were conducted to assess cytotoxicity and hTERT gene expression inhibition effects, respectively. MTT results showed that IC50 values for 24, 48, and 72 h after treatment were 47, 44, and 43 μM, respectively. Culturing cells with disogenin treatment caused down-regulation of hTERT expression (Mohammad et al. 2013).

20.2.1.7 Inhibition of human breast cancer MDA-MB-231 cells

Disogenin caused a marked inhibition of cell migration in MDA-MB-231 cell by transwell assay. In addition, disogenin significantly impacted MDA-MB-231 cell migratory behavior under real-time observation. Disogenin significantly inhibited actin polymerization, Vav2 phosphorylation and Cdc42 activation, which might be, at least in part, attributed to the anti-metastatic potential of disogenin (He et al. 2014).

20.2.1.8 Human hepatocellular carcinoma

Disogenin significantly inhibited the growth of Bel-7402, SMMC-7721, and HepG2 HCC cells in a concentration-dependent manner. Disogenin treatment for 24 h induced G2/M cell cycle arrest and apoptosis of hepatoma cells. Disogenin inhibited Akt phosphorylation and upregulated p21 and p27 expression, but did not alter the expression of p53, suggesting disogenin-induced upregulation of p21 and p57 is p53-independent in HCC cells. Disogenin induced HCC cell apoptosis by activating caspase cascades-3, -8, and -9 (Li et al. 2015).

20.2.2 Antidiabetic

In this study, researchers determined the effects of fenugreek on adipocyte size and inflammation in adipose tissues in diabetic obese KK-Ay mice, and identified the active substance in fenugreek. Treatment of KK-Ay mice with a high fat diet supplemented with 2% fenugreek ameliorated diabetes. Moreover, fenugreek miniaturized the adipocytes and increased the mRNA expression levels of differentiation-related genes in adipose tissues.

Fenugreek also inhibited macrophage infiltration into adipose tissues and decreased the mRNA expression levels of inflammatory genes. In addition, disogenin was identified as a major aglycone of saponins in fenugreek to promote adipocyte differentiation and to inhibit expressions of several molecular candidates associated with inflammation in 3T3-L1 cells (Uemura et al. 2010).

After one week of STZ injection, fasting glucose level was measured in blood taken from the tail vein every 30 min for 150 min after injection of disogenin or dioscorea (3 mg/kg). On another day, muscle was resected 150 min after disogenin or dioscorea injections. Serum DHEA level increased significantly 120 min after disogenin or dioscorea injections; concomitantly, the blood glucose level decreased significantly. Moreover, GLUT4 translocation, as well as phosphorylation of Akt and PKC ζ/λ, increased significantly by disogenin or dioscorea administration. However, these effects of disogenin and dioscorea were blocked by a 5α-reductase inhibitor that inhibits synthesizing dehydrotestosterone from testosterone (Sato et al. 2014).

20.2.3 Anti-atherosclerotic

A study was designed to evaluate the potential effects of disogenin on macrophage cholesterol metabolism and the development of aortic atherosclerosis, and to explore its underlying mechanisms. The present study demonstrated that Dgn enhances ATP-binding cassette transporter A1 dependent cholesterol efflux and inhibits aortic atherosclerosis progression by suppressing macrophage miR-19b expression (Lv et al. 2015).

20.2.4 Anti-inflammatory

Disogenin potently and concentration-dependently inhibited the extracellular and intracellular superoxide anion generation in Formyl-Met-Leu-Phe (FMLP)-activated neutrophils, with IC50 values of 0.50 ± 0.08 μM and 0.66 ± 0.13 μM, respectively. Disogenin exhibited inhibitory effects on superoxide anion production through the blockade of cAMP, PKA, cPLA2, PAK, Akt, and MAPKs signaling pathways (Lin et al. 2014).

20.2.5 Antioxidant

This present study evaluated the dose-dependent effect of disogenin on high phosphate induced vascular calcification in adenine-induced chronic renal failure rats. High phosphate environment causes elevated calcium accumulation with related histological changes and alkaline phosphatase activity in aorta.

Further disogenin downregulated the activity of enzymatic antioxidants and elevates the level of lipid peroxidative markers. Moreover, the renal failure leads to reduced nitric oxide production. But, treatment with disogenin at a dose of 10, 20, and 40 mg/kg given via oral gavages causes reversion of all the above events in a dose-dependent manner (Manivannan et al. 2013).

The overall results of a study proved that disogenin attenuated CRF-induced impairment in acetylcholine induced endothelium-dependent and sodium nitroprusside induced endothelium-independent vascular relaxation. Moreover, it elevated the GSH and restores the eNOS mRNA expression level. CRF-induced dyslipidemia and ACE activity was also inhibited by disogenin treatment (Manivannan et al. 2013a).

20.2.6 Cardioprotective

A study was aimed to evaluate the effects of disogenin on myocardial ischaemia-reperfusion injury and the potential involvement of mitochondrial K_{ATP} (mitoK_{ATP}) channel and nitric oxide (NO) system blockades in this field. The results showed that disogenin may have cardioprotective effects against myocardial reperfusion injury through activating the mitoK_{ATP} channels (Badalzadeh et al. 2014).

20.2.7 The proapoptotic effect

Disogenin, caused an inhibition of the growth of fibroblast-like synoviocytes from human rheumatoid arthritis, with apoptosis induction associated with cyclooxygenase-2 (COX-2) up-regulation. Celecoxib (Fig. 20.3), a selective COX-2 inhibitor, provoked a large decrease in disogenin-induced apoptosis even in the presence of exogenous prostaglandin E2, whereas interleukin-1 β, a COX-2 inducer, strongly increased disogenin induced apoptosis of these synoviocytes (Bertrand et al. 2004).

20.2.8 Effects on the adrenal gland

Adult female Sprague Dawley rats were divided into three equal groups (n = 54, 250–300 gm BW). Rats in group I served as the control group, animals in group II were ovariectomized, and supplemented with tricalcium phosphate drug delivery system loaded with 500 mg disogenin. The results indicated that ovariectomized animals had a significant increase in body weight and spleen weights. Slight increases in wet adrenal weights were observed in the ovariectomized group compared to the control animals (Benghuzzi et al. 2003).

20.3 Dioscin (Glycoside form of disogenin)

Dioscin (Fig. 20.4), a typical steroid saponin, is isolated from *Dioscorea nipponica* Makino and *Dioscorea zingiberensis* Wright.

20.4 Pharmacology

20.4.1 Anticancer

20.4.1.1 Apoptosis of human ovarian cancer cells

A study investigated the anti-tumor effects of dioscin from *Paris chinensis* and correlated mechanisms regarding apoptosis in human ovarian cancer SKOV3 cells. Dioscin had an anti-proliferation effect on human ovarian cancer SKOV3 cells in a dose- and time-dependent manner. After treatment with dioscin, the apoptotic rate significantly increased, and accompanied with the increased levels of caspase-3 and cytochrome C protein in SKOV3 cells (Gao et al. 2011).

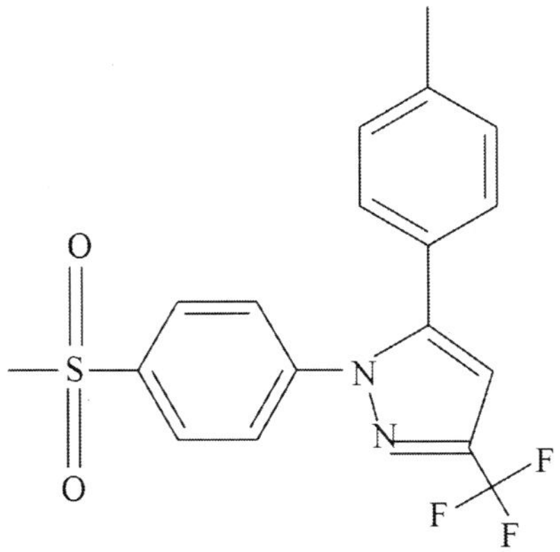

Fig. 20.3: Structure of celecoxib.

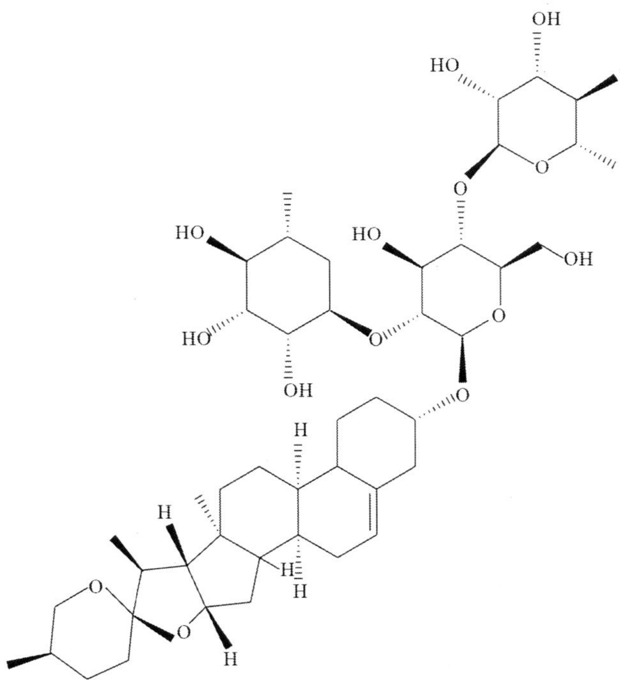

Fig. 20.4: Structure of dioscin.

20.4.1.2 Apoptosis induction in Huh7 Cells

In this study, the unique effects of dioscin on autophagy of hepatoma cells were investigated. Results found that dioscin induced caspase-3- and -9-dependent cell apoptosis in a dose-dependent manner. Moreover, inhibition of ERK1/2 phosphorylation significantly abolished the dioscin-induced apoptosis (Hsieh et al. 2012).

20.4.1.3 Induction of apoptosis in human lung cancer cell lines

Results from 4′-6-diamidino-2-phenylindole and annexin-V/PI double-staining assay showed that caspase-3- and caspase-8-dependent, and dose-dependent apoptoses were detected after a 24-h dioscin treatment. Blockade of autophagy with bafilomycin A1 or 3-methyladenine sensitized the A549 and H1299 cells to apoptosis. Treatment of A549 and H1299 cells with dioscin caused a dose-dependent increase in ERK1/2 and JNK1/2 activity, accompanied with a decreased PI3K expression and decreased phosphorylation of Akt and mTOR (Hsieh et al. 2013).

20.4.1.4 Apoptosis induction glioblastoma multiforme

In this study, dioscin significantly inhibited proliferation of C6 glioma cells and caused reactive oxygen species (ROS) generation and Ca^{2+} release. ROS accumulation affected levels of malondialdehyde, nitric oxide, glutathione disulfide and glutathione, and caused cell apoptosis. Simultaneously, dioscin down-regulated protein expression of Bcl-2, Bcl-xl, up-regulated expression of Bak, Bax, Bid and cleaved poly (ADP-ribose) polymerase (Lv et al. 2013).

20.4.1.5 Anti-cancer effects on three kinds of human lung cancer cell lines

Dioscin inhibited the proliferation of human A549, NCI-H446, and NCI-H460 cancer cells. Furthermore, dioscin caused mitochondrial structure changes and blocked cell cycle at S phase based on transmission electron microscope and flow cytometry analysis. In addition, dioscin treatment caused the release of cytochrome c from mitochondria into cytosol. The activities of Caspase-3 and -9 in dioscin-treated groups were significantly increased compared with the control group (Wei et al. 2013).

20.4.1.6 Cytotoxicity in human gastric carcinoma cells

Radioimmunoassays showed that the tumor necrosis factor (TNF)-α concentration in cells treated with dioscin significantly increased compared with untreated cells. Expression of Bid, bcl-2 and bcl-xl was markedly downregulated, and the expression of Bak and Bax was upregulated. In addition, cytochrome c was released from the mitochondria into the cytosol, which indicates activation of the mitochondrial pathway by dioscin. Increased mRNA expression of p53 was also found in dioscin-treated SGC-7901 cells, and the activation of caspase-3 and -8 was also observed (Hu et al. 2013).

20.4.1.7 Multidrug-resistant human leukemia

Dioscin, significantly inhibited MDR1 mRNA and protein expression and MDR1 promoter and nuclear factor κ-B (NF-κB) activity in K562/ADR cells. MDR1 mRNA and protein suppression resulted in the subsequent recovery of intracellular drug accumulation. Dioscin reversed adriamycin-induced multidrug resistance by down-regulating MDR1 expression by a mechanism that involves the inhibition of the NF-κB signaling pathway (Wang et al. 2013).

20.4.1.8 Apoptosis induction of human LNCaP prostate carcinoma cells

Dioscin (1, 2, and 4 μmol/L) could significantly inhibit the viability of LNCaP cells in a time- and concentration-dependent manner. Flow cytometry revealed that the apoptosis rate was increased after treatment of LNCaP cells with dioscin for 24 h, indicating that apoptosis was an important mechanism by which dioscin inhibited cancer (Chen et al. 2014).

20.4.1.9 Cytotoxicity in colon cancer cells

The present work was undertaken to study activity of dioscin on colorectal cancer. Dioscin increased the levels of NO and inducible NO synthase. Dioscin caused mitochondrial damage and G2/M cell cycle arrest through transmission electron microscopy and flow cytometry analysis, respectively (Chen et al. 2014).

20.4.1.10 Effect on apoptosis in pancreatic cancer MiaPaCa-2 cells

The aim of this study was to observe the effects of dioscin on apoptosis and on expression of PRDX1 in pancreatic cancer MiaPaCa-2 cells *in vitro*. Dioscin considerably inhibited the proliferation of MiaPaCa-2 cells *in vitro*. Decrease of PRDX1 expression was observed after dioscin treatment. Moreover, after PRDX1 over expression, dioscin treatment no longer induced high levels of ROS and apoptosis, and the apoptotic rate was decreased to (21.3 ± 5.9)% (Zhao et al. 2014).

20.4.2 Promotion of osteoblasts proliferation

Dioscin (0.25 μg/ml, 0.5 μg/ml, and 1.0 μg/ml) promoted MC3T3-E1 cells and MG-63 cells proliferation and differentiation dose dependently. Western blot analysis results showed that estrogen receptor α (ER-α), estrogen receptor β (ER-β), β-catenin, and Bcl-2 protein expression increased after MC3T3-E1 cells were treated with dioscin. Dioscin promotes osteoblasts proliferation and differentiation via Lrp5 and ER pathway (Zhang et al. 2014).

20.4.3 Effects against liver fibrosis

Dioscin effectively inhibited the cell viabilities of HSC-T6, LX-2, and primary rat hepatic stellate cells, but not hepatocytes. Furthermore, dioscin markedly increased peroxisome proliferator activated receptor-γ (PPAR-γ) expression and significantly reduced a-smooth muscle actin (α-SMA), transforming growth factor-β1 (TGF-β1), collagen α1 (I) (COL1A1), and collagen α1 (III) (COL3A1) levels *in vitro*.

Dioscin facilitated matrix degradation, and exhibited hepatoprotective effects through the attenuation of oxidative stress and inflammation, in addition to exerting anti-fibrotic effects (Zhang et al. 2015).

20.4.4 Action on hepatic ischemia-reperfusion

Seventy percent partial hepatic warm ischemia was induced in Wistar rats for 60 min followed by succedent reperfusion. In the prophylactic test, dioscin was administered intragastrically to the rats at doses of 20, 40, and 60 mg/kg once daily for seven consecutive days before I/R.

> Dioscin significantly decreased serum alanine aminotransferase and aspartate aminotransferase activities, increased survival rate of rats, and improved I/R-induced hepatocyte abnormality. In addition, dioscin obviously increased the levels of SOD, CAT, GSH-Px, GSH, decreased the levels of MDA, TNOS, iNOS, NO, and prevented DNA fragmentation caused by I/R injury (Tao et al. 2014).

20.4.5 Action on cerebral ischemia/reperfusion injury

In this work, an *in vitro* oxygen-glucose deprivation and reoxygenation (OGD/R) model and an *in vivo* middle cerebral artery occlusion (MCAO) model were used. The results indicated that dioscin clearly protected PC12 cells and primary cortical neurons against OGD/R insult and significantly prevented cerebral I/R injury (Tao et al. 2015).

20.4.6 Anti-obesity

High-fat diet-induced C57BL/6J mice and ob/ob mice were used as the experimental models. Dioscin significantly attenuated oxidative damage, suppressed inflammation, inhibited triglyceride and cholesterol synthesis, promoted fatty acid β-oxidation, down-regulated MAPK phosphorylation levels, and induced autophagy to alleviate fatty liver conditions. Dioscin prevented diet induced obesity and NAFLD by increasing energy expenditure (Liu et al. 2015).

20.4.7 Antifungal

A study investigated the antifungal effect of dioscin against different fungal strains and its antifungal mechanism(s) in *Candida albicans* cells. The ability of dioscin to disrupt the plasma membrane potential, using 3,3'-dipropylthiadicarbocyanine iodide [DiSC(3)(5)] and bis-(1,3-dibarbituric acid)-trimethine oxanol [DiBAC(4)(3)] was also investigated. The results suggest that dioscin exerts a considerable antifungal activity by disrupting the structure in membrane after invading into the fungal membrane, resulting in fungal cell death (Cho et al. 2013).

20.4.8 Hepatoprotective

20.4.8.1 Acetaminophen-induced hepatotoxicity

In the *in vivo* experiments, mice were orally administrated dioscin for five days and then given acetaminophen. Dioscin showed a remarkable protective effect against acetaminophen-induced hepatotoxicity by adjusting mitochondrial function. Following administration of dioscin, APAP-induced hepatotoxicity in mice was significantly attenuated (Zhao et al. 2012).

20.4.8.2 CCl4-induced hepatotoxicity

Dioscin significantly inhibited ($p < 0.01$) the increases of serum ALT and AST activities compared with the CCl(4)-treated animals. The hepatic lipid peroxidation formation and, concentrations of TNF-α and IL-6 were also decreased. Liver histopathologic studies and a DNA laddering assay indicated that dioscin protected hepatocytes against CCl(4)-induced apoptosis and necrosis (Lu et al. 2012).

20.4.8.3 Ethanol-induced hepatotoxicity

Ethanol-induced acute and chronic liver damage rat models were used, and the results showed that dioscin significantly alleviated liver steatosis, reduced the levels of alanine aminotransferase, aspartate aminotransferase, total triglyceride, total cholesterol and malondialdehyde, and increased the levels of high-density lipoprotein, superoxide dismutase, glutathione, and glutathione peroxidise (Xu et al. 2014).

20.4.9 Antiviral

In this study, dioscin's antiviral effects were tested against several viruses including adenovirus, vesicular stomatitis virus, and hepatitis B virus. By time-of-addition assay, dioscin not only blocked the initial stage of adenovirus infection, but also affected the host cell's response for viral infection. Over expression of CAR in 293 cells pretreated with dioscin restored the infectivity of adenovirus (Liu et al. 2013).

20.4.10 Tyrosinase inhibition

A methanol extract of *Smilax china* was partitioned into hexane, ethyl acetate, and water. Of the three fractions, ethyl acetate extract showed the strongest inhibition of tyrosinase activity with l-tyrosine or l-DOPA as a substrate. Two compounds were isolated from a final active fraction by activity-guided column chromatography. These compounds were identified as dioscin and oxyresveratrol (Fig. 20.5). Dioscin showed little inhibition activity of tyrosinase, whereas oxyresveratrol, a known tyrosinase inhibitor, showed a strong tyrosinase inhibitory activity (Liang et al. 2012).

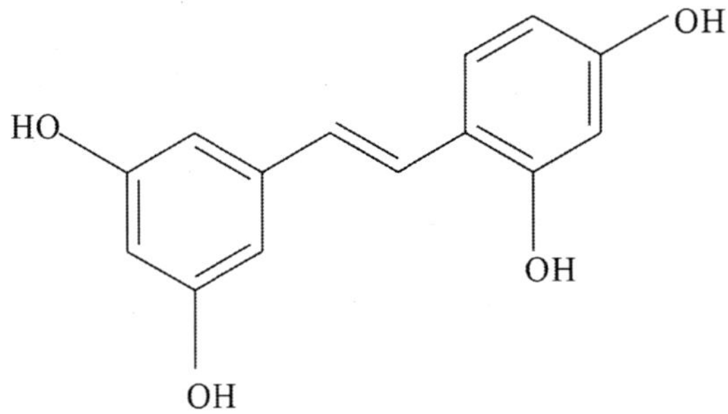

Fig. 20.5: Structure of oxyresveratrol.

20.4.11 ADME

The stability was determined in simulated gastric and intestinal fluids (SGF, pH 1.2 and SIF, pH 6.8), and intestinal transport was evaluated in Caco-2 model. Phase I and phase II metabolic stability was determined in human liver microsomes and S9 fractions, respectively. Quantitative analysis of dioscin and disogenin was performed by UPLC-MS system. Dioscin degraded up to 28.3% in SGF and 12.4% in SIF, which could be accounted for by its conversion to disogenin (24.2% in SGF and 2.4% in SIF). Dioscin was stable in both HLM and S9 fractions. No phase I metabolism was detected (Manda et al. 2013).

20.4.12 Pharmacokinetics

The dose-dependent pharmacokinetics of dioscin was characterized after intravenous administrations (0.064, 0.16, 0.4, and 1.0 mg/kg) to rats. There was significant decrease in clearance with increasing dose (4.67 +/– 0.09 ml/min/kg (0.064 mg/kg) versus 3.49 +/– 0.23 ml/min/kg (1.0 mg/kg), $P < 0.05$), and the plot of reciprocal clearance values versus the doses was linear ($r = 0.909$, $P < 0.05$). After an I.V. dose of 1 mg/kg, simultaneous oral gavage of activated charcoal did not change the pharmacokinetic parameters indicating enterohepatic recycling of dioscin is not important in rat. The absolute oral bioavailability was very low (0.2%) (Li et al. 2005).

20.4.13 Subchronic toxicity

The rats were divided into four groups and dioscin was administered orally at doses of 0, 75, 150, and 300 mg/kg/day, respectively. The results showed that dioscin had no subchronic toxicity in female rats and had slight subchronic toxicity in male rats. Compared with the control group, body weight gain was significantly decreased in male rats (Xu et al. 2012).

20.5 Protodioscin

Protodioscin is a steroidal saponin compound found in a number of plant species, most notably in the *Tribulus*, *Trigonella*, and *Dioscorea* families (Hibasami et al. 2003;

Dong et al. 2004). It is best known as the putative active component of the herbal aphrodisiac plant *Tribulus terrestris* (Dinchev et al. 2008).

20.5.1 Proerectile effect

Extracts from *T. terrestris* standardised for protodioscin content have been demonstrated to produce proerectile effects in isolated tissues and aphrodisiac action in several animal species (Gauthaman et al. 2002).

> Protodioscin achieves proerectile effect primarily through causing an increase in androgen receptor immunoreactivity, meaning it increases the concentration of androgen receptors in cells, causing the organism to become more sensitive to androgens like testosterone and DHT (Gauthaman and Adaikan 2005).

20.5.2 Antiproliferative

Some metabolites including methyl protodioscin showed potent antiproliferative activities against HepG2, NCI-H460, MCF-7, and HeLa cell lines *in vitro* (He et al. 2010). Treatment of methyl protodioscin resulted in G2/M arrest and apoptosis in HepG2 cells. These effects were attributed to down-regulation of Cyclin B1 and the signaling pathways leading to up-regulation of Bax and down-regulation of BCL2, suggesting that methyl protodioscin may be a novel anti-mitotic agent (Wang et al. 2006).

20.5.3 Induction of apoptosis

Methyl protodioscin induced a biphasic alteration (i.e., an early hyperpolarization, followed by depolarization) in mitochondrial membrane potential of K562 cells. The transient decline of intracellular Ca2+ concentration was observed at early stage. The generation of reactive oxygen species was also detected. The anti-apoptotic Bcl-x(L) transiently increased and then decreased. And the pro-apoptotic Bax was markedly up-regulated (Liu et al. 2005).

20.5.4 Intestinal inflammation

Methyl protodioscin increased the percentage of survival from high-dose dextran sulfate sodium-(4%) treated mice, and accelerated mucosal healing and epithelial proliferation in low-dose dextran sulfate sodium-(2.5%) treated mice characterized by marked reduction in NF-κB activation, pro-inflammatory cytokines expression, and bacterial translocation. Consistently, methyl protodioscin protected colonic mucosa from C. rodentium-induced colonic inflammation and bacterial colonization. *In vitro* studies showed that MPD significantly increased crypt formation and restored intestinal barrier dysfunction induced by pro-inflammatory cytokines (Zhang et al. 2015).

20.5.5 Pharmacokinetics

The dose-dependent pharmacokinetics of methyl protodioscin were characterized after i.v. injection (20, 40, and 120 mg/kg of MPD) to rats. A good linearity ($r = 0.9989$, $P < 0.05$) was found in the regression analysis of the AUC0-t-dose. The plasma concentrations of methyl protodioscin declined rapidly with an elimination half-life (t1/2) from 25.56 to 29.32 min. The methyl protodioscin kinetics was in line with one-compartment model after i.v. injection. 23.43% and 32.86% of methyl protodioscin was recovered in urine and bile, respectively (Cao et al. 2010).

20.6 Hecogenin

Hecogenin (Fig. 20.6) is a crystalline steroid sapogenin obtained from *Hechtia texensis*, *Agave attenuata* Salm-Dyck (Bedour et al. 1979), *Agave macroacantha* (Eskander et al. 2010), *Agave sisalana* Perr. (Botura et al. 2011), *Agave americana* L. (Xu and Zhou 1982), *Polygonum chinensis* L. (Tsai et al. 1998).

20.6.1 Pharmacology

20.6.1.1 Anticancer

20.6.1.1.1 Induction of apoptosis

The results demonstrated hecogenin and tigogenin-induced apoptosis through activation of p38 without affecting the JNK and ERK pathways. Indeed, pre-treatment with a p38 inhibitor decreased saponin-induced apoptosis with a significant decrease in DNA fragmentation. Furthermore, the rate of apoptosis induced by hecogenin or tigogenin was associated with over expression of COX-2 correlated with overproduction of endogenous PGE2. These new results provide strong evidence that a family of structurally similar plant steroids is capable of inducing apoptosis in human RA FLS with different rates and different signalling pathways (Liagre et al. 2007).

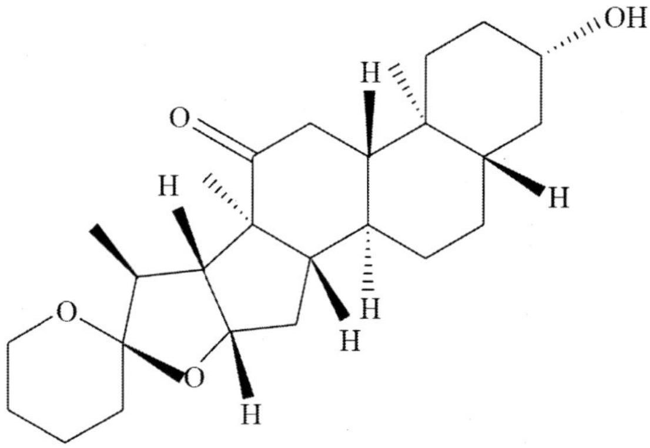

Fig. 20.6: Structure of hecogenin.

20.6.1.1.2 Inhibition of a 549 human lung cancer cell

Hecogenin acetate significantly inhibited increase in intracellular reactive species production induced by H_2O_2. In addition, hecogenin acetate blocked ERK1/2 phosphorylation and inhibited the increase in MMP-2 caused by H_2O_2. Treatment with hecogenin acetate induced G_0/G_1-phase arrest at two concentrations (75 and 100 μM, 74% and 84.3% respectively), and increased the staining of senescence-associated β-galactosidase positive cells (Gasparotto et al. 2014).

20.6.1.2 Anti-hypergesic

Acute pre-treatment with Hecogenin acetate (5, 10, or 20 mg/kg; i.p.) inhibited the development of mechanical hyperalgesia induced by carrageenan, TNF-α, dopamine, and PGE2. Additionally, the immunofluorescence data demonstrated that acute pre-treatment with HA, at all doses tested, significantly inhibited Fos-like expression in the spinal cord dorsal horn normally observed after carrageenan-inflammation (Quintans et al. 2014).

20.6.1.3 Gastro protective

Hecogenin (3.1, 7.5, 15, 30, 60, and 90 mg/kg, p.o.) acutely administered, before ethanol or indomethacin, exhibited a potent gastro protective effect. The hecogenin pre-treatment normalized GSH levels and significantly reduced lipid peroxidation and nitrite levels in the stomach, as evaluated by the ethanol-induced gastric lesion model. The drug alone increased COX-2 expression and this effect was further enhanced in the presence of ethanol (Cerqueira et al. 2012).

20.6.1.4 Larvicidal

The present study was undertaken to evaluate the effect of hecogenin in *Aedes aegypti* mortality. The results confirm that the hecogenin have larvicidal activity against *A. aegypti*. As mechanism of action, it is possible that the acetate mimics the insect growth hormone, stopping its development and causing them to die (Oliveira et al. 2014).

20.7 Yamogenin

Yamogenin is a diastereomer of disogenin (Fig. 20.7).

20.7.1 Sources

Balanites orbicularis (Hardman and Wood 1971a), *Balanites pedicellaris* (Hardman and Wood 1971b), *Solanum spirale* (Quyen et al. 1987), *Trigonella foenum-graecum* (Puri et al. 1976), *Dioscores colettii* (Minghe and Yanyong 1983), *Dioscorea multiflora* (da Costa and Mukherjee 1984).

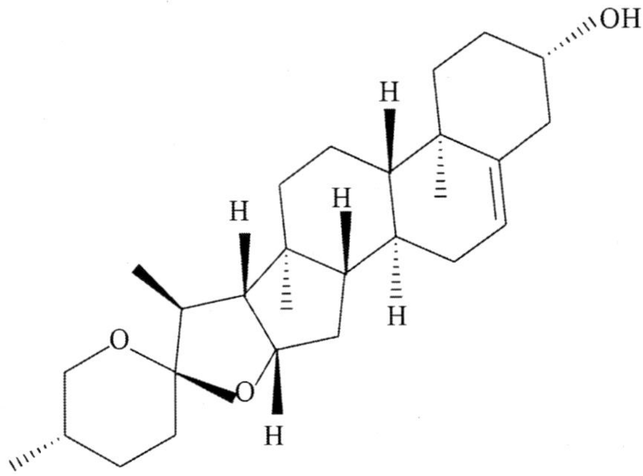

Fig. 20.7: Structure of yamogenin.

20.7.2 Pharmacology

Anti-hyperlipidemic

It was demonstrated that yamogenin inhibited triacylglyceride accumulation in HepG2 hepatocytes and suppressed the mRNA expression of fatty acid synthesis-related genes such as fatty acid synthase and sterol response element-binding protein-1c. Indeed, yamogenin also antagonized the activation of the liver X receptor in luciferase ligand assay similar to disogenin. However, yamogenin could not exert such effects in the presence of T0901713, a potent agonist of liver X receptor (Moriwaki et al. 2014).

20.8 Ruscogenin

Ruscogenin (Fig. 20.8) is a steroidal glycoside extracted from roots of *Ophiopogon japonicas* and *Ruscus aculeatus*.

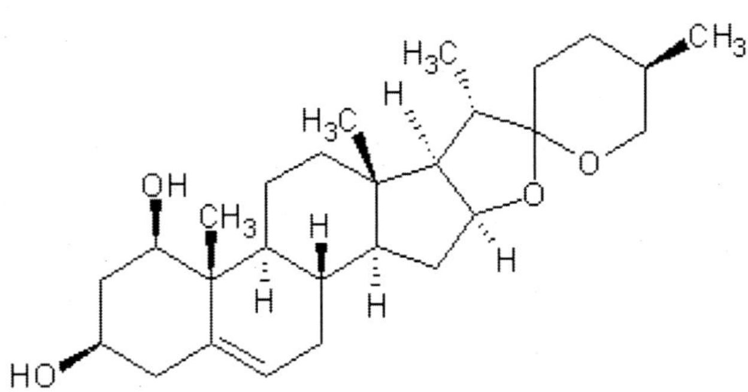

Fig. 20.8: Structure of ruscogenin.

20.8.1 Pharmacology

20.8.1.1 Anti-inflammatory

Ruscogenin significantly suppressed zymosan A-evoked peritoneal total leukocyte migration in mice in a dose-dependent manner, while it had no obvious effect on PGE(2) content in peritoneal exudant. Ruscogenin also inhibited TNF-alpha-induced over expression of ICAM-1 both at the mRNA and protein levels and suppressed NF-kappaB activation considerably by decreasing NF-kappaB p65 translocation and DNA binding activity (Huang et al. 2008).

The present study was performed to compare the anti-inflammatory activities *in vitro* of RUS-2HS (Succinylated Derivative) and ruscogenin. Both compounds reduced tumor necrosis factor-a (TNF-a)-induced adhesion of human pro-myelocytic leukemia cells (HL-60) to endothelial ECV304 cells with IC50 values of 6.90 nM and 7.45 nM, respectively.

RUS-2HS (Succinylated Derivative) and ruscogenin also inhibited overexpression of ICAM-1 in ECV304 cells at the mRNA level as evaluated by real-time PCR and at the protein level evaluated by flow cytometry with similar potency. Such data demonstrate that the functional groups of ruscogenin were not blocked by derivation, suggesting further use of the ruscogenin affinity column for target investigation. Meanwhile, RUS-2HS was found to have remarkable anti-inflammatory activity for the first time (Huang et al. 2008a).

20.8.1.2 Cerebroprotective

Adult male mice (C57BL/6 strain) were pretreated with ruscogenin and then subjected to transient middle cerebral artery occlusion (MCAO)/reperfusion. After 1 h MCAO and 24 h reperfusion, neurological deficit, infarct sizes, and brain water content were measured.

Ruscogenin markedly decreased the infarct size, improved neurological deficits and reduced brain water content after MCAO. The activation of NF-κB Signaling pathway was observed after 1 h of ischemia and 1 h of reperfusion, and ruscogenin significantly inhibited NF-κB p65 expression, phosphorylation and translocation from cytosol to nucleus at this time point in a dose-dependent manner (Guan et al. 2013).

20.8.1.3 Diabetic neuropathy

Ruscogenin treatment was found to markedly improve histological architecture in the diabetic kidney. Renal NF-κB activity, as wells as protein expression and infiltration of macrophages were increased in diabetic kidneys, accompanied by an increase in protein content of intercellular adhesion molecule-1 and monocyte chemoattractant protein-1 in kidney tissues (Hung et al. 2014).

20.8.1.4 Nonalcoholic steatohepatitis

Ruscogenin (10.0 μmol/l) had inhibitory effects on PA-induced triglyceride accumulation and inflammatory markers in HepG2 cells. Male golden hamsters were randomly divided into five groups and fed a normal diet, a high-fat diet (HFD), or a HFD supplemented with ruscogenin (0.3, 1.0, or 3.0 mg/kg/day) by gavage once daily for 8 weeks.

> Ruscogenin alleviated dyslipidemia, liver steatosis, and necroinflammation and reversed plasma markers of metabolic syndrome in HFD-fed hamsters. Hepatic mRNA levels involved in fatty acid oxidation were increased in ruscogenin-treated HFD-fed hamsters (Lu et al. 2014).

20.8.1.5 Pulmonary arterial hypertension

Ruscogenin had favorable effects on hemodynamics and pulmonary vascular remodeling, preventing the development of PAH 3 weeks after MCT. In addition, ruscogenin resulted in markedly reduced expression of inflammatory cytokine and leukocyte infiltration via the inhibition of nuclear factor (NF)-κB activity in rat lungs. Ruscogenin also attenuated MCT-induced endothelial cell apoptosis in the remodeled pulmonary arterioles and rescued destruction of endothelial cell membrane proteins such as eNOS, caveolin-1, and CD31 (Bi et al. 2013).

20.9 Tigogenin

Tigogenin (Fig. 20.9) is obtained from *Agave sisalana* (Zullo et al. 1989), *Costus speciosus* Koen (Surendra et al. 2013), *Digitalis lanata* (Lack et al. 1963), *Yucca gloriosa* (Kemertelidze and Pkheidze 1972).

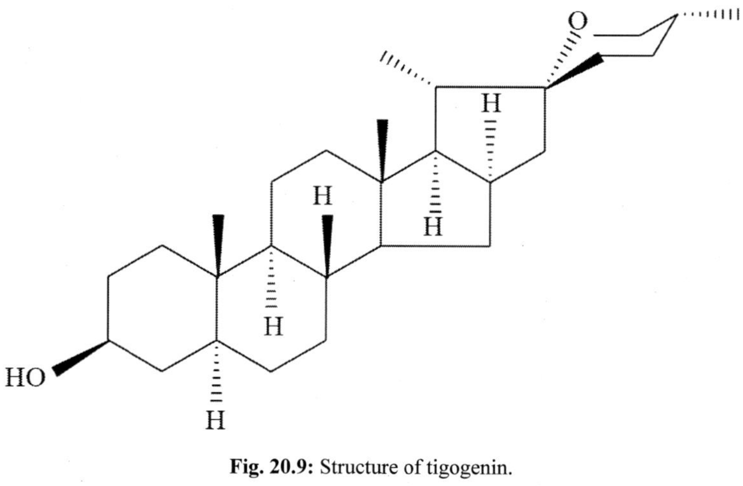

Fig. 20.9: Structure of tigogenin.

20.9.1 *Pharmacology*

20.9.1.1 Antidiabetic

The Streptozotocin induced diabetic rats were treated with tigogenin at a dose of 10, 20, and 30 mg/kg body mass for 15 days. After 15 days, the blood samples were collected from each group of rats and the blood glucose level was estimated.

A significant increase in blood glucose level was observed in diabetic rats. After treatment with tigogenin, the blood glucose level was found to be normal. In diabetic rats, the amount of glycogen, and Hexokinase activities were decreased, but they were normal in the tigogenin treated rats (Sunitha et al. 2012).

20.9.1.2 Induction of apoptosis

The results demonstrated hecogenin and tigogenin-induced apoptosis through activation of p38 without affecting the JNK and ERK pathways. Indeed, pre-treatment with a p38 inhibitor decreased saponin-induced apoptosis with a significant decrease in DNA fragmentation.

Furthermore, the rate of apoptosis induced by hecogenin or tigogenin was associated with over expression of COX-2 correlated with overproduction of endogenous PGE2. These new results provide strong evidence that a family of structurally similar plant steroids is capable of inducing apoptosis in human RA FLS with different rates and different signalling pathways (Liagre et al. 2007).

20.9.1.3 Inhibition of adipocytic differentiation and induction of osteoblastic differentiation

Tigogenin may modulate differentiation of mouse bone marrow stromal cells to cause a lineage shift away from the adipocytes and toward the osteoblasts, which is at least mediated by inhibition of PPARγ and via p38 MAPK pathway. Tigogenin is a potential drug preventing the development of osteoporosis and the related disorders (Zhou et al. 2007a).

20.10 Steroidal Saponins of *Asparagus acutifolius* L.

Three furostanol and four spirostanol saponins were isolated from the roots of *Asparagus acutifolius* L. spirostanol saponins demonstrated antifungal activity against the human pathogenic yeasts *Candida albicans*, *C. glabrata*, and *C. tropicalis* (Sautour et al. 2007).

20.11 Steroidal Saponins of *Asparagus adscendens*

From the methanol extract of the fruits of *Asparagus adscendens* sitosterol-β-D-glucoside, two spirostanol glycosides (asparanin A and B) and two furostanol glycosides (asparoside A and B) have been isolated (Sharma et al. 1982).

20.12 Steroidal Saponins of *Asparagus africanus* L.

(25R)-3 beta-hydroxy-5 beta-spirostan-12-one 3-O-{beta-D-glucopyranosyl-(1--> 2)-[alpha-1-arabinopyranosyl-(1-->6)]-beta- D-glucopyranoside}, (25R)-5 beta-spirostan-3beta-ol3-O-{beta-D-glucopyranosyl-(1-->2)-[alpha-L-arabinopyranosyl-(1-->6)]-beta-D-glucopyranoside} and 26-O-beta-D-glucopyranosyl]-22 alpha-methoxy-(25R)-furostan-3 beta,26-diol 3-O-{beta-D-glucopyranosyl-(1-->2)-[beta-D-glucopyranoside} have been reported (Debella et al. 1999).

20.13 Steroidal Saponins of *Asparagus dumosus* Baker in J. Linn.

A new steroidal saponin, dumoside, has been reported (Ahmad et al. 1999).

20.14 Steroidal Saponins of *Asparagus filicinus* Buch.–Ham

Two new steroidal saponins with a new aglycone moiety, aspafiliosides E and F, were isolated from the roots of *Asparagus filicinus* Buch.–Ham (Zhou and Chen 2008). Four new steroidal saponins, filiasparosides A–D (1–4), together with known aspafiliosides A and B were isolated from the roots of *A. filicinus*. Filiasparoside C showed the most potent cytotoxicity, with EC_{50} values of 2.3 and 3.0 μg/mL toward A549 and MCF-7 cell lines, respectively (Zhou et al. 2007b).

20.15 Steroidal Saponins of *Asparagus officinalis* L.

Two oligofurostanosides, saponins 1 and 2 were isolated from *A. officinalis*. The two compounds were shown to inhibit the growth of human leukemia HL-60 cells in culture and macromolecular synthesis in dose-dependent manner (Shao et al. 1999).

20.16 Steroidal Saponins of *Asparagus racemosus* Willd.

Five steroidal saponins, shatavarins VI–X, together with five known saponins, shatavarin I (or asparoside B), shatavarin IV (or asparinin B), shatavarin V, immunoside and schidigerasaponin D5 (or asparanin A), have been isolated from the roots of *A. racemosus* (Hayes et al. 2008). Shatavaroside A and shatavaroside B together with a known saponin, filiasparoside C, have been isolated from the roots of *A. racemosus* (Sharma et al. 2009).

20.17 Steroidal Saponins of *Camassia cusickii* S. Watson

In murine leukemic L1210 cells, both compounds showed cytotoxicity with an EC_{50} value of 0.06 μM. The morphological observation revealed that TGHS-1 and TGHS-2 induced shrinkage in cell soma and chromatin condensation, suggesting apoptotic cell death (Candra et al. 2001).

20.18 Steroidal Saponins of *Chlorophytum borivilianum* L.

Stigmasterol and saponin named as furostanol and Chlorophytoside-I (3b, 5a, 22R, 25R)-26-(β-Dglucopyranosyloxy)-22-hydroxy-furostan-12-one-3yl O-β-D-galactopyranosyl (1-4) glucopyranoside has been isolated (Qiu et al. 2000; Acharya et al. 2009).

20.19 Steroidal Saponins of *Chlorophytum orchidastrum* Lindl.

Six new spirostane-type saponins, named orchidastrosides A–F, and chloromaloside D were isolated from an ethanol extract of the roots of *C. orchidastrum* (Acharya et al. 2010).

20.20 Steroidal Saponins of *Chlorophytum malayense* Ridl.

Four new steroidal saponins, chloromalosides A–D, were isolated from the rhizomes of *C. malayense* (Li et al. 1990).

20.21 Steroidal Saponins of *Dracaena angustifolia* Roxb.

Two new steroidal saponins, named drangustosides A-B, together with eight known compounds were isolated and characterized from the MeOH extract of *D. angustifolia*. Compounds 1 and 2 showed anti-inflammatory activity by superoxide generation and elastase release by human neutrophils in response to fMLP/CB (Huang et al. 2013).

20.22 Steroidal Saponins from *Dracaena cambodiana* Pierre ex Gagnepain

Six new steroidal saponins, cambodianosides A–F, together with seven known ones, were isolated from the dragon's blood of *D. cambodiana*. The cytotoxicities of all the isolated compounds were evaluated *in vitro* against three human cancer cell lines, and compounds 7, 8, and 11 showed significant inhibitory activities (Shen et al. 2014).

20.23 Steroidal Saponins from *Dracaena cochinchinensis* (Loureiro) S.C. Chen in F.T. Wang & Tang

A phytochemical investigation on the fresh stem of *D. cochinchinensis* yielded 18 steroidal saponins. Fourteen of which are new compounds, designated as 25(R,S)-dracaenosides E–H, M, O–Q, dracaenosides I–L, R, and 25(S)-dracaenoside N (Zheng et al. 2004).

20.24 Steroidal Saponins from *Fritillaria pallidiflora* Schrenk.

Three new steroidal saponins, pallidiflosides A, B, C, have been isolated from the dry bulbs of *F. pallidiflora* (Shen et al. 2011).

20.25 Steroidal Saponins of *Polygonatum sibiricum* Mill.

Four new steroidal saponins, named neosibiricosides A–D, were isolated from the rhizomes of *Polygonatum sibiricum*, along with two known spirostanol glycosides. The cytotoxic activity of the isolated compounds was evaluated with human MCF-7 breast cancer cells (Ahn et al. 2006).

20.26 Steroidal Saponins from *Polygonatum zanlanscianense* Mill.

Four new steroidal saponins, polygonatosides A–D, were isolated from the rhizomes of *P. zanlanscianense*, together with six known spirostanols and a known megastigmane glycoside (Jin et al. 2004).

20.27 Steroidal Saponins of *Smilacina atropurpurea* (Franch.) F.T. Wang & Tang

Atropurosides A–G, seven new steroidal were isolated from the rhizomes of *S. atropurpurea* atropurosides B and F were fungicidal against *Candida albicans*, *Candida glabrata*, *Cryptococcus neoformans*, and *Aspergillus fumigatus* with minimum fungicidal concentrations (MFCs) < or = 20 microg/ml, while dioscin was selectively active against *C. albicans* and *C. glabrata* (MFC < or = 5.0 microg/ml) (Zhang et al. 2006).

20.28 Steroidal Saponins of *Smilacina japonica* A. Gray.

Three new steroidal saponins, japonicoside A, japonicoside B, and japonicoside C were isolated from the dried rhizomes and roots of *S. japonica*. The cytotoxicity of isolated compounds was evaluated *in vitro* for cytotoxic properties against human hepatocellular carcinoma cells (SMMC-7221) and human colorectal adenocarcinoma cells (DLD-1), respectively (Liu et al. 2012).

20.29 Steroidal Saponins of *Smilax officinalis* L.

Sarsasapogenin 3-O-beta-D-glucopyranosyl-(1-->4)-[alpha-L-arabinopyranosyl-(1-->6)-beta- D-glucopyranoside, neotigogenin 3-O-beta-D-glucopyranosyl-(1-->4)-[alpha-L-arabinopyranosyl-(1-->6)]-beta- D-glucopyranoside and 25S-spirostan-6 beta-ol 3-O-beta-D-glucopyranosyl-(1-->4)-[alpha-L-arabinopyranosyl-(1-->6)]-beta-D-glucopyranoside have been reported (Bernardo et al. 1996).

20.30 Steroidal Saponins of *Smilax lebrunii* H. Léveillé

Two new steroidal saponins, (25 R)-spirostan-3 beta-ol-6-one-3-O-[alpha-L-arabinopyranosyl (1–6)]-beta-D-glucopyranoside and (25 R)-spirostan-3 beta-ol-6-one-3-O-[beta-D-glucopyranosyl(1–4)] [alpha-L-arabinopyranosyl(1–6)]-beta-glucopyranoside, were isolated from the rhizomes of *S. lebrunii* (Jia and Ju 1992).

20.31 Steroidal Saponins of *Smilax menispermoidea* A. de Candolle in A. de Candolle & C. de Candolle

(25S)spirost-5-en-3 beta,17 alpha-triol-3-O-[alpha-L-rhamnopyranosyl(1–2)] [alhpa-L-rhamnopyranosyl(1–4)]-beta-D-glucopyranoside, dioscin, methyl protodioscin, and pseudoprotodioscin have been reported (Ju and Jia 1992).

20.32 Steroidal Saponins of *Tribulus terrestris* L.

Presence of chlorgenin and gitogenin with disogenin has been reported (Gheorghiu et al. 1968). Campesterol, β-sitosterol, stigmasterol disogenin, and neotigogenin have been reported from roots of *T. terrestris* (Tomova et al. 1973). The same authors (7) have further reported a new saponin–terrestroside F along with saponins C and G from aerial parts; saponin C and G proved to be mixture of two tigogenin and disogenin glycosides each containing glucose rhamnose and astragalin (Tomova et al. 1974). A study reported trillin, gracillin and dioscin (Perepelista and Kintya 1975). Hecogenin has been reported (Tomova et al. 1977).

20.33 Steroidal Saponins of *Trillium tschonoskii* Maxim

TTB2, a steroidal saponin isolated from n-BuOH extracts had anti-proliferative and morphological influence, leading to the loss of mitochondrial membrane potential in a dose-dependent manner. A significant increase in the level of intracellular ROS and an accumulation of cells in the G2/M phases of the cell cycle were also observed in treated cells (Huang and Zou 2011).

20.34 Steroidal Saponins of *Yucca schidigera* Roezl.

Eight steroidal saponins have been isolated from *Y. schidigera*. These included three novel furostanol glycosides including 3-O-beta-D-glucopyranosyl-(1-->2)-[beta-D-xylopyranosyl-(1-->3)]-beta-D-glucopyranosyl-5 beta(25R)-furostan-3 beta,22 alpha,26-triol 26-O-beta-D-glucopyranoside, 3-O-beta-D-glcopyranosyl-(1-->2)-[beta-D-xylopyranosyl-(1-->3)]-beta-D-glucopyranosyl-5 beta(25R)-furost-20(22)-en-3 beta,26-diol-12-one 26-O-beta-D-glucopyranoside, 3-O-beta-D-glcopyranosyl-(1-->2)-beta-D-glucopyranosyl-5 beta(25R)-furostan-3 beta,22 alpha,26-triol 26-O-beta-D-glucopyranoside, and five known spirostanol glycosides (Oleszek et al. 2001).

20.35 Steroidal Saponins of *Yucca smalliana* Fern.

The purified steroidal saponin, yuccalan showed antifungal activities against both *Rhizoctonia solani* and *Fusarium oxysporum* (Jin et al. 2007).

20.36 Steroidal Saponins of *Yucca elephantipes* Regel.

Ten steroidal saponins with *cis*-fused A/B ring, including a smilagenin glycoside, elephanoside A, and the five furostanol bisdesmosides, elephanosides B–F, were isolated from the stems of *Y. elephantipes* (Zhang et al. 2008).

20.37 Antifungal activity of C-27 Steroidal Saponins

The antifungal activity of the steroidal saponins was associated with their aglycone moieties and the number and structure of monosaccharide units in their sugar chains. Within the 10 active saponins, four tigogenin saponins with a sugar moiety of four or five monosaccharide units exhibited significant activity against *C. neoformans* and *A. fumigatus*, comparable to the positive control amphotericin B (Yang et al. 2006).

Further Readings

Acharya D, Mitaine-Offer AC, Kaushik N, Miyamoto T, Paululat T, Mirjolet JF, Duchamp O and Lacaille-Dubois MA. Cytotoxic spirostane-type saponins from the roots of *Chlorophytum borivilianum*. *J Nat Prod* 2009; **72**: 177–81.

Acharya D, Mitaine-Offer AC, Kaushik N, Miyamoto T, Paululat T, Mirjolet JF, Duchamp O, Lacaille-Dubois MA. Steroidal saponins from *Chlorophytum orchidastrum*. *J Nat Prod* 2010; **73**: 7–11.

Ahmad VU, Khaliq-uz-Zaman SM, Shameel S, Perveen S, Ali Z. Steroidal saponins from *Asparagus dumosus*. *Phytochemistry* 1999; **50**: 481–4.

Ahn MJ, Kim CY, Yoon KD, Ryu MY, Cheong JH, Chin YW, Kim J. Steroidal saponins from the rhizomes of *Polygonatum sibiricum*. *J Nat Prod* 2006; **69**: 360–4.

Badalzadeh R, Yousefi B, Tajaddini A, Ahmadian N. Diosgenin-induced protection against myocardial ischaemia-reperfusion injury is mediated by mitochondrial KATP channels in a rat model. *Perfusion* 2014. pii: 0267659114566064. [Epub ahead of print].

Bedour MS, Elgamal MHA, El-Tawil BAH. Steroid sapogenins. Part XV. The constituents of *Agave utahensis* var. *nevadensis*, *A. lophanta* and *A. parasana*. *Planta Med* 1979; **36**: 180–181.

Benghuzzi H, Tucci M, Eckie R, Hughes J. The effects of sustained delivery of diosgenin on the adrenal gland of female rats. *Biomed Sci Instrum* 2003; **39**: 335–40.

Bernardo RR, Pinto AV, Parente JP. Steroidal saponins from *Smilax officinalis*. *Phytochemistry* 1996; **43**: 465–9.

Bertrand L, Pascale V-S, Cecile C, Charissoux JL, Beneytout JL. Diosgenin, a plant steroid, induces apoptosis in human rheumatoid arthritis synoviocytes with cyclooxygenase-2 overexpression. *Arthritis Res Ther* 2004; **6**: 373–383.

Bi LQ, Zhu R, Kong H, Wu SL, Li N, Zuo XR, Zhou SM, Kou JP, Yu BY, Wang H, Xie WP. Ruscogenin attenuates monocrotaline-induced pulmonary hypertension in rats. *Int Immunopharmacol* 2013; **16**: 7–16.

Botura MB, Silva GD, Lima HG, Oliveira JV, Souza TS, Santos JD, Branco A, Moreira EL, Almeida MA, Batatinha MJ. *In vivo* anthelmintic activity of an aqueous extract from sisal waste (Agave sisalana Perr.) against gastrointestinal nematodes in goats. *Vet Parasitol* 2011; **177:** 104–10.

Candra E, Matsunaga K, Fujiwara H, Mimaki Y, Sashida Y, Yamakuni T, Ohizumi Y. Two steroidal saponins from *Camassia cusickii* induce L1210 cell death through the apoptotic mechanism. *Canadian J Physiol Pharmacol* 2001; **79**: 953–958.

Cao X, Yao Z, Shao M, Chen H, Ye W, Yao X. Pharmacokinetics of methyl protodioscin in rats. *Pharmazie* 2010; **65**: 359–62.

Chen H, Xu L, Yin L, Xu Y, Han X, Qi Y, Zhao Y, Liu K, Peng J. iTRAQ-based proteomic analysis of dioscin on human HCT-116 colon cancer cells. *Proteomics* 2014; **14**: 51–73.

Chen J, Li HM, Zhang XN, Xiong CM, Ruan JL. Dioscin-induced apoptosis of human LNCaP prostate carcinoma cells through activation of caspase-3 and modulation of Bcl-2 protein family. *J Huazhong Univ Sci Technolog Med Sci* 2014; **34**: 125–30.

Cho J, Choi H, Lee J, Kim MS, Sohn HY, Lee DG. The antifungal activity and membrane-disruptive action of dioscin extracted from *Dioscorea nipponica*. *Biochim Biophys Acta* 2013; **1828**: 1153–8.

Corbiere C, Liagre B, Terro F, Beneytout JL. Induction of antiproliferative effect by diosgenin through activation of p53, release of apoptosis-inducing factor (AIF) and modulation of caspase-3 activity in different human cancer cells. *Cell Res* 2004; **14**: 188–96.

da Costa FF, Mukherjee R. Diosgenin and Yamogenin from *Dioscorea multiflora*. *J Nat Prod* 1984; **47**: 909–910.

Das S, Dey KK, Dey G, Pal, Majumder A, MaitiChoudhury S, kundu SC, Mandal M. Antineoplastic and apoptotic potential of traditional medicines thymoquinone and diosgenin in squamous cell carcinoma. *PLoS One* 2012; **7**: e46641.

Debella A, Haslinger E, Kunert O, Michl G, Abebe D. Steroidal saponins from *Asparagus africanus*. *Phytochemistry* 1999; **51**: 1069–75.

Dinchev D, Janda B, Evstatieva L, Oleszek W, Aslani MR, Kostova I. Distribution of steroidal saponins in *Tribulus terrestris* from different geographical regions. *Phytochemistry* 2008; **69**: 176–86.

Dong M, Feng XZ, Wang BX, Ikejima T, Wu, LJ. Steroidal saponins from *Dioscorea panthaica* and their cytotoxic activity. *Die Pharmazie* 2004; **59**: 294–6.

Eskander J, Lavaud C, Harakat D. Steroidal saponins from the leaves of *Agave macroacantha*. *Fitoterapia* 2010; **81**: 371–4.

Gao LL, Li FR, Jiao P, Yao ST, Sang H, Si YH. Apoptosis of human ovarian cancer cells induced by Paris chinensis dioscin via a Ca(2+)-mediated mitochondrion pathway. *Asian Pac J Cancer Prev* 2011; **12**: 1361–6.

Gasparotto J, Somensi N, Kunzler A, Girardi CS, de Bittencourt Pasquali MA, Ramos VM, Simoes-Pires A, Quintans-Junior LJ, Branco A, Moreira JC, Gelain DP. Hecogenin acetate inhibits reactive oxygen species production and induces cell cycle arrest and senescence in the a549 human lung cancer cell line. *Anticancer Agents Med Chem* 2014; **14**: 1128–35.

Gauthaman K, Adaikan PG, Prasad RN. Aphrodisiac properties of *Tribulus terrestris* extract (Protodioscin) in normal and castrated rats. *Life Sciences* 2002; **71**: 1385–96.

Gauthaman K, Adaikan PG. Effect of *Tribulus terrestris* on nicotinamide adenine dinucleotide phosphate-diaphorase activity and androgen receptors in rat brain. *J Ethnopharmacol* 2005; **96**: 127–32.

Gheorghiu A, lonesui M. presence of chlorogenin with diosgenin and gitogenin in *T. terrestris*. *Annls Pharm Fr* 1968; **26**: 795.

Guan T, Liu Q, Qian Y, Yang H, Kong J, Kou J, Yu B. Ruscogenin reduces cerebral ischemic injury via NF-κB-mediated inflammatory pathway in the mouse model of experimental stroke. *Eur J Pharmacol* 2013; **714**: 303–11.

Hardman R, Wood CN. The effect of ripening and aqueous incubation on the yield of diosgenin and yamogenin from the fruits of Balanites pedicellaris. *Planta Med* 1971a; **20**: 350–6.

Hardman R, Wood CN. The ripe fruits of *Balanites orbicularis* as a new source of diosgenin and yamogenin. *Phytochemistry* 1971b; **10**: 887–889.

Hayes PY, Jahidin AH, Lehmann R, Penman K, Kitching W, De Voss JJ. Steroidal saponins from the roots of *Asparagus racemosus*. *Phytochemistry* 2008; **69**: 796–804.

He X, Qiao A, Wang X, Liu B, Jiang M, Su L, Yao X. Structural identification of methyl protodioscin metabolites in rats' urine and their antiproliferative activities against human tumor cell lines. *Steroids* 2006; **71**: 828–33.

He Z, Chen H, Li G, Zhu H, Gao Y, Zhang L, Sun J. Diosgenin inhibits the migration of human breast cancer MDA-MB-231 cells by suppressing Vav2 activity. *Phytomedicine* 2014; **21**: 871–6.

Hibasami H, Moteki H, Ishikawa K, Katsuzaki H, Imai K, Yoshioka K, Ishii Y, Komiya T. Protodioscin isolated from fenugreek (*Trigonella foenum graecum* L.) induces cell death and morphological change indicative of apoptosis in leukemic cell line H-60, but not in gastric cancer cell line KATO III. *Int J Mol Med* 2003; **11**: 23–6.

Hsieh MJ, Yang SF, Hsieh YS, Chen TY, Chiou HL. Autophagy inhibition enhances apoptosis induced by dioscin in huh7 cells. *Evid Based Complement Alternat Med* 2012; **2012**: 134512.

Hsieh MJ, Tsai TL, Hsieh YS, Wang CJ, Chiou HL. Dioscin-induced autophagy mitigates cell apoptosis through modulation of PI3K/Akt and ERK and JNK signaling pathways in human lung cancer cell lines. *Arch Toxicol* 2013; **87**: 1927–37.

Hu M, Xu L, Yin L, Qi Y, Li H, Xu Y, Han X, Peng J, Wan X. Cytotoxicity of dioscin in human gastric carcinoma cells through death receptor and mitochondrial pathways. *J Appl Toxicol* 2013; **33**: 712–22.

Huang HC, Lin MK, Hwang SY, Hwang TL, Kuo YH, Chang CI, Ou CY, Kuo YH. Two anti-inflammatory steroidal saponins from *Dracaena angustifolia* Roxb. *Molecules* 2013; **18**: 8752–63.

Huang W, Zou K. Cytotoxicity of a plant steroidal saponin on human lung cancer cells. *Asian Pac J Cancer Prev* 2011; **12**: 513–7.

Huang YL, Kou JP, Ma L, Song JX, Yu BY. Possible mechanism of the anti-inflammatory activity of ruscogenin: role of intercellular adhesion molecule-1 and nuclear factor-kappaB. *J Pharmacol Sci* 2008; **108**: 198–205.

Hung-Jen Lu, Thing-Fong Tzeng, Shorong-Shii Liou, Sheng Da Lin, Ming-Chang Wu, I-Min Liu. Ruscogenin ameliorates diabetic nephropathy by its anti-inflammatory and anti-fibrotic effects in streptozotocin-induced diabetic rat. *BMC Complementary and Alternative Medicine* 2014; **14**: 110.

Jagadeesan J, Nandakumar N, Rengarajan T, Balasubramanian MP. Diosgenin, a steroidal saponin, exhibits anticancer activity by attenuating lipid peroxidation via enhancing antioxidant defense system during NMU-induced breast carcinoma. *J Environ Pathol Toxicol Oncol* 2012; **31**: 121–9.

Jia ZH, Ju Y. Steroidal saponins from *Smilax lebrunii*. *Phytochemistry* 1992; **31**: 3173–5.

Jin JM, Zhang YJ, Li HZ, Yang CR. Cytotoxic steroidal saponins from Polygonatum zanlanscianense. *J Nat Prod* 2004; **67**: 1992–5.

Jin YL, Kuk JH, Oh KT, Kim YJ, Piao XL, Park RD. A new steroidal saponin, yuccalan, from the leaves of *Yucca smalliana*. *Arch Pharm Res* 2007; **30**: 543–6.

Ju Y, Jia ZH. Steroidal saponins from the rhizomes of *Smilax menispermoidea*. *Phytochemistry* 1992; **31**: 1349–51.

Kemertelidze ÉP, Pkheidze TA. Tigogenin from *Yucca gloriosa*, a possible raw material for the synthesis of steroid hormonal preparations. *Pharm Chem J* 1972; **6**: 795–797.

Lack R, Newman BC, Shoppee CW. The chemical constituents of the leaves of *Digitalis lanata*. *Australian J Chem* 1963; **16**: 896–899.

Li K, Tang Y, Fawcett JP, Gu J, Zhong D. Characterization of the pharmacokinetics of dioscin in rat. *Steroids* 2005; **70**: 525–30.

Li Y, Wang X, Cheng S, Du J, Deng Z, Zhang Y, Liu Q, Gao J, Cheng B, Ling C. Diosgenin induces G2/M cell cycle arrest and apoptosis in human hepatocellular carcinoma cells. *Oncol Rep* 2015; **33**: 693–8.

Liagre B, Vergne-Salle P, Leger DY, Beneytout JL. Inhibition of human rheumatoid arthritis synovial cell survival by hecogenin and tigogenin is associated with increased apoptosis, p38 mitogen-activated protein kinase activity and up regulation of cyclooxygenase-2. *Int J Mol Med* 2007; **20**: 451–60.

Liang C, Lim JH, Kim SH, Kim DS. Dioscin: a synergistic tyrosinase inhibitor from the roots of *Smilax china*. *Food Chem* 2012; **134**: 1146–8.

Lin Y, Jia R, Liu Y, Gao Y, Zeng X, Kou J, Yu B. Diosgenin inhibits superoxide generation in FMLP-activated mouse neutrophils via multiple pathways. *Free Radic Res* 2014; **48**: 1485–93.

Liu C, Wang Y, Wu C, Pei R, Song J, Chen S, Chen X. Dioscin's antiviral effect *in vitro*. *Virus Res* 2013; **172**: 9–14.

Liu M, Xu L, Yin L, Qi Y, Xu Y, Han X, Zhao Y, Sun H, Yao J, Lin Y, Liu K, Peng J. Potent effects of dioscin against obesity in mice. *Sci Rep* 2015; **5**: 7973.

Liu MJ, Yue PY, Wang Z, Wong RN. Methyl protodioscin induces G2/M arrest and apoptosis in K562 cells with the hyperpolarization of mitochondria. *Cancer Lett* 2005; **224**: 229–41.

Liu X, Zhang H, Niu XF, Xin W, Qi L. Steroidal saponins from *Smilacina japonica*. *Fitoterapia* 2012; **83**: 812–6.

Lu B, Xu Y, Xu L, Cong X, Yin L, Li H, Peng J. Mechanism investigation of dioscin against CCl4-induced acute liver damage in mice. *Environ Toxicol Pharmacol* 2012; **34**: 127–35.

Lu HJ, Tzeng TF, Liou SS, Chang CJ, Yang C, Wu MC, Liu IM. Ruscogenin ameliorates experimental nonalcoholic steatohepatitis via suppressing lipogenesis and inflammatory pathway. *Biomed Res Int* 2014; **2014**: 652–680.

Lv L, Zheng L, Dong D, Xu L, Yin L, Xu Y, Qi Y, Han X, Peng J. Dioscin, a natural steroid saponin, induces apoptosis and DNA damage through reactive oxygen species: a potential new drug for treatment of glioblastoma multiforme. *Food Chem Toxicol* 2013; **59**: 657–69.

Lv YC, Yang J, Yao F, Xie W, Tang YY, Ouyang XP, He PP, Tan YL, Li L, Zhang M, Liu D, Cayabyab FS, Zheng XL, Tang CK. Diosgenin inhibits atherosclerosis via suppressing the MiR-19b-induced downregulation of ATP-binding cassette transporter A1. *Atherosclerosis*. 2015; **240**: 80–9.

Manda VK, Avula B, Ali Z, Wong YH, Smillie TJ, Khan IA, Khan SI. Characterization of in vitro ADME properties of diosgenin and dioscin from *Dioscorea villosa*. *Planta Med* 2013; **79**: 1421–8.

Manivannan J, Barathkumar TR, Sivasubramanian J, Arunagiri P, Raja B, Balamurugan E. Diosgenin attenuates vascular calcification in chronic renal failure rats. *Mol Cell Biochem* 2013; **378**: 9–18.

Manivannan J, Balamurugan E, Silambarasan T, Raja B. Diosgenin improves vascular function by increasing aortic eNOS expression, normalize dyslipidemia and ACE activity in chronic renal failure rats. *Mol Cell Biochem* 2013a; **384**: 113–20.

Minghe Y, Yanyong C. Steroidal sapogenins in *Dioscorea collettii*. *Planta Med* 1983; **49**: 38–42.

Mohammad RY, Somayyeh G, Gholamreza H, Majid M, Yousef R. Diosgenin inhibits hTERT gene expression in the A549 lung cancer cell line. *Asian Pac J Cancer Prev* 2013; **14**: 6945–8.

Moriwaki S, Murakami H, Takahashi N, Uemura T, Taketani K, Hoshino S, Tsuge N, Narukami T, Goto T, Kawada T. Yamogenin in fenugreek inhibits lipid accumulation through the suppression of gene expression in fatty acid synthesis in hepatocytes. *Biosci Biotechnol Biochem* 2014; **78**: 1231–6.

Oleszek W, Sitek M, Stochmal A, Piacente S, Pizza C, Cheeke P. Steroidal saponins of *Yucca schidigera* Roezl. *J Agric Food Chem* 2001; **49**: 4392–6.

Oliveira L, Lacerda D, Nunes F. Effects of hecogenin on Larvicidal activity against *Aedes aegypti* mosquito, the dengue vector. *BMC Proceedings* 2014; **8**: 33.

Perepelista ED, Kintya PK. Chemical study of steroid glycosides of *T. terrestris* IV steroid saponins. *Khim Prir Soedin* 1975; **11**: 260.

Puri HS, Jefferies TM, Hardman R. Diosgenin and yamogenin levels in some Indian plant samples. *Planta Med* 1976; **30**: 118–21.

Qiu SX, Li XC and Xiong Y. Isolation and characterization of cytotoxic saponin Chloromaloside-A from *Chlorophytum borivilianum*. *Planta Med* 2000; **66**: 587–90.

Quintans JS, Barreto RS, de Lucca W Jr, Villarreal CF, Kaneto CM, Soares MB, Branco A, Almeida JR, Taranto AG, Antoniolli AR, Freitas RM, Quintans LJ Jr. Evidence for the involvement of spinal cord-inhibitory and cytokines-modulatory mechanisms in the anti-hyperalgesic effect of hecogenin acetate, a steroidal sapogenin-acetylated, in mice. *Molecules* 2014; **19**: 8303–16.

Quyen le T, Khoi NH, Suong NN, Schreiber K, Ripperger H. Steroid alkaloids and yamogenin from *Solanum spirale*. *Planta Med* 1987; **53**: 292–3.

Raju J, Patlolla JM, Swamy MV, Rao CV. Diosgenin, a steroid saponin of Trigonella foenum graecum (Fenugreek), inhibits azoxymethane-induced aberrant crypt foci formation in F344 rats and induces apoptosis in HT-29 human colon cancer cells. *Cancer Epidemiol Biomarkers Prev* 2004; **13**: 1392–8.

Raju J, Bird RP. Diosgenin, a naturally occurring steroid [corrected] saponin suppresses 3-hydroxy-3-methylglutaryl CoA reductase expression and induces apoptosis in HCT-116 human colon carcinoma cells. *Cancer Lett* 2007; **255**: 194–204.

Santos Cerqueira G, dos Santos e Silva G, Rios Vasconcelos E, Fragoso de Freitas AP, Arcanjo Moura B, Silveira Macedo D, Lopes Souto A, Barbosa Filho JM, de Almeida Leal LK, de Castro Brito GA, Souccar C, de Barros Viana GS. Effects of hecogenin and its possible mechanism of action on experimental models of gastric ulcer in mice. *Eur J Pharmacol* 2012; **683**: 260–9.

Sato K, Fujita S, Iemitsu M. Acute administration of diosgenin or dioscorea improves hyperglycemia with increases muscular steroidogenesis in STZ-induced type 1 diabetic rats. *J Steroid Biochem Mol Biol* 2014; **143**: 152–9.

Sautour M, Miyamoto T, Lacaille-Dubois MA. Steroidal saponins from *Asparagus acutifolius*. *Phytochemistry* 2007; **68**: 2554–62.

Shao Y, Poobrasert O, Kennelly Edward J, Chin C, Ho C, Huang M, Garrison SA, Cordell GA. Cytotoxic activity of steroidal saponins from *Asparagus officinalis*. Acta Hort. (ISHS) 1999; **479**: 277–282.

Sharma SC, Chand R, Sati OP. Steroidal saponins of *Asparagus adscendens*. *Phytochemistry* 1982; **21**: 2075–2078.

Sharma U, Saini R, Kumar N, Singh B. Steroidal saponins from *Asparagus racemosus*. *Chem Pharm Bull* (Tokyo) 2009; **57**: 890–3.

Shen HY, Zuo WJ, Wang H, Zhao YX, Guo ZK, Luo Y, Li XN, Dai HF, Mei WL. Steroidal saponins from dragon's blood of Dracaena cambodiana. *Fitoterapia* 2014; **94**: 94–101.

Shen S, Chen CJ, Bu R, Ga L, Li GY, Tan Y, Li X, Wang JH. Three new steroidal saponins from *Fritillaria pallidiflora*. *J Asian Nat Prod Res* 2011; **13**: 1014–22.

Sunitha Kumari K, Immanuel G, Dhanya BS, Raj A. Evaluation of the effect of tigogenin on the activities of certain key enzymes of carbohydrate metabolism in streptozotocin induced diabetic rats. *Int J Biol Med Res* 2012; **3**: 1242–1247.

Surendra Kumar M, Aswathy TN, Suhail CN, Astalakshmi N. Evaluation of *Costus speciosus* Koen aqueous extract for larvicidal activity. *Der Pharmacia Lettre* 2013; **5**: 283–285.

Tao X, Wan X, Xu Y, Xu L, Qi Y, Yin L, Han X, Lin Y, Peng J. Dioscin attenuates hepatic ischemia-reperfusion injury in rats through inhibition of oxidative-nitrative stress, inflammation and apoptosis. *Transplantation* 2014; **98**: 604–11.

Tao X, Sun X, Yin L, Han X, Xu L, Qi Y, Xu Y, Li H, Lin Y, Liu K, Peng J. Dioscin ameliorates cerebral ischemia/reperfusion injury through the downregulation of TLR4 signaling via HMGB-1 inhibition. *Free Radic Biol Med* 2015; **84**: 103–115.

Tomova MP, Panova DL, Vulfson NS. Phytosterol from *T. terrestris*. *Dokl Bolg Akad Nauk* 1973; **26**: 379.

Tomova MP, Panova DL, Vulfson NS. Steroid Saponins and sapongenins IV. Saponins from *T. terrestris*. *Planta Med* 1974; **25**: 231.

Tomova MP, Bochova D, Zaiku CG, Vulfson NS. Steroid Saponins and sapongenin V. Hecogenin from T. terrestris. *Planta Med* 1977; **32**: 223.

Tsai PL, Wang JP, Chang CW, Kuo SC, Chao PD. Constituents and bioactive principles of Polygonum chinensis. *Phytochemistry* 1998; **49**: 1663–6.

Uemura T, Hirai S, Mizoguchi N, Goto T, Lee JY, Taketani K, Nakano Y, Shono J, Hoshino S, Tsuge N, Narukami T, Takahashi N, Kawada T. Diosgenin present in fenugreek improves glucose metabolism by promoting adipocyte differentiation and inhibiting inflammation in adipose tissues. *Mol Nutr Food Res* 2010; **54**: 1596–608.

Wang G, Chen H, Huang M, Wang N, Zhang J, Zhang Y, Bai G, Fong WF, Yang M, Yao X. Methyl protodioscin induces G2/M cell cycle arrest and apoptosis in HepG2 liver cancer cells. *Cancer Lett* 2006; **241**: 102–9.

Wang L, Meng Q, Wang C, Liu Q, Peng J, Huo X, Sun H, Ma X, Liu K. Dioscin restores the activity of the anticancer agent adriamycin in multidrug-resistant human leukemia K562/adriamycin cells by down-regulating MDR1 via a mechanism involving NF-κB signaling inhibition. *J Nat Prod* 2013; **76**: 909–14.

Wei Y, Xu Y, Han X, Qi Y, Xu L, Xu Y, Yin L, Sun H, Liu K, Peng J. Anti-cancer effects of dioscin on three kinds of human lung cancer cell lines through inducing DNA damage and activating mitochondrial signal pathway. *Food Chem Toxicol* 2013; **59**: 118–28.

Xing-Cong Li, De-Zu Wang, Chong-Ren Yang. Steroidal saponins from *Chlorophytum malayense*. *Phytochemistry* 1990; **29**: 3893–3898.

Xu LX, Zhou TH. Colorimetric method for the determination of hecogenin in *Agave americana* L. residue. *Yao Xue Xue Bao* 1982; **17**: 609–14.

Xu T, Zhang S, Zheng L, Yin L, Xu L, Peng J. A 90-day subchronic toxicological assessment of dioscin, a natural steroid saponin, in Sprague-Dawley rats. *Food Chem Toxicol* 2012; **50**: 1279–87.

Xu T, Zheng L, Xu L, Yin L, Qi Y, Xu Y, Han X, Peng J. Protective effects of dioscin against alcohol-induced liver injury. *Arch Toxicol* 2014; **88**: 739–53.

Ya-Lin Huang, Jun-Ping Kou, Ji-Hua Liu, Nan Liu, Bo-Yang Yu. Comparison of anti-inflammatory activities of ruscogenin, a major steroidal sapogenin from Radix Ophiopogon japonicus, and its succinylated derivative, RUS-2HS. *Drug Development Research* 2008a; **69**: 196–202.

Yang CR, Zhang Y, Jacob MR, Khan SI, Zhang YJ, Li XC. Antifungal activity of C-27 steroidal saponins. *Antimicrob Agents Chemother* 2006; **50**: 1710–4.

Zhang C, Peng J, Wu S, Jin Y, Xia F, Wang C, Liu K, Sun H, Liu M. Dioscin promotes osteoblastic proliferation and differentiation via Lrp5 and ER pathway in mouse and human osteoblast-like cell lines. *J Biomed Sci* 2014; **17**: 21–30.

Zhang X, Han X, Yin L, Xu L, Qi Y, Xu Y, Sun H, Lin Y, Liu K, Peng J. Potent effects of dioscin against liver fibrosis. *Sci Rep* 2015; **5**: 9713.

Zhang Y, Li HZ, Zhang YJ, Jacob MR, Khan SI, Li XC, Yang CR. Atropurosides A-G, new steroidal saponins from Smilacina atropurpurea. *Steroids* 2006; **71**: 712–9.

Zhang Y, Zhang YJ, Jacob MR, Li XC, Yang CR. Steroidal saponins from the stem of *Yucca elephantipes*. *Phytochemistry* 2008; **69**: 264–70.

Zhang R, Gilbert S, Yao X, Vallance J, Steinbrecher K, Moriggl R, Zhang D, Eluri M, Chen H, Cao H, Shroyer N, Denson L, Han X. Natural compound methyl protodioscin protects against intestinal inflammation through modulation of intestinal immune responses. *Pharmacol Res Persp* 2015. DOI: 10.1002/prp2.118.

Zhao X, Cong X, Zheng L, Xu L, Yin L, Peng J. Dioscin, a natural steroid saponin, shows remarkable protective effect against acetaminophen-induced liver damage *in vitro* and *in vivo*. *Toxicol Lett* 2012; **214**: 69–80.

Zhao X, Ren H, Gao S, Hao J. Effects of dioscin on apoptosis in pancreatic cancer MiaPaCa-2 cells and its mechanism. *Zhonghua Zhong Liu Za Zhi* 2014; **36**: 5–10.

Zheng QA, Zhang YJ, Li HZ, Yang CR. Steroidal saponins from fresh stem of Dracaena cochinchinensis. *Steroids* 2004; **69**: 111-9.

Zhou H, Yang X, Wang N, Zhang Y, Cai G. Tigogenin inhibits adipocytic differentiation and induces osteoblastic differentiation in mouse bone marrow stromal cells. *Mol Cell Endocrinol* 2007a; **270**: 17–22.

Zhou L-B, Chen T-H, Bastow KF , Shibano M, Lee K-H, Chen D-F. Filiasparosides A–D, cytotoxic steroidal saponins from the roots of *Asparagus filicinus*. *J Nat Prod* 2007b; **70**: 1263–1267.

Zhou Li-Bo, Chen Dao-Feng. Steroidal saponins from the roots of *Asparagus filicinus*. *Steroids* 2008; **73**: 83–87.

Zullo MAT, Azzini A, Salgado ALD, Ciaramello D. Steroidal sapogenins in sisal. *Bragantia* 1989; **48**: http://dx.doi.org/10.1590/S0006-87051989000100003.

Chapter 21

Steroidal Alkaloids

21.1 Introduction

The steroidal alkaloids represent an important class of alkaloids that essentially afford a close structural relationship to sterols, i.e., they contain a perhydro-1, 2-cyclopentanophenanthrene nucleus. Steroid alkaloids invariably occur in the plant kingdom as glucosidal combination with carbohydrate moieties.

21.2 Classification of Steroidal Alkaloids

Two important classes of steroid alkaloids are

- The *Solanum* type—one example is solanidne. This steroid alkaloid is the nucleus (i.e., aglycone) for two important glycoalkaloids, solanine and chaconine, found in potatoes. Other plants in the Solanum family including various nightshades, Jerusalem cherries, and tomatoes also contain solanum-type glycoalkaloids. Glycoalkaloids are glycosides of alkaloids.
- The *Veratrum* type—There are more than 50 Veratrum alkaloids including veratramine, cyclopamine, cycloposine, jervine, and muldamine occurring in plants of the Veratrum spp. *Zigadenus venenosus* produces several veratrum-type of steroid alkaloids including, zygacine.

21.3 Solanum-type Alkaloids

Solanum-type alkaloids are found in plants in the form of glycosides of alkaloids. Glycosides are ethers that join a noncarbohydrate moiety, the aglycone, by a ester bond to a carbohydrate moiety. In solanum-type glycoalkaloids, the aglycone is a steroid alkaloid. Solanine and chaconine cause poisoning in potatoes. They have the same aglycone, solanidine, but the structure of their carbohydrate sidechains is different. Tomatine is a glycoalkaloid found in tomatoes. Its aglycone is tomatidine.

21.4 Steroidal Alkaloids found in Solanaceae

α-Chaconine
Solamargine
Solanidine
Solanine
Solanocapsine
Solasodamine
Solasodine
Solasonine
Tomatine

21.5 Steroidal Alkaloids of *Veratrum nigrum* L.

21 steroidal alkaloids (5 protoverine-type alkaloids, 14 germine-type alkaloids, and 2 zygadenin-type alkaloids) have been selectively identified. Xingangermine and deacetyl xinganveratrine were found to be novel steroidal alkaloids (Li et al. 2007).

21.6 Steroidal Alkaloids of *Veratrum lobelianum* Bernh.

Twelve steroidal alkaloids including veralosinine, veralosine, teinemine, and (+/–)-15-O-(2-Methylbutyroyl)germine were isolated from four populations of *Veratrum lobelianum* and *Veratrum nigrum* (Christov et al. 2010).

21.7 Steroidal Alkaloids of *Sarcococca saligna* D. Don

Two new pregnane-type steroidal alkaloids, saligcinnamide, and N(a)-methyl epipachysamine-D along with epipachysamine D were isolated from the EtOH extracts of the roots and stems of *Sarcococca saligna*. Two derivatives of saligcinnamide, dihydrosaligcinnarnide, and dihydrosaligcinnamin were tested for antibacterial activity (Rahman et al. 1998).

21.8 Steroidal Alkaloids of *Sarcococca hookeriana*

Hookerianamide, hookerianamide-E, hookerianamide-F, and hookerianamide-G have been reported (Choudhary et al. 2005).

21.9 Steroidal Alkaloids of *Buxus macowanii* Oliv.

Chemical investigation of the crude methanolic extract of B. macowanii resulted in the isolation of five new steroidal alkaloids, 31-hydroxybuxatrienone, macowanioxazine, 16α-hydroxymacowanitriene, macowanitriene, macowamine, along with five known steroidalbases, Nb-demethylpapillotrienine, moenjodaramine, irehine, buxbodine B and buxmicrophylline C. All isolates exhibited moderate to weak anti-AChE activity with IC50 values in the range of 10.8–98 μM (Lam et al. 2015).

21.10 Steroidal Alkaloids of *Buxus hyrcana* Pojark.

Two new steroidal alkaloids, (+)-O6-buxafurandiene (1), and (+)-7-deoxy-O6-buxafurandiene along with four known steroidal bases, (+)-benzoylbuxidienine, (+)-buxapapillinine, (+)-buxaquamarine, and (+)-irehine have been reported. All compounds exhibited acetylcholinesterase enzyme inhibitory activity (Babar et al. 2006).

21.11 Steroidal Alkaloids of *Fritillaria puqiensis* G.D. Yu et G.Y. Chen

Six new steroidal alkaloids, namely puqienines C–E, puqiedine, 3alpha-puqiedin-7-ol, and puqietinedione, along with two known puqiedinone and peimisine have been isolated from the bulbs of *F. puqiensis* (Jiang et al. 2006).

21.12 Steroidal Alkaloids of *Fritillaria unibracteata*

Two new steroidal alkaloids peimisine-3-O-β-D-glucopyranoside and puqiedinone-3-O-β-D-glucopyranoside, together with three known compounds peimisine, puqiedinone , and puqiedine, were isolated and characterized from the bulbs of *F. unibracteata* peimisine-3-O-β-D-glucopyranoside showed moderate protection effect on neurotoxicity of PC12 cell lines induced by rotenone (Zhang et al. 2011).

21.13 Steroidal Alkaloids of *Fritillaria shuchengensis* S.C. Chen et S.F. Yin

A novel steroidal alkaloid, suchengbeisine, along with two known steroidal alkaloids, N-oxide of verticinone and zhebeininoside, has been isolated from the bulbs of *F. shuchengensis* (Huang et al. 2013).

21.14 Steroidal Alkaloids of *Fritillaria cirrhosa*

Imperialine and peimisine have been reported (Wang et al. 2014).

21.15 Steroidal Alkaloids of *Fritillaria imperialis*

Two new cevanine steroidal alkaloids, impericine and forticine, along with known bases delavine, persicanidine A, and imperialine were isolated. The alkaloids showed anti-acetylcholinesterase and anti-butyrylcholinesterase inhibitory activity (Rahman et al. 2002).

21.16 Steroidal Alkaloids of *Fritillaria lichuanensis* P. Li et C.P. Yang (Liliaceae)

Two new steroidal alkaloids, lichuanine and lichuanisinine, were isolated from the bulbs of *F. lichuanensis* (Pi et al. 2006).

21.17 Dendrogenin A and B

Dendrogenin A and B were found to have potent proliferative effects in neural stem cells. Additionally, they induce neuronal outgrowth from neurospheres during *in vitro* cultivation (Khalifa et al. 2014).

21.18 Neuroprotective Effect of Steroidal Alkaloids

Extracts from *Lycopersicon esculentum* Mill. leaves and their isolated steroidal alkaloids (tomatine and tomatidine) afford neuroprotective effect against glutamate-induced toxicity in SH-SY5Y neuroblastoma cells (Taveira et al. 2014).

21.19 Steroidal Alkaloids Isolated from *Solanum paniculatum* L. (Solanaceae)

Steroidal alkaloids isolated from *S. paniculatum* strongly protected cells against mitomycin C aneugenic and/or clastogenic activities as well as modulated mitomycin C cytotoxic action (Vieira et al. 2013).

21.20 Steroidal Alkaloids of *Holarrhena antidysenterica*

Chemical investigations on the stem bark of *H. antidysenterica* resulted in the isolation of a new steroidal alkaloid designated as holadysenterine, together with conessine, isoconessimine, and kurchessine (Kumar et al. 2007).

21.21 α-chaconine

α-chaconine (Fig. 21.1) is a steroidal glycoalkaloid that occurs in plants of the Solanaceae family. α-Chaconine is a natural toxicant produced in green potatoes. It is responsible for the bitter taste of the potatoes. Tubers of potato produce α-chaconine in response to stress, providing the potato with fungicidal and insecticidal properties.

21.21.1 Pharmacology

21.21.1.1 Anti-angiogenic

α-chaconine inhibited proliferation of bovine aortic endothelial cells in a dose-dependent manner. When treated with non-toxic doses of α-chaconine, cell

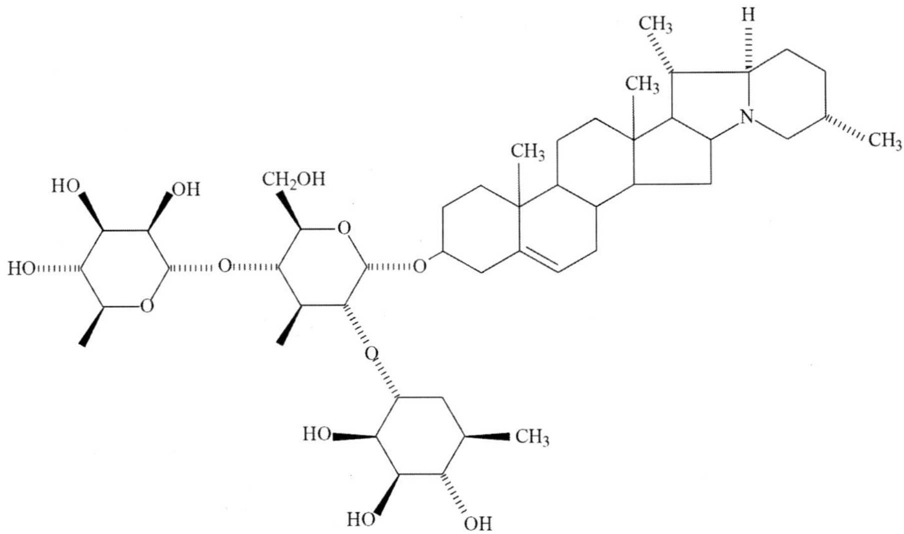

Fig. 21.1: Structure of α-chaconine.

migration, invasion, and tube formation were markedly suppressed. Results suggests that α-chaconine could inhibit NF-kappaB activity (Lu et al. 2010).

21.21.1.2 Chemotherapeutic action

α-chaconine could inhibit phosphorylation of c-Jun N-terminal kinase and Akt, whereas it did not affected phosphorylation of extracellular signal regulating kinase and p38. In addition, α-chaconine significantly decreased the nuclear level of nuclear factor kappa B and the binding ability of NF-kappaB (Shih et al. 2007).

21.21.1.3 Cytotoxic and apoptotic action

The aim of this study was to screen potato glycoalkaloids for cytotoxicity in a range of cancerous cell lines. IC50 values (human cancerous cell lines including Jurkat T lymphocytes, U937 leukemic monocyte lymphoma cells, Caco-2 epithelial colorectal cells, HepG2 liver cells, HFFF2 foetal foreskin fibroblasts, and MCF-7 breast cells).

> α-chaconine was the most cytotoxic glycoalkaloid across all cell lines. When compared to therapeutic drugs, α-chaconine has similar inhibitory effects to tamoxifen (Fig. 21.2) on cell viability (Kenny et al. 2013).

21.21.1.4 Induction of apoptosis of HT-29 human colon cancer cells

In a study it was found that caspase-3 activity and the active form of caspase-3 were increased 12 h after α-chaconine treatment. Caspase inhibitors, N-Ac-DEVD-CHO and Z-VAD-fmk, prevented α-chaconine-induced apoptosis, whereas alpha-

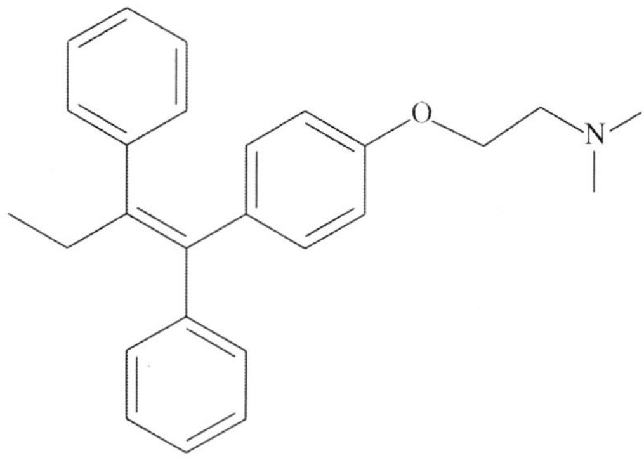

Fig. 21.2: Structure of tamoxifen.

chaconine-induced apoptosis was potentiated by PD98059, an extracellular signal-regulated kinase inhibitor (Yang et al. 2006).

21.21.2 Toxicology

The LD50 values of chaconine in the mouse and rabbit are also similar and suggest that compounds other than these are probably responsible for the predominant toxic effects of certain hybrid potatoes in man and animals. The failure of chaconine failed to produce a significant teratological effect in the chick embryo (Nishie et al. 1975).

A study evaluated the effect of several potato glycoalkaloids and aglycones in the frog embryo teratogenesis assay-Xenopus with and without metabolic activation by Aroclor 1254-induced rat liver microsomes. The data suggest that the glycoalkaloid alpha-chaconine is teratogenic and more embryotoxic than α-solanine, in terms of the median lethal concentration (LC50) after 96 hr of exposure, the concentration inducing gross terata in 50% of the surviving frog embryos (96-hr EC50, malformation), and the minimum concentration needed to inhibit the growth of the embryos (Friedman et al. 1991).

21.22 Conessine

Conessine (Fig. 21.3) is a steroid alkaloid found in *Holarrhena floribunda* (L.) Wall. (Duez et al. 1987), *Holarrhena antidysenterica* (L.) Wall. (Dohnal et al. 1990; Kumar et al. 2007) and *Funtumia elastica* (Preuss) Stapf. (Zirihi et al. 2005).

21.22.1 Pharmacology

21.22.1.1 Antimalarial

Holarrhetine (Fig. 21.4), conessine, holarrhesine, and isoconessimine (Fig. 21.5) exhibited *in vitro* antiplasmodial activity against the chloroquine-resistant strain FcB1 of *Plasmodium falciparum* with IC50 values ranging from 0.97 to 3.39 microM.

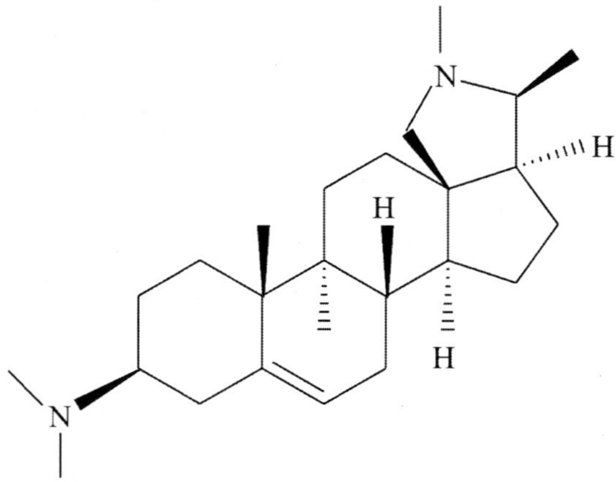

Fig. 21.3: Structure of conessine.

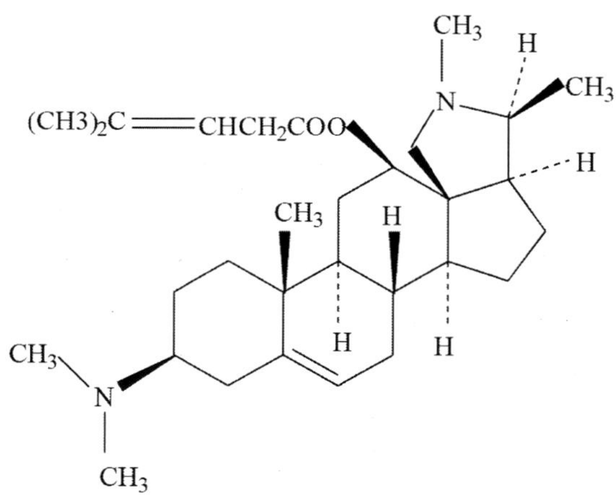

Fig. 21.4: Structure of holarrhetine.

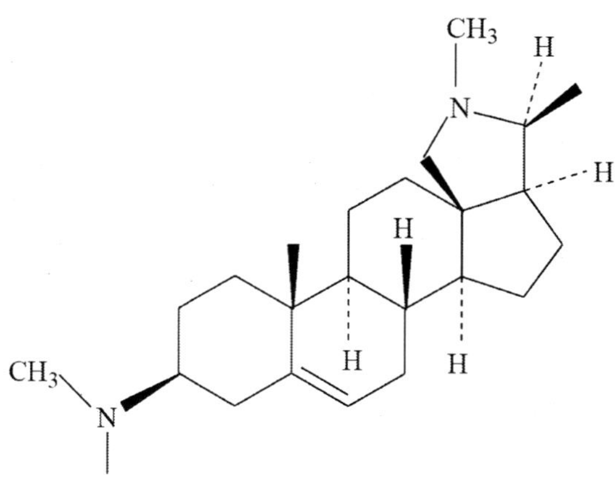

Fig. 21.5: Structure of isoconessimine.

They showed weak cytotoxicity against a rat cell line L-6 with IC50 values ranging from 5.13 to 36.55 microM (Zirihi et al. 2005).

> Conessine isolated from the bark of *H. antidysenterica*, showed *in vitro* anti-plasmodial activity with its IC_{50} value 1.9 μg/ml and 1.3 μg/ml using schizont maturation and parasite lactate dehydrogenase assay respectively. Conessine showed cytotoxicity IC_{50} = 14 μg/ml against L6 cells of rat skeletal muscle myoblast. Conessine significantly reduced parasitaemia (at 10 mg/kg exhibited 88.95% parasite inhibition) in *P. berghei*-infected mice (Dua et al. 2013).

21.22.1.2 Antimicrobial

The methanolic extract of the bark of *H. pubescens* and conessine were found to possess significant activity against some of the bacteria tested. The alkaloidal fraction and conessine also exhibited marginal activity against some of the fungi tested. The minimum inhibitory concentration value of conessine was determined against various bacteria, and the highest activity was seen against *Micrococcus luteus* ATCC 9341 (MIC: 15.6 μg per disc) (Siddiqui et al. 2012).

21.22.1.3 Antihistaminic

Conessine was discovered to bind to histamine H3 receptors in a radioligand-based high-throughput screen. Conessine displayed high affinity at both rat and human H3 receptors (pKi = 7.61 and 8.27) and generally high selectivity against other sites, including histamine receptors H1, H2, and H4 (Zhao et al. 2008).

21.22.1.4 AChE inhibitory

The total alkaloidal extract from the seeds of *H. antidysenterica* was found to have potent AChE inhibitory activity with an IC(50) value of 6.1 μg/mL. Conessine, conessimin, conarrhimin, and conimin, showed strong AChE inhibiting activity with IC(50) values ranging from 4 to 28 μM. The most active inhibitor conessimin with an IC(50) value of 4 Mm (Yang et al. 2012).

21.23 Malouetine

A steroidal alkaloid obtained from *Malouetia bequaertiana* E. Woodson (Janot et al. 1960).

21.24 Muldamine

A phytosterol alkaloid isolated from *Veratrum californicum* (Keeler 1971).

21.25 Pingbeinine

A steroidal alkaloid isolated from *Fritillaria ussuriensis* Maxim (Xu et al. 1990).

21.26 Solamargine

Solamargine (Fig. 21.6) is a glycoalkaloid derived from solasodine. It is isolated from *Solanum americanum* Miller (Al Chami et al. 2003), *Solanum erianthum* (Chou et al. 2012), *Solanum heteracanthum* (Manase et al. 2012), *Solanum incanum* (Hsu et al. 1996), *Solanum nigrum* Linn. (Sani et al. 2015), *Solanum palinacanthum*, and *Solanum lycocarpum* (Moreira et al. 2013).

21.26.1 Pharmacology

21.26.1.1 Anticancer

The appearance in solamargine-treated cells of chromatin condensation, DNA fragmentation, and a sub-G(1) peak in a DNA histogram suggests that solamargine induces cell death by apoptosis. The IC(50) values of solamargine for control, G(0)/G(1)-, M-, and G(2)/M-synchronized Hep3B cells were 5.0, > 10, 3.7, and 3.1 microg/mL, implying that cells in the G(2)/M phases are relatively susceptible to solamargine-mediated apoptosis (Kuo et al. 2000).

Cytotoxicity: Solamargine possessed a potent cytotoxicity to human hepatocyte (Hep3B) and normal skin fibroblast. The inhibition curves of solamargine to the both cells were essentially overlapped, suggesting a parallel effect for the cell death. Since TNF Receptor I has been involved in apoptosis, the overexpression of TNF receptor I may be related with the mechanism of cytotoxicity of solamargine (Hsu et al. 1996).

Nonsmall-cell lung cancer: After treatment with solamargine, the expression of HER2 mRNA was correlated with the expression of topoisomerase II α (TOP2A) mRNA. The combinatory use of low concentrations of solamargine with low-toxic topoisomerase II inhibitor epirubicin (Fig. 21.7) accelerated apoptotic cell death. Therefore, the downregulation of the HER2 and TOP2A expression by solamargine with epirubicin may partially explain the solamargine and epirubicin cytotoxicity synergy effect in nonsmall-cell lung cancer (Liang et al. 2007).

Apoptosis induction in cisplatin-resistant breast cancer: The combined treatment of solamargine and cisplatin significantly reduced Bcl-2 and Bcl-xL expressions, and enhanced Bax, cytochrome c, caspase-9, and -3 expressions in breast cancer cells. Thus, the combined use of solamargine and cisplatin (Fig. 21.8) may be effective in cisplatin-resistant breast cancer (Shiu et al. 2007).

Human hepatocellular carcinoma cells: A study investigated the effects of solamargine on tumor migration and invasion in aggressive human hepatocellular carcinoma cells. A wound healing migration assay and Boyden chamber invasion assay showed that solamargine significantly inhibited *in vitro* migration and invasion of HepG2 cells. At the highest dose, solamargine decreased cell migration and invasion by more than 70% and 72% in HepG2 cells, respectively. Western blotting and gelatin zymography results showed that solamargine reduced expression and function of MMP-2 and MMP-9 proteins (Sani et al. 2015).

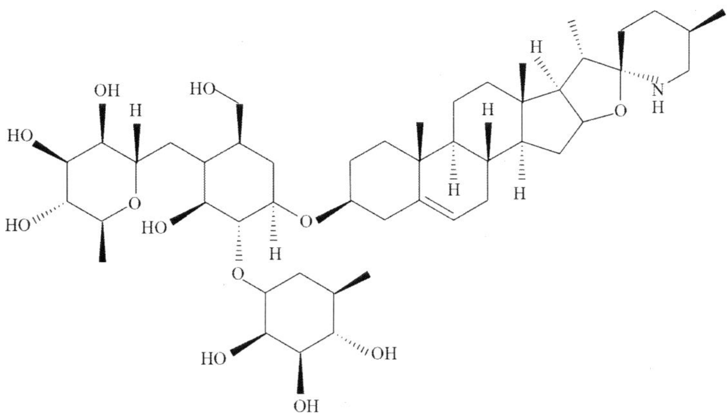

Fig. 21.6: Structure of solamargine.

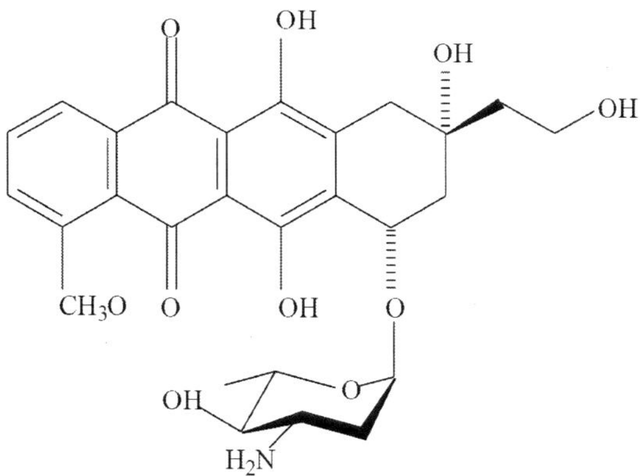

Fig. 21.7: Structure of epirubicin.

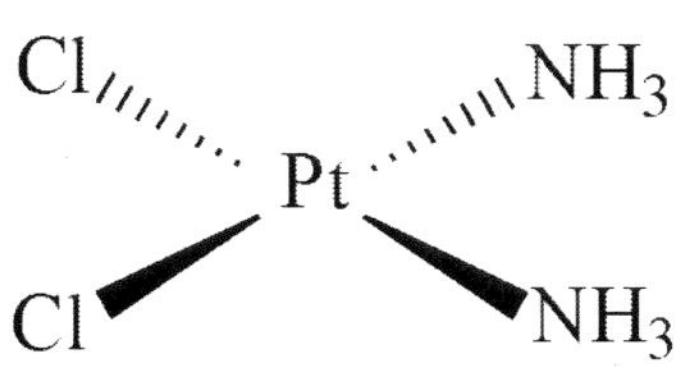

Fig. 21.8: Structure of cisplatin.

21.26.1.2 Trypanocidal

Extracts of *Solanum palinacanthum* and *Solanum lycocarpum* and solamargine were analysed for trypanocidal activity by using MTT colorimetric assay. Extracts of *S. palinacanthum* showed to be more active (IC50 = 175.9 μg.ml^{-1}) than *S. lycocarpum* (IC50 = 194.7 μg.ml^{-1}). Solamargine presented a strong activity (IC50 = 15.3 μg.ml^{-1}), which can explain the better activity of the both extracts (Moreira et al. 2013).

21.26.1.3 Toxicity

Lethality studies in rats showed a dose-mortality relationship with a LD(50) of 42 mg/kg body weight intraperitoneally. The chronic and subchronic toxicity investigations indicated that the size of the glycoalkaloid dose was more important than the total glycoalkaloid intake (Al Chami et al. 2003).

21.27 Solanidine

Solanidine (Fig. 21.9) is obtained from *Solanum tuberosum* Linn. (Attoumbré et al. 2013).

21.27.1 Pharmacology

21.27.1.1 Inhibition of acetyl cholinesterase

Solanidine, tomatidine, and demissidine (Fig. 21.10) had lower anticholinesterase (4.2 to 15.4%) than the seven glycoalkaloids α-chaconine, β_2-chaconine, α-solanine, dehydrocommersonine, commersonine, and demissine, except for tomatine (Bushway et al. 1987).

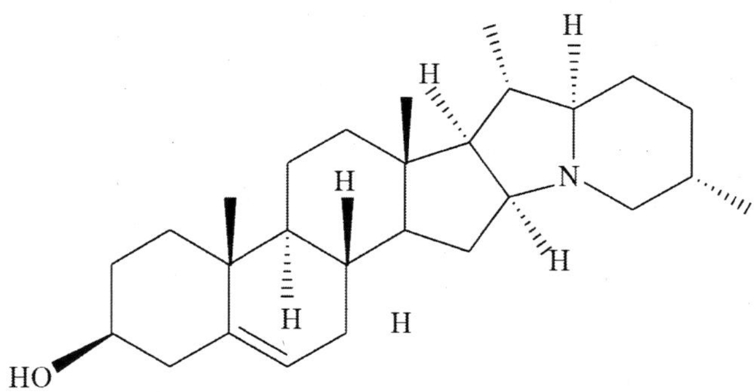

Fig. 21.9: Structure of solanidine.

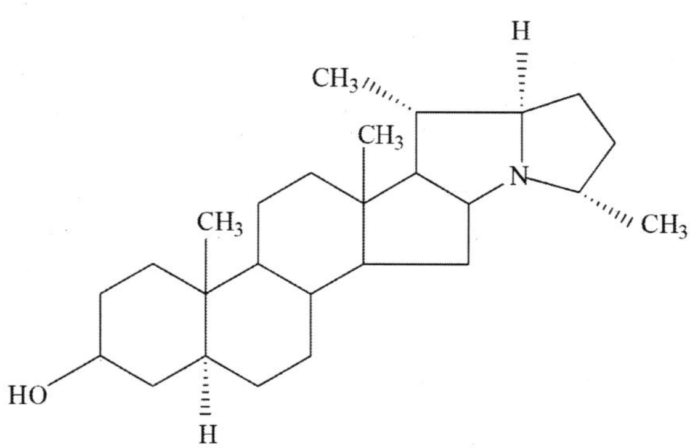

Fig. 21.10: Structure of demissidine.

21.27.1.2 Kinetics

There was a significant correlation between serum solanidine concentration and normal dietary intake of potato by the individual concerned. When two subjects abstained from potato and its products serum solanidine fell markedly and became minimal after the second week onwards (Harvey et al. 1985).

When [3H] solanidine was administered to normal human subjects by i.v. injection, the tritium concentration in the erythrocytes was 2–5 times greater than in the plasma. Three phases in the clearance of tritium from the plasma were identified having half-times of 2–5 min, 120–300 min, and 70–105 h. Rates of excretion of 3H in urine and faeces were low: 24 h after administration, 1–4% of the dose of 3H had been excreted in urine and 1–3% in faeces (Claringbold et al. 1982).

21.28 Solanine

Solanine is a glyco-alkaloid which is found in the blackcurrant-like fruits of *Solanum nigrum* L. and the redcurrant-like fruits of *Solanum dulcamara* L.

21.28.1 Anticancer

21.28.1.1 Inhibition of human melanoma cell migration and invasion

A study examined the effect of α-solanine on metastasis *in vitro.* α-solanine inhibited proliferation of human melanoma cell line A2058 in a dose-dependent manner. α-solanine inhibited migration and invasion of A2058 cells by reducing matrix metalloproteinase-2/9 activities (Lu et al. 2010b).

21.28.1.2 The protective and therapeutic effects on mice breast cancer

5 mg/kg solanine has been chosen for assessing its protective and therapeutic effects in mice breast cancer. The average tumor size and weight were significantly lower in solanine-treated animals than its respective control ones ($P < 0.05$). Solanine exerts a significant chemoprotective and chemotherapeutic effects on an animal model of breast cancer through apoptosis induction, cell proliferation, and angiogenesis inhibition (Mohsenikia et al. 2013).

21.28.1.3 Synergistic cytotoxicity induction by α-solanine and α-chaconine

A study confirmed α-solanine and α-chaconine synergistic cytotoxic effects on C6 rat glioma cells by three different cell viability tests, namely WST-1 (water-soluble tetrazolium) assay sensitive to intracellular NADH concentration, menadione-catalysed chemiluminescent assay depending on both NAD(P)H concentration and NAD(P)H:quinone reductase activity, and LDH (lactate dehydrogenase) assay sensitive to the release of LDH from damaged cells. The maximum cytotoxic effect was observed at a ratio of 1:1 between α-solanine and α-chaconine at micromolar concentrations (Yamashoji and Matsuda 2013).

21.28.1.4 Inhibition of invasion of human prostate cancer cell

The inhibition of the human prostate cancer cell PC-3 cell invasion by α-solanine may be, at least in part, through blocking epithelial-mesenchymal transition and matrix metalloproteinase-2 expression. α-Solanine also reduces ERK and PI3K/Akt signaling pathways and regulates expression of microRNA-21 (miR-21) and miR-138 (Shen et al. 2014).

21.28.1.5 Induction of mitochondria-mediated apoptosis in human pancreatic cancer cells

In pancreatic cancer cells and nu/nu nude mice model, solanine inhibited cancer cells grew through caspase-3 dependent mitochondrial apoptosis. Mechanically, solanine promotes the opening of mitochondrial membrane permeability transition pore by down regulating the Bcl-2/Bax ratio; thereafter, Cytochrome c and Smac are released from mitochondria into cytosol to process the caspase-3 zymogen into an activated form (Sun et al. 2014).

21.29 Solanocapsine

Solanocapsine (Fig. 21.11) is a toxic steroidal alkaloid from *Solanum pseudocapsicum* Linn. (Barger and Fraenkel-Conrat 1936).

21.30 Solasodine

Solasodine (Fig. 21.12) is found in *Solanum dulcamara* L. (Kumar et al. 2009), *Solanum sisymbriifolium* Linn. (Chauhan et al. 2011), and *Solanum trilobatum* Linn. (Pandurangan et al. 2010). Solasonine and solamargine are glycoalkaloid derivatives of solasodine.

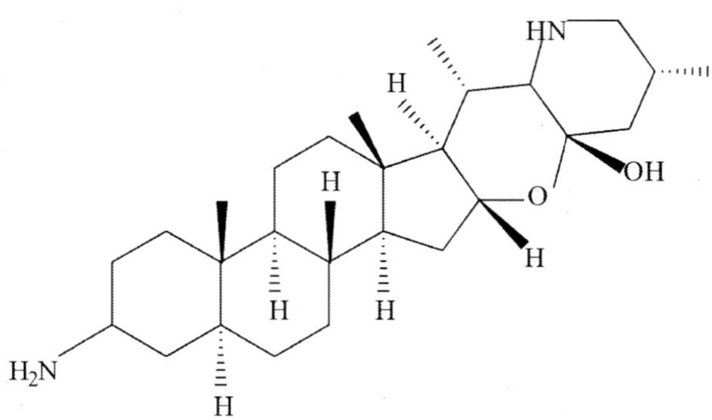

Fig. 21.11: Structure of solanocapsine.

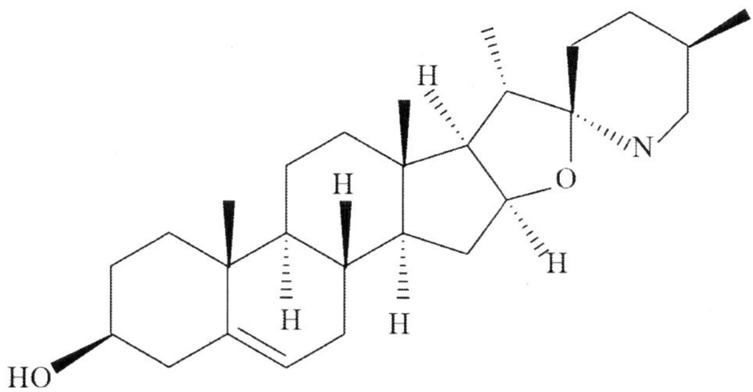

Fig. 21.12: Structure of solasodine.

21.30.1 Pharmacology

21.30.1.1 Induction of neurogenesis in vitro and in vivo

Solasodine treatment in rats resulted in a dramatic increase in expression of the cholesterol- and drug-binding translocator protein in ependymal cells, suggesting a possible role played by neurosteroid production in solasodine-induced neurogenesis.

In GAD65-GFP mice that express the green fluorescent protein under the control of the glutamic acid decarboxylase 65-kDa promoter, solasodine treatment increased the number of GABAergic progenitors and neuroblasts generated in the subventricular zone and present in the olfactory migratory tract (Lecanu et al. 2011).

21.30.1.2 Neuroprotective

Prior administration of solasodine (100 and 200 mg/kg, p.o.) significantly heightened superoxide dismutase, catalase, and glutathionetotal thiols, whereas reduced lipid peroxidation and nitric oxide levels in the brain. Brain coronal sectioning and histopathology studies revealed a marked reversal of I/R-provoked neuronal damage in the solasodine treatment groups (Sharma et al. 2014).

21.30.1.3 Anticonvulsant

Intraperitoneal injection of solasodine (25 mg/kg) significantly delayed ($p < 0.01$) latency of hind limb tonic extensor phase in the PCT-induced convulsions. In the MES model, solasodine significantly reduced ($p < 0.001$) duration of HLTE at 25, 50, and 100 mg/kg, i.p. in a dose-dependent manner (Chauhan et al. 2011).

21.30.1.4 Antinociceptive

Solasodine, when used at doses of 2, 4, and 8 mg/kg, this steroidal alkaloid caused a significant and dose-dependent decrease in the nociception induced by an intraperitoneal injection of acetic acid ($p < 0.001$). It also led to a significant reduction of the painful sensation caused by formalin in both phases of the formalin test ($p < 0.001$) (Pandurangan et al. 2010).

21.30.1.5 Antibacterial

Solanine (from unripe fruits), solasodine (from flowers), and beta-solamarine (from roots) of *Solanum dulcamara* Linn. inhibited the growth of *Escherichia coli* and Staphylococcus aureus. However, no significant activity was observed against *Enterobacter aerogenes* (Kumar et al. 2009).

21.31 Solasonine

Solasonine (Fig. 21.13) is a glycoalkaloid derived from solasodine. Solauricidine is the poisonous aglycone chemical compound of the glycoalkaloid solauricine. It closely resembles an isomer of solasonine. Solauricine is found in *Solanum mauritianum* Scop.

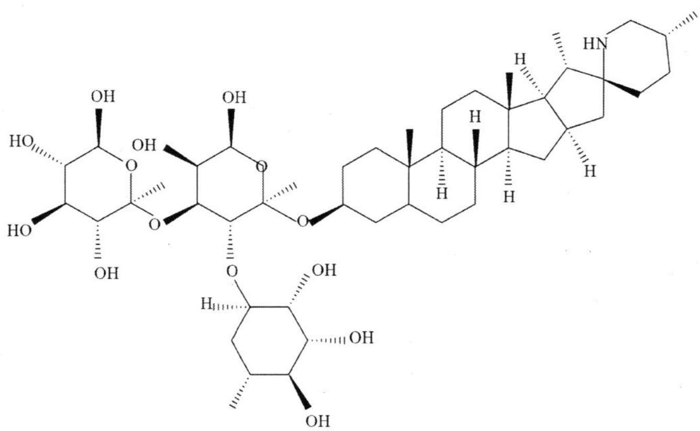

Fig. 21.13: Structure of solasonine.

21.32 Tomatine

Tomatine is a glycoalkaloid found in the stems and leaves of *Lycopersicon esculentum* (Kozukue et al. 2004). Its aglycone is tomatidine. It is a strong fungitoxic. Anticarcinogenic, cardioprotective, and other health benefits of tomato compounds lycopene, α-tomatine, and tomatidine has been discussed (Friedman 2013). Dehydrotomatine is another glycoalkaloid found in tomato (Kozukue et al. 2004).

21.32.1 Pharmacology

21.32.1.1 Anticarcinogenic

α-tomatine induced activation of caspase-3, -8, and -9, suggesting that both intrinsic and extrinsic apoptosis pathways are involved. Furthermore, nuclear factor-kappa

B (NF-κB) nuclear translocation was inhibited, which in turn resulted in significant decreased in NF-κB/p50 and NF-κB/p65 in the nuclear fraction of the treated cells compared to the control untreated cells (Lee et al. 2011).

Cell viability experiments showed that α-tomatine had significant cytotoxic effects on the human leukemia cancer cell lines HL60 and K562, and the cells were found to be in the Annexin V-positive/propidium iodide-negative phase of cell death. In addition, α-tomatine induced both HL60 and K562 cell apoptosis in a cell cycle- and caspase-independent manner (Chao et al. 2012).

Compared to untreated controls, the high-tomatine green tomato extracts strongly inhibited the following human cancer cell lines: breast (MCF-7), colon (HT-29), gastric (AGS), and hepatoma (liver) (HepG2), as well as normal human liver cells (Chang). There was little inhibition of the cells by the three low-tomatine red tomato extracts (Friedman et al. 2013).

Combination alpha-tomatine (1 mg/kg) and doxorubicin (2 mg/kg) had a synergistic effect and significantly prolonged the survival of the mice. Neither alpha-tomatine nor doxorubicin influenced the infiltration of tumours with CD3+ lymphocytes; nor were we able to find an *in vivo* modulation of the key molecules of two regulatory pathways reported *in vitro* as the principal anti-cancer mechanisms of alpha-tomatine.

Combined treatment with a sub-toxic dose of α-tomatine and paclitaxel (Fig. 21.14) significantly decreased cell viability with concomitant increase in the percentage of apoptotic PC-3 cells. The combined treatment, however, had no cytotoxic effect on the non-neoplastic prostate RWPE-1 cells. Apoptosis of PC-3 cells was accompanied by the inhibition of PI3K/Akt pro-survival signaling, an increase in the expression of the pro-apoptotic protein BAD but a decrease in the expressions of anti-apoptotic proteins, Bcl-2 and Bcl-xL (Lee et al. 2013).

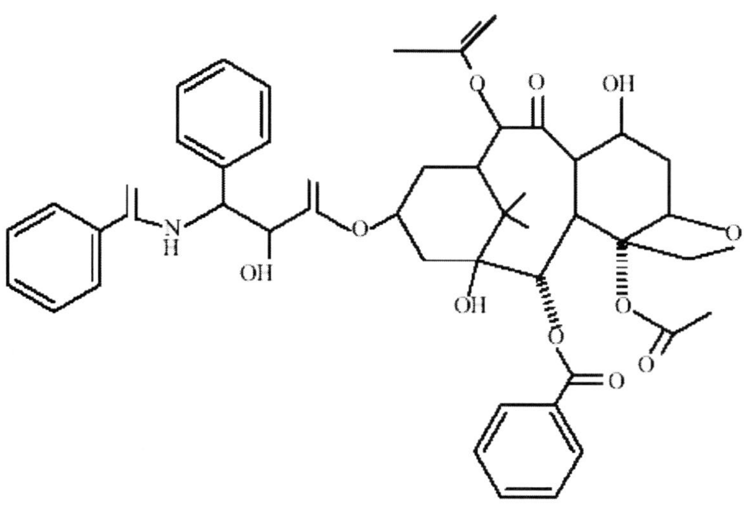

Fig. 21.14: Structure of paclitaxel.

21.32.1.2 Hypolipidemic

Tomatine has an effect on cholesterol absorption and on other aspects of lipid metabolism in the rat similar to that of cholestyramine, with the notable exception that tomatine increased sterol excretion while cholestyramine increased bile acid excretion (Cayen 1971).

21.32.2 Toxicology

After addition of alpha-solanine (80 microgram/ml) and tomatine (40 microgram/ml) to the culture medium, the cells ceased beating within a few minutes. At a concentration of 40 microgram/ml alpha-solanine and 20 microgram/ml tomatine, both compounds caused a pronounced increase of the contraction frequency, lasting for at least 2 h. K-strophantin (Fig. 21.15), a reference heart glycoside, caused arrhythmic beating at 20 microgram/ml and complete cessation of contractions at 160 microgram/ml (Bergers and Alink 1980).

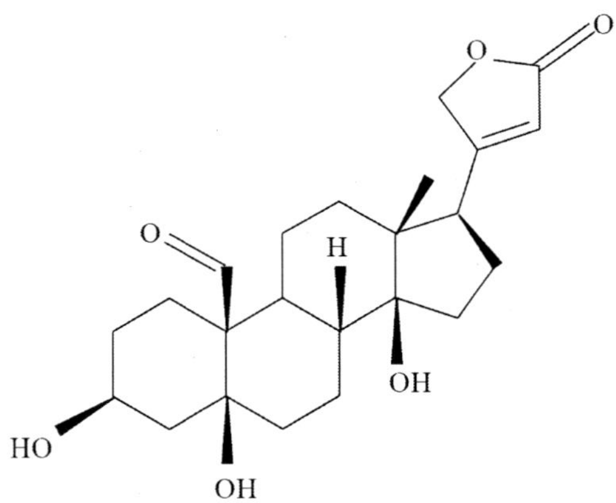

Fig. 21.15: Structure of K-strophantin.

21.33 The *Veratrum* Type Alkaloids

The Veratrum alkaloids represent the most important and medicinally significant class of steroidal alkaloids. In general, the majority of Veratrum alkaloids may be classified into two categories solely based on their characteristic structural features, namely:

(i) Cevaratrum alkaloids, and
(ii) Jeveratrum alkaloids

Cevaratrum alkaloids are important group of alkaloids. Main alkaloids are protoveratrines; veratridine, cevadine and germine.

Jeveratrum alkaloids are represented by the structure of veratramine, jervine, and pseudojervine.

21.34 Protoveratrines

Protoveratrines A and B, collectively known as protoveratrines, are obtained from the rhizome of *Veratrum album* L. and *Veratrum viride* Ait. (Nash and Brooker 1953). Protoveratrine A and protoveratrine B are tetraesters of the alkamine, protoverine. Protoveratrine A and protoveratrine B are used in hypertension (Hoobler et al. 1952).

21.34.1 Hypotensive, bradyerotic, and emetic

Comparative studies of intravenously administered protoveratrine A and protoveratrine B in hypertensive maan indicate that the alkaloids have qualitatively similar hypotensive, bradyerotic, and emetic actions, but quantitatively different hypotensive and emetic potency (Abreu et al. 1954; Winer 1960).

21.34.2 Promotion of apnea expression

After administration of 0.2, 0.5, or 1 mg/kg sc of protoveratrines, cardiopulmonary changes lasting at least 6 h were observed in all three behavioral states [heart period increased up to 23% in wakefulness, 21% in non-rapid-eye-movement (non-REM) sleep, and 20% in REM sleep; $P < 0.005$ for each] (Trbovic et al. 1985).

21.34.3 Anti-hemorrhagic

In a rat model of severe hypotension and respiratory depression induced by step-wise bleeding, protoveratrines cause a prompt and sustained improvement of cardiovascular and respiratory functions, both in anesthetized and in conscious animals, seemingly through a magnification of the reflex response originated by the chemoreceptors of aortic and carotid bodies (Bertolini et al. 1990).

21.35 Veratridine

Veratridine (Fig. 21.16) is obtained from the seeds of *Schoenocaulon officinale* (Schelecht. and Cham.) A. Gray and from the rhizome of *Veratrum album* L.

21.35.1 Effects of veratridine on Na and Ca currents

Single depolarizing prepulses that induced complete inactivation of the fast Na channels, prevented development of the slow current. The low concentration of veratridine (1 microM) did not affect the Ca current, while 100 microM veratridine reversibly suppressed the Ca current and shifted its peak current-voltage relation towards more negative potentials (Nánási et al. 1994).

21.35.2 Increasing the survival of retinal ganglion cells

Veratridine (3.0 μM), promoted a two-fold increase in the survival of retinal ganglion cells kept in culture for 48 h. This effect was dose-dependent and was blocked by 1.0

μM tetrodotoxin (a classical voltage-dependent Na+ channel blocker) and 30.0 μM flunarizine (a Na+ and Ca2+ channel blocker) (Pereira and Araujo 1997).

21.36 Veratrine (Cevadine)

Veratrine (Fig. 21.17) is obtained from the seeds of *Schoenocaulon officinale* (Schelecht. And Cham.) A. Gray (*Sabadilla officinarum* Brandt.).

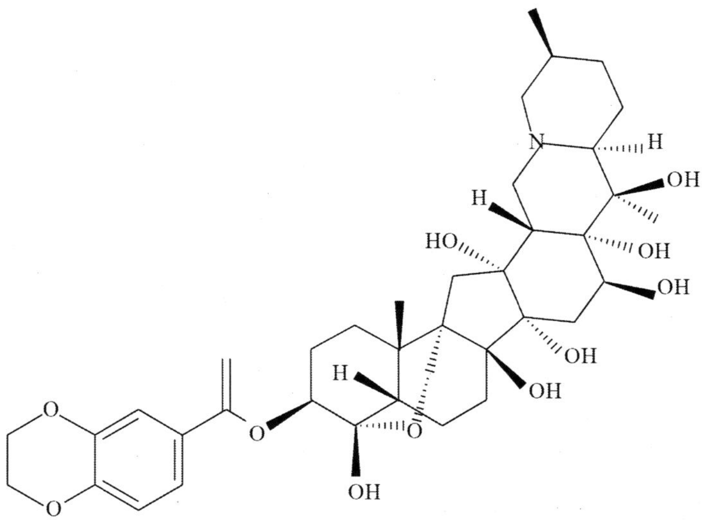

Fig. 21.16: Structure of veratridine.

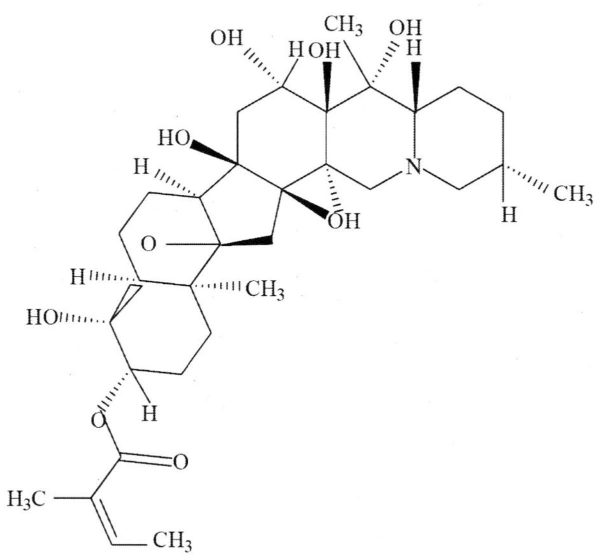

Fig. 21.17: Structure of veratrine.

21.36.1 Effect on the isolated rat aorta

Veratrine (1 x m10(–6) to 1 x 10(–4) g/ml) could induce contraction on the isolated rat aorta in a concentration-related manner. Paeoniflorin had no effect on the isolated rat aorta. The inhibition of paeoniflorin on veratrine was more potent on rat isolated aorta with endothelium than without endothelium (Tsai et al. 1999).

21.36.2 Effect on isolated mouse vas deferens

Paeoniflorin had no effect on isolated mouse vas deferens. Veratrine (1 x 10(–5) approximately 1 x 10(–3) g/ml) could directly induce contraction of isolated rat and mouse vas deferens. Veratrine (1 x 10(–5) g/ml)-induced contractions could be decreased by pretreatment with ryanodine (1 x 10(–5) M) in both the epididymal and the prostatic portions (Chen et al. 2002).

21.37 Germine

Germine is present in *Zygadenus venenosus* and *Veratrum viride* Ait.

21.38 Veratramine

Veratramine (Fig. 21.18) is a hypotensive alkaloid isolated from the rhizomes of *Veratrum nigrum* L.

21.38.1 Hypotensive effect and toxicology

There was a dose-dependent reduction in blood pressure and heart rate after a single ingestion (1.0 to 4.0 mg/kg, intragastric administration) of total alkaloids. A single oral ingestion (0.56 to 2.24 mg/kg) of veratramine, the major component of total alkaloids, dose-dependently decreased blood pressure and heart rate, suggesting that veratramine was involved in the hypotensive effect of VTA in spontaneously hypertensive rats (Wang et al. 2008).

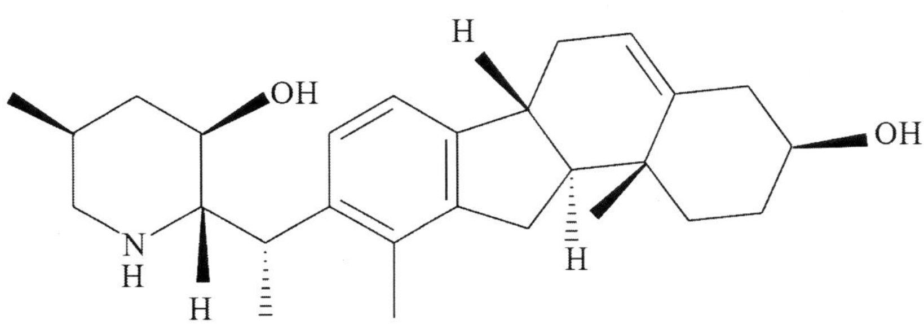

Fig. 21.18: Structure of veratramine.

21.39 Antitumor and Antiplatelet Alkaloids from *Veratrum dahuricum*

The ethanol extract and total alkaloids of *V. dahuricum* were evaluated for their antitumor and antiplatelet activities. Cyclopamine, veratramine, and germine significantly inhibited the hedgehog pathway in NIH/3T3 cells. Veratroylgermine was found to produce the strongest inhibition against the platelet aggregation induced by arachidonic acid, with inhibition rate of 92.0% at 100 microM (Tang et al. 2010).

21.40 Steroidal Alkaloids of *Veratrum lobelianum* Bernh. and *Veratrum nigrum* L.

Veralosine, teinemine, and (+/–)-15-O-(2-Methylbutyroyl)germine were isolated. The antiproliferative activities of veranigrine, veralosinine, and neogermitrine have shown that they are a perspective for further studies (Christov et al. 2010).

21.41 Cycloposine

Cycloposine (Fig. 21.19) is a polar, glycosidic alkaloid, isolated from *Veratrum californicum* (Durand) (Keeler 1968). Cycloposine is the glycoside of cyclopamine (Riet-Correa 2011). Jervine and cyclopamine are also identified in *V. californicum.*

21.42 Cyclopamine (11-deoxojervine)

Cyclopamine (Fig. 21.20) has teratogen potential as it causes usually fatal birth defects. Cyclopamine can result in holoprosencephaly (prevents the fetal brain from dividing into two lobes), (holoprosencephaly), and cyclopia (the development of a single eye). Cyclopamine causes holoprosencephaly and cyclopia by inhibiting the hedgehog signaling pathway.

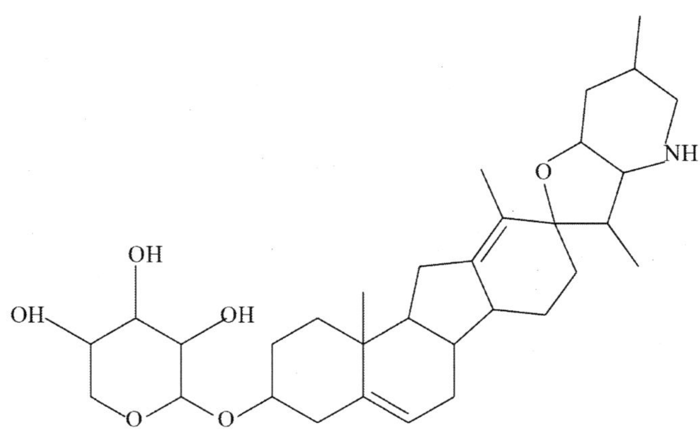

Fig. 21.19: Structure of cycloposine.

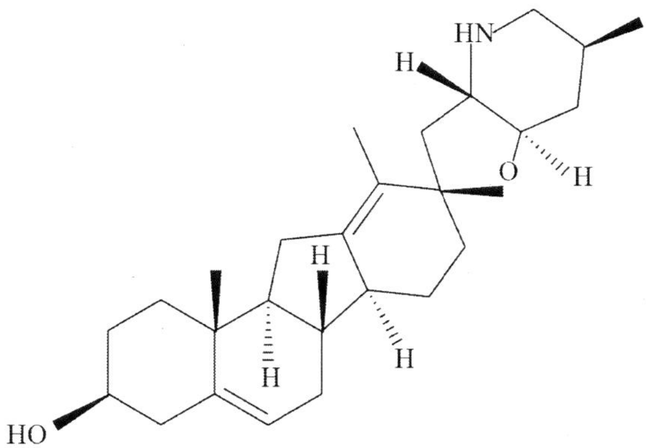

Fig. 21.20: Structure of cyclopamine.

21.42.1 Anticancer

21.42.1.1 Trichoepithelioma

Application of a cream preparation of cyclopamine onto skin tumors in patients who were scheduled for the excision of these tumors four basal cell carcinomas and a trichoepithelioma in four unrelated patients. All of the cyclopamine-treated tumors regressed rapidly (Tabs and Avci 2004).

21.42.1.2 Head and neck squamous cell carcinoma

The hedgehog signaling pathway is involved in tumorigenesis in a variety of human malignancies. Cyclopamine concentration-dependently inhibited HNSCC *ex vivo* [(IC50) at about 500 nM]. In binary combinations, cyclopamine additively enhanced the suppressive effects of cisplatin and docetaxel on HNSCC colony formation (Mozet et al. 2013).

21.42.1.3 Breast cancer cell

Cyclopamine blocks Hedgehog signaling by antagonizing Smo function, which induces tumor apoptosis. Here, we show that the combined use of cyclopamine and paclitaxel was able to induce breast cancer cell apoptosis both *in vivo* and *in vitro* (Chai et al. 2013).

21.42.1.4 Anti-psoriatic

Psoriatic skin lesions were subjected to cyclopamine under conditions. All treated lesions in different patients having plaque and guttate forms of psoriasis regressed rapidly. Differentiation of the epidermal cells of lesional skin and disappearances of infiltrating inflammatory cells were evident within a day. Lesions were cleared commonly on days 3–4 (Tabs and Avci 2004a).

21.42.1.5 Teratogenicity

Cyclopamine produces cyclopia and holoprosencephaly when administered to gastrulation-stage amniote embryos. Cyclopamine-induced malformations in chick embryos are associated with interruption of Sonic hedgehog-mediated dorsoventral patterning of the neural tube and somites. These findings suggest that cyclopamine-induced teratogenesis is due to a more direct antagonism of Shh signal transduction (Incardona et al. 1998).

Further Readings

Abreu BE, Richards WM, Weaver LC. Cardiovascular, emetic and pharmaeodynamic properties of certain veratrum alkaloids. *J Pharmacol Exper Therap* 1954; **112**: 73.

Al Chami L, Méndez R, Chataing B, O'Callaghan J, Usubillaga A, LaCruz L. Toxicological effects of alpha-solamargine in experimental animals. *Phytother Res* 2003; **17**: 254–8.

Atta-Ur-Rahman, Akhtar MN, Choudhary MI, Tsuda Y, Sener B, Khalid A, Parvez M. New steroidal alkaloids from *Fritillaria imperialis* and their cholinesterase inhibiting activities. *Chem Pharm Bull* (Tokyo) 2002; **50**: 1013–6.

Attoumbré J, Giordanengo P, Baltora-Rosset S. Solanidine isolation from *Solanum tuberosum* by centrifugal partition chromatography. *J Sep Sci* 2013; **36**: 2379–85.

Babar ZU, Ata A, Meshkatalsadat MH. New bioactive steroidal alkaloids from Buxus hyrcana. *Steroids* 2006; **71**: 1045–51.

Barger G, Fraenkel-Conrat HL. Alkaloids from *Solanum pseudocapsicum* L. *J Chem Soc* 1936; **1**: 1537–1542.

Bergers WW, Alink GM. Toxic effect of the glycoalkaloids solanine and tomatine on cultured neonatal rat heart cells. *Toxicol Lett* 1980; **6**: 29–32.

Bertolini A, Ferrari W, Guarini S, Tagliavini S. Circulatory and respiratory consequences of massive hemorrhage are reversed by protoveratrines. *Experientia* 1990; **46**: 704–8.

Bushway RJ, Savage SA, Ferguson BS. Inhibition of acetyl cholinesterase by solanaceous glycoalkaloids and alkaloids. *American Potato J* 1987; **64**: 409–413.

Cayen AI N. Effect of dietary tomatine on cholesterol metabolism in the rat. *J Lipid Res* 1971; **12**: 482–90.

Chai F, Zhou J, Chen C, Xie S, Chen X, Su P, Shi J. The Hedgehog inhibitor cyclopamine antagonizes chemoresistance of breast cancer cells. *Onco Targets Ther* 2013; **6**: 1643–1647.

Chao MW, Chen CH, Chang YL, Teng CM, Pan SL. α-Tomatine-mediated anti-cancer activity *in vitro* and *in vivo* through cell cycle- and caspase-independent pathways. *PLoS One* 2012; **7**: e44093.

Chauhan K, Sheth N, Ranpariya V, Parmar S. Anticonvulsant activity of solasodine isolated from *Solanum sisymbriifolium* fruits in rodents. *Pharm Biol* 2011; **49**: 194–9.

Chen YF, Lin YT, Tan TW, Tsai HY. Effects of veratrine and paeoniflorin on isolated mouse vas deferens. *Phytomedicine* 2002; **9**: 296–301.

Chou SC, Huang TJ, Lin EH, Huang CH, Chou CH. Antihepatitis B virus constituents of *Solanum erianthum*. *Nat Prod Commun* 2012; **7**: 153–6.

Choudhary MI, Devkota KP, Nawaz SA, Ranjit R, Atta-ur-Rahman. Cholinesterase inhibitory pregnane-type steroidal alkaloids from *Sarcococca hookeriana*. *Steroids* 2005; **70**: 295–303.

Christov V, Mikhova B, Ivanova A, Serly J, Molnar J, Selenge D, Solongo A, Kostova N, Gerelt-Od Y, Dimitrov D. Steroidal alkaloids of *Veratrum lobelianum* Bernh. and V*eratrum nigrum* L. *Z Naturforsch C* 2010; **65**: 195–200.

Claringbold WD, Few JD, Renwick JH. Kinetics and retention of solanidine in man. *Xenobiotica* 1982; **12**: 293–302.

Dohnal B, Miedzobrodzki J, Włodarczyk B. Alkaloids in the tissue culture of *Holarrhena antidysenterica*. *Acta Pol Pharm* 1990; **47**: 71–3.

Dua VK, Verma G, Singh B, Rajan A, Bagai U, Agarwal DD, Gupta NC, Kumar S, Rastogi A. Anti-malarial property of steroidal alkaloid conessine isolated from the bark of *Holarrhena antidysenterica*. *Malar J* 2013; **12**: 194.

Duez P, Chamart S, Lejoly J, Hanocq M, Zeba B, Sawadogo M, Guissou P, Molle L. Changes in conessine in stem bark of *Holarrhena floribunda* in Burkina Faso. *Ann Pharm Fr* 1987; **45**: 307–13.

Friedman M, Rayburn JR, Bantle JA. Developmental toxicology of potato alkaloids in the frog embryo teratogenesis assay—Xenopus (FETAX). *Food Chem Toxicol* 1991; **29**: 537–47.

Friedman M, Levin CE, Lee SU, Kim HJ, Lee IS, Byun JO, Kozukue N. Tomatine-containing green tomato extracts inhibit growth of human breast, colon, liver, and stomach cancer cells. *J Agric Food Chem* 2009; **57**: 5727–33.

Friedman M. Anticarcinogenic, cardioprotective, and other health benefits of tomato compounds lycopene, α-tomatine, and tomatidine in pure form and in fresh and processed tomatoes. *J Agric Food Chem* 2013; **61**: 9534–50.

Harvey MH, McMillan M, Morgan MR, Chan HW. Solanidine is present in sera of healthy individuals and in amounts dependent on their dietary potato consumption. *Hum Toxicol* 1985; **4**: 187–94.

Hoobler SW, Corley RW, Kabza RW, Loyke HF. Treatment of hypertension with oral protoveratrine. *Ann Int Med* 1952; **37**: 465.

Hsu SH, Tsai TR, Lin CN, Yen MH, Kuo KW. Solamargine purified from Solanum incanum Chinese herb triggers gene expression of human TNFR I which may lead to cell apoptosis. *Biochem Biophys Res Commun* 1996; **229**: 1–5.

Huang S, Zhou XL, Wen J, Wang CJ, Wang HY, Shan LH, Weng J. A novel steroidal alkaloid from *Fritillaria shuchengensis*. *J Nat Med* 2013; **67**: 647–51.

Incardona JP, Gaffield W, Kapur RP, Roelink H. The teratogenic Veratrum alkaloid cyclopamine inhibits sonic hedgehog signal transduction. *Development* 1998; **125**: 3553–62.

Janot MM, Laine F, Goutarel R. Steroid alkaloids. V. Alkaloids of *Malouetia bequaertiana* E. Woodson (Apocynaceae): funtuphyllamine B and malouetine. Preliminary communication. *Ann Pharm Fr* 1960; **18**: 673–7.

Jiang Y, Li P, Li HJ, Yu H. New steroidal alkaloids from the bulbs of *Fritillaria puqiensis*. *Steroids* 2006; **71**: 843–8.

Keeler R, Binns W. Teratogenic compounds of *Veratrum californicum* (Durand). V. Comparison of cyclopian effects of steroidal alkaloids from the plant and structurally related compounds from other sources. *Teratology* 1968; **1**: 5–10.

Keeler RF. Teratogenic compounds of *Veratrum californicum* (Durand). 13. Structure of muldamine. *Steroids* 1971; **18**: 741–52.

Kenny OM, Brunton NP, Rai DK, Collins SG, Jones PW, Maguire AR, O'Brien NM. Cytotoxic and apoptotic potential of potato glycoalkaloids in a number of cancer cell lines. *J Agric Sci Appl* 2013; **2**: 184–192.

Khalifa SA, de Medina P, Erlandsson A, El-Seedi HR, Silvente-Poirot S, Poirot M. The novel steroidal alkaloids dendrogenin A and B promote proliferation of adult neural stem cells. *Biochem Biophys Res Commun* 2014; **446**: 681–6.

Kozukue N, Han JS, Lee KR, Friedman M. Dehydrotomatine and alpha-tomatine content in tomato fruits and vegetative plant tissues. *J Agric Food Chem* 2004; **52**: 2079–83.

Kumar N, Singh B, Bhandari P, Gupta AP, Kaul VK. Steroidal alkaloids from *Holarrhena antidysenterica* (L.) Wall. *Chem Pharm Bull* (Tokyo) 2007; **55**: 912–4.

Kumar P, Sharma B, Bakshi N. Biological activity of alkaloids from *Solanum dulcamara* L. *Nat Prod Res* 2009; **23**: 719–23.

Kuo KW, Hsu SH, Li YP, Lin WL, Liu LF, Chang LC, Lin CC, Lin CN, Sheu HM. Anticancer activity evaluation of the solanum glycoalkaloid solamargine. Triggering apoptosis in human hepatoma cells. *Biochem Pharmacol* 2000; **60**: 1865–73.

Lam CW, Wakeman A, James A, Ata A, Gengan RM, Ross SA. Bioactive steroidal alkaloids from *Buxus macowanii* Oliv. *Steroids*. 2015; **95**: 73–9.

Lecanu L, Hashim AI, McCourty A, Giscos-Douriez I, Dinca I, Yao W, Vicini S, Szabo G, Erdélyi F, Greeson J, Papadopoulos V. The naturally occurring steroid solasodine induces neurogenesis *in vitro* and *in vivo*. *Neuroscience* 2011; **183**: 251–64.

Lee ST, Wong PF, Cheah SC, Mustafa MR. Alpha-tomatine induces apoptosis and inhibits nuclear factor-kappa B activation on human prostatic adenocarcinoma PC-3 cells. *PLoS One* 2011; **6**: e18915.

Lee ST, Wong PF, Hooper JD, Mustafa MR. Alpha-tomatine synergises with paclitaxel to enhance apoptosis of androgen-independent human prostate cancer PC-3 cells *in vitro* and *in vivo*. *Phytomedicine* 2013; **20**: 1297–305.

Li HL, Tang J, Liu RH, Lin M, Wang B, Lv YF, Huang HQ, Zhang C, Zhang WD. Characterization and identification of steroidal alkaloids in the Chinese herb *Veratrum nigrum* L. by high-performance liquid chromatography/electrospray ionization with multi-stage mass spectrometry. *Rapid Commun Mass Spectrom* 2007; **21**: 869–79.

Liang CH, Shiu LY, Chang LC, Sheu HM, Kuo KW. Solamargine upregulation of Fas, downregulation of HER2, and enhancement of cytotoxicity using epirubicin in NSCLC cells. *Mol Nutr Food Res* 2007; **51**: 999–1005.

Lu MK, Chen PH, Shih YW, Chang YT, Huang ET, Liu CR, Chen PS. alpha-Chaconine inhibits angiogenesis *in vitro* by reducing matrix metalloproteinase-2. *Biol Pharm Bull* 2010; **33**: 622–30.

Lu MK, Shih YW, Chang Chien TT, Fang LH, Huang HC, Chen PS. α-Solanine inhibits human melanoma cell migration and invasion by reducing matrix metalloproteinase-2/9 activities. *Biol Pharm Bull* 2010; **33**: 1685–91.

Manase MJ, Mitaine-Offer AC, Pertuit D, Miyamoto T, Tanaka C, Delemasure S, Dutartre P, Mirjolet JF, Duchamp O, Lacaille-Dubois MA. *Solanum incanum* and *S. heteracanthum* as sources of biologically active steroid glycosides: confirmation of their synonymy. *Fitoterapia* 2012; **83**: 1115–9.

Mohsenikia M, Alizadeh AM, Khodayari S, Khodayari H, Kouhpayeh SA, Karimi A, Zamani M, Azizian S, Mohagheghi MA. The protective and therapeutic effects of alpha-solanine on mice breast cancer. *Eur J Pharmacol* 2013; **718**: 1–9.

Moreira RR, Martins GZ, Magalhães NO, Almeida AE, Pietro RC, Silva FA, Cicarelli RM. *In vitro* trypanocidal activity of solamargine and extracts from *Solanum palinacanthum* and *Solanum lycocarpum* of Brazilian cerrado. *An Acad Bras Cienc* 2013; **85**: 903–7.

Mozet C, Stoehr M, Dimitrova K, Dietz A, Wichmann G. Hedgehog targeting by cyclopamine suppresses head and neck squamous cell carcinoma and enhances chemotherapeutic effects. *Anticancer Res* 2013; **33**: 2415–24.

Nánási PP, Varró A, Lathrop DA, Bryant SH. Effects of veratridine on Na and Ca currents in frog skeletal muscle. *Gen Pharmacol* 1994; **25**: 1661–6.

Nash HA, Brooker RM. Hypotensive alkaloids from *Veratrum album*, protoveratrine A, protoveratrine B and germitetrine B. *J Am Chem Soc* 1953; **75**: 1942.

Nishie K, Norred WP, Swain AP. Pharmacology and toxicology of chaconine and tomatine. *Res Commun Chem Pathol Pharmacol* 1975; **12**: 657–68.

Pandurangan A, Khosa RL, Hemalatha S. Antinociceptive activity of steroid alkaloids isolated from Solanum trilobatum Linn. *J Asian Nat Prod Res* 2010; **12**: 691–5.

Pereira SPF, Araujo EG. Veratridine increases the survival of retinal ganglion cells *in vitro*. *Braz J Med Biol Res* 1997; **30**: 1467–70.

Pi HF, Ruan HL, Zhan YH, Wu JZ. Two new steroidal alkaloids from bulbs of *Fritillaria lichuanensis*. *J Asian Nat Prod Res* 2006; **8**: 253–7.

Rahman A, Anjum S, Farooq A, Khan MR, Parveen Z, Choudhary MI. Antibacterial steroidal alkaloids from *Sarcococca saligna*. *J Nat Prod* 1998; **61**: 202–6.

Riet-Correa F. Poisoning by Plants, Mycotoxins and Related Toxins. 2011, 244 pp.

Sani IK, Marashi SH, Kalalinia F. Solamargine inhibits migration and invasion of human hepatocellular carcinoma cells through down-regulation of matrix metalloproteinases 2 and 9 expression and activity. *Toxicol In Vitro* 2015; **29**: 893–900.

Sharma T, Airao V, Panara N, Vaishnav D, Ranpariya V, Sheth N, Parmar S. Solasodine protects rat brain against ischemia/reperfusion injury through its antioxidant activity. *Eur J Pharmacol* 2014; **725**: 40–6.

Shen KH, Liao AC, Hung JH, Lee WJ, Hu KC, Lin PT, Liao RF, Chen PS. α-Solanine inhibits invasion of human prostate cancer cell by suppressing epithelial-mesenchymal transition and MMPs expression. *Molecules* 2014; **19**: 11896–914.

Shih YW, Chen PS, Wu CH, Jeng YF, Wang CJ. Alpha-chaconine-reduced metastasis involves a PI3K/Akt signaling pathway with downregulation of NF-kappaB in human lung adenocarcinoma A549 cells. *J Agric Food Chem* 2007; **55**: 11035–43.

Shiu LY, Chang LC, Liang CH, Huang YS, Sheu HM, Kuo KW. Solamargine induces apoptosis and sensitizes breast cancer cells to cisplatin. *Food Chem Toxicol* 2007; **45**: 2155–64.

Siddiqui BS, Ali ST, Rizwani GH, Begum S, Tauseef S, Ahmad A. Antimicrobial activity of the methanolic bark extract of *Holarrhena pubescens* (Buch. Ham), its fractions and the pure compound conessine. *Nat Prod Res* 2012; **26**: 987–92.

Sun H, Lv C, Yang L, Wang Y, Zhang Q, Yu S, Kong H, Wang M, Xie J, Zhang C, Zhou M. Solanine induces mitochondria-mediated apoptosis in human pancreatic cancer cells. *Biomed Res Int* 2014; **2014**: 805926.

Tabs S, Avci O. Induction of the differentiation and apoptosis of tumor cells in vivo with efficiency and selectivity. *Eur J Dermatol* 2004; **14**: 96–102.

Tang J, Li HL, Shen YH, Jin HZ, Yan SK, Liu XH, Zeng HW, Liu RH, Tan YX, Zhang WD. Antitumor and antiplatelet activity of alkaloids from *Veratrum dahuricum*. *Phytother Res* 2010; **24**: 821–6.

Taş, Avci O. Rapid clearance of psoriatic skin lesions induced by topical cyclopamine. A preliminary proof of concept study. *Dermatology* 2004a; **209**: 126–31.

Taveira M, Sousa C, Valentão P, Ferreres F, Teixeira JP, Andrade PB. Neuroprotective effect of steroidal alkaloids on glutamate-induced toxicity by preserving mitochondrial membrane potential and reducing oxidative stress. *J Steroid Biochem Mol Biol* 2014; **140**: 106–15.

Tomsik P, Micuda S, Sucha L, Cermakova E, Suba P, Zivny P, Mazurova Y, Knizek J, Niang M, Rezacova M. The anticancer activity of alpha-tomatine against mammary adenocarcinoma in mice. *Biomed Pap Med Fac Univ Palacky Olomouc Czech Repub* 2013; **157**: 153–61.

Trbovic SM, Radulovacki M, Carley DW. Protoveratrines A and B increase sleep apnea index in Sprague-Dawley rats. *J Appl Physiol* (1985) 1997; **83**: 1602–6.

Tsai HY, Lin YT, Chen CF, Tsai CH, Chen YF. Effects of veratrine and paeoniflorin on the isolated rat aorta. *J Ethnopharmacol* 1999; **66**: 249–55.

Vieira PM, Marinho LP, Ferri SC, Chen-Chen L. Protective effects of steroidal alkaloids isolated from *Solanum paniculatum* L. against mitomycin cytotoxic and genotoxic actions. *An Acad Bras Cienc* 2013; **85**: 553–60.

Wang D, Wang S, Du Q, Wang N, Liu S, Wang X, Jiang J. Optimization of extraction and enrichment of steroidal alkaloids from bulbs of cultivated *Fritillaria cirrhosa*. *Biomed Res Int* 2014; **2014**: 258402.

Wang L, Li W, Liu Y. Hypotensive effect and toxicology of total alkaloids and veratramine from roots and rhizomes of *Veratrum nigrum* L. in spontaneously hypertensive rats. *Pharmazie* 2008; **63**: 606–10.

Winer BM. Comparative studies of Protoveratrine A and Protoveratrine B intravenously in hypertensive man. *Circulation* 1960; **XXII**: 1074–9.

Xu D-M , Xu M-L, Wang S-Q, Hung E-X, Wen X-G, Arihara S, Shoji N. Two new steroidal alkaloids from *Fritillaria ussuriensis*. *J Nat Prod* 1990; **53**: 549–552.

Yamashoji S, Matsuda T. Synergistic cytotoxicity induced by α-solanine and α-chaconine. *Food Chem* 2013 15; **141**: 669–74.

Yang SA, Paek SH, Kozukue N, Lee KR, Kim JA. Alpha-chaconine, a potato glycoalkaloid, induces apoptosis of HT-29 human colon cancer cells through caspase-3 activation and inhibition of ERK 1/2 phosphorylation. *Food Chem Toxicol* 2006; **44**: 839–46.

Yang ZD, Duan DZ, Xue WW, Yao XJ, Li S. Steroidal alkaloids from *Holarrhena antidysenterica* as acetylcholinesterase inhibitors and the investigation for structure-activity relationships. *Life Sci* 2012; **90**: 929–33.

Zhang QJ, Zheng ZF, Yu DQ. Steroidal alkaloids from the bulbs of *Fritillaria unibracteata*. *J Asian Nat Prod Res* 2011; **13**: 1098–103.

Zhao C, Sun M, Bennani YL, Gopalakrishnan SM, Witte DG, Miller TR, Krueger KM, Browman KE, Thiffault C, Wetter J, Marsh KC, Hancock AA, Esbenshade TA, Cowart MD. The alkaloid conessine and analogues as potent histamine H3 receptor antagonists. *J Med Chem* 2008; **51**: 5423–30.

Zirihi GN, Grellier P, Guédé-Guina F, Bodo B, Mambu L. Isolation, characterization and antiplasmodial activity of steroidal alkaloids from *Funtumia elastica* (Preuss) Stapf. *Bioorg Med Chem Lett* 2005; **15**: 2637–40.

Chapter 22

Guggulsterones

22.1 Guggulsterone

Guggulsterone (Figs. 22.1, 22.2) is a plant steroid found in the resin of the guggul plant, *Commiphoramukul*. Guggulsterone can exist as either of two stereoisomers, *E*-guggulsterone and *Z*-guggulsterone.

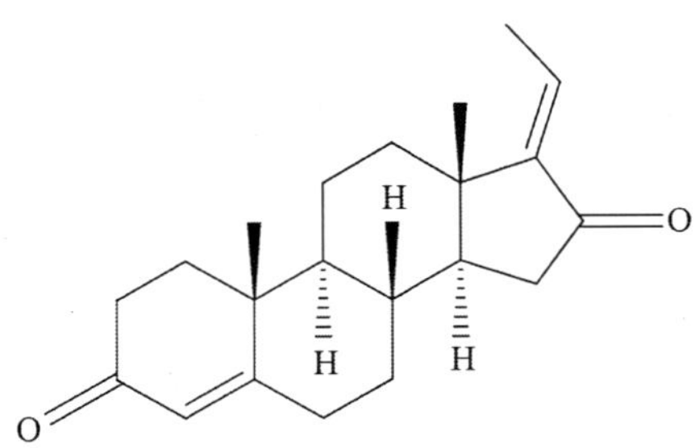

Fig. 22.1: Structure of *E*-Guggulsterone.

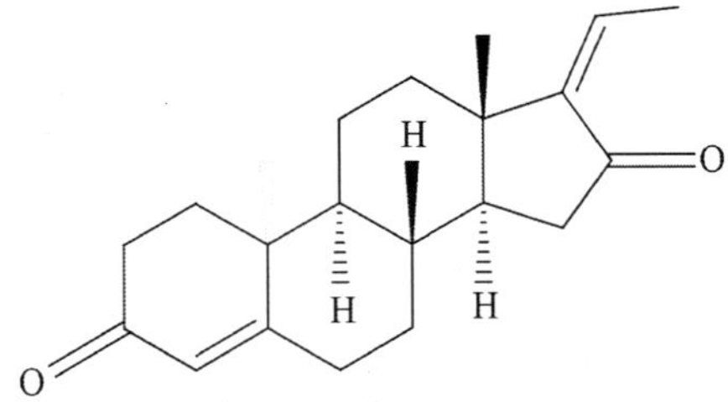

Fig. 22.2: Structure of *Z*-Guggulsterone.

22.2 Pharmacology

22.2.1 The appetite regulatory effect

Guggulsterones at the dose of 400 mg/kg body weight was able to significantly reduce food intake and limit body weight gain over a period of 15 days. It also significantly decreased the plasma ghrelin, glucose, triglyceride levels and increased plasma leptin, serotonin, dopamine levels, but did not show much effect on cholecystokinin levels (Mithila and Khanum 2014).

22.2.2 Anticancer

22.2.2.1 Acute myeloid leukaemia

Cis-guggulsterone, trans-guggulsterone, and 16-dehydroprogesterone (Fig. 22.3) inhibited the proliferation of HL60 and U937 cells, with IC50s ranging from 3.6 to 10.9 micromol/L after treatment for six days. This study is the first to show that guggulsterones and 16-dehydroprogesterone exert antileukemic effects via the induction of apoptosis and differentiation (Samudio et al. 2005).

22.2.2.2 Head and neck squamous cell carcinoma

Treatment of human head and neck squamous cell carcinoma cell lines with guggulsterone demonstrated dose-dependent decreases in cell viability with EC50s ranging from 5 to 8 mM. Guggulsterone induced apoptosis and cell cycle arrest, inhibited invasion, and enhanced the efficacy of erlotinib, cetuximab, and cisplatin in head and neck squamous cell carcinoma cell lines (Leeman-Neill et al. 2009).

Guggulsterone treatment not only inhibited proliferation, but also induced apoptosis by abrogating the effects of smokeless tobacco/nicotine on PI3K/Akt pathway in head and neck cancer cells. These findings provide a rationale for designing future studies to evaluate the chemopreventive potential of guggulsteronein smokeless tobacco/nicotine associated with head and neck cancer (Macha et al. 2011).

22.2.2.3 Melanoma

Guggulsterone dose-dependently inhibited isobutylmethylxanthine (Fig. 22.4)-induced melanogenesis and cellular tyrosinase activity with no cytotoxicity. Guggulsterone also inhibited α-melanocyte stimulating hormone-or forskolin-induced increases in melanogenesis, suggesting an action on the cAMP-dependent melanogenic pathway (Koo et al. 2012).

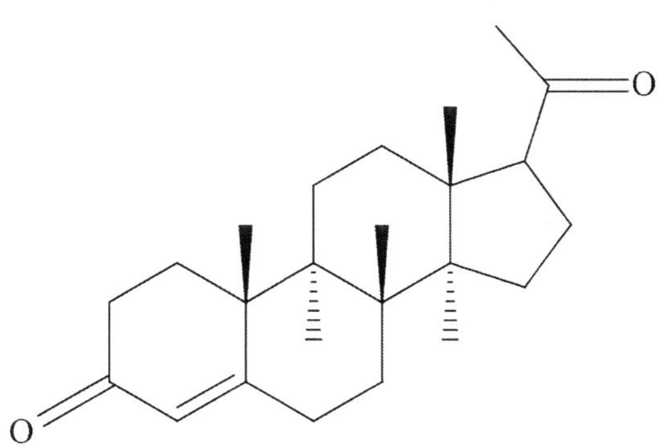

Fig. 22.3: Structure of dehydroprogesterone.

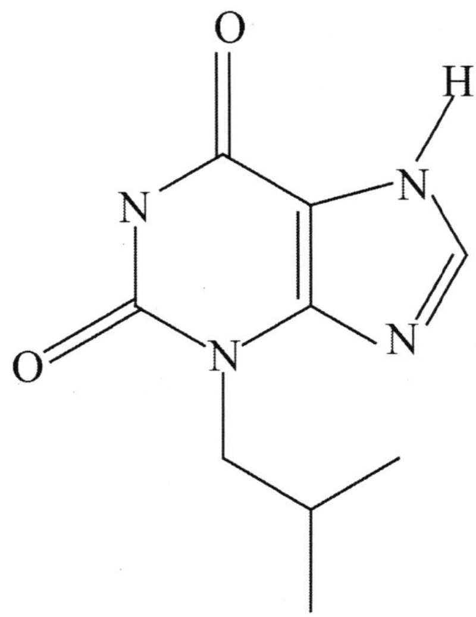

Fig. 22.4: Structure of isobutylmethylxanthine.

22.2.2.4 Glioblastoma

Guggulsterone inhibited Ras and NFκB activity and sensitized cells to SANT-1 induced apoptosis via intrinsic apoptotic mechanism. Inhibition of either Ras or NFκB activity was sufficient to sensitize cells to SANT-1. Guggulsterone induced ERK activation also contributed to Caspase-9 activation (Dixit et al. 2013).

22.2.2.5 Esophageal adenocarcinoma

A combination of amiloride (Fig. 22.5) and guggulsterone showed more than additive effects in suppressing esophageal cancer cell growth *in vitro* and in nude mouse xenografts. The study suggested that inhibition of expression of gastric acid-inducing gene Na+/H+ exchanger-1 or activity or combination of amiloride and guggulsterone could be useful in control of esophageal adenocarcinoma (Guan et al. 2014).

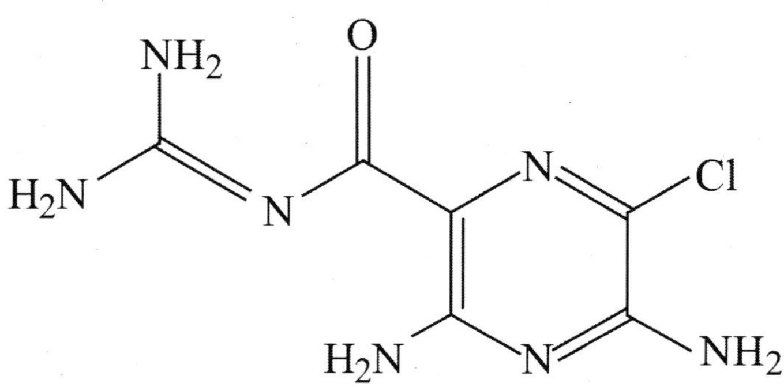

Fig. 22.5: Structure of amiloride.

22.2.2.6 Prostate cancer

The present study builds upon the novel observations and now reveals a novel mechanism of guggulsterone-anticancer activity that ATP citrate lyase-regulated Akt inactivation is involved in guggulsterone-mediated inhibition of prostate cancer growth. Oral gavage of guggulsterone significantly retarded the growth of PC-3 xenografts in athymic mice without causing weight loss and any other side effects. The Gug-induced apoptosis was associated with remarkably down-regulation of Akt and ATP citrate lyase in both cancer cells and xenografts tumor tissue of Gug-treated group (Gao et al. 2014).

22.2.2.7 Breast cancer

The guggulsterone and bexarotene (Fig. 22.6) combination reduced cellular levels of BCRP to 20% of control cells by inducing its association and secretion with exosomes. Exogenous C6 ceramide also induced secretion of breast cancer resistance protein-associated exosomes, while siRNA-mediated knockdown or GW4869-mediated inhibition of neutral sphingomyelinase, an enzyme generating ceramide, restored cellular breast cancer resistance protein (Kong et al. 2015).

22.2.2.8 Gall bladder cancer

Gall bladder cancer cells treated with a combination of guggulsterone and gemcitabine demonstrated significant inhibition of cell proliferation and invasion when compared to treatment with gemcitabine alone. In addition, NF-κB p65 activation decreased significantly in cells treated with a combination of guggulsterone and gemcitabine when compared to treatment with gemcitabine alone (Yang et al. 2012).

22.2.2.9 Pancreatic cancer

In vitro, the combination treatment of gemcitabine (Fig. 22.7) and guggulsterone resulted in more growth inhibition and apoptosis through the down-regulation of nuclear factor κB activity with suppression of Akt and BcL-2 and through the activation of c-Jun NH(2)-terminal kinase and Bax in pancreatic cancer cell lines.

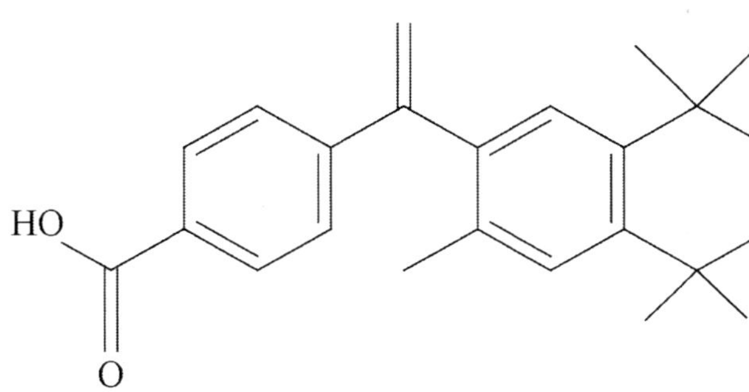

Fig. 22.6: Structure of bexarotene.

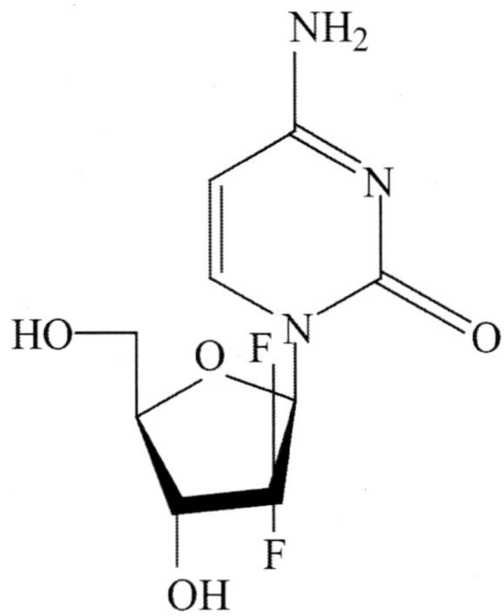

Fig. 22.7: Structure of gemcitabine.

In vivo, the combination therapy augmented tumor growth inhibition through the same mechanisms in tumor tissue (Ahn et al. 2012).

22.2.2.10 Reversal of multidrug resistance by guggulsterone

Co-administration of guggulsterone (10 μM) resulted in a significant increase in chemosensitivity of MCF-7/DOX cells to doxorubicin (Fig. 22.8), compared with doxorubicin treatment alone ($p < 0.01$). However, guggulsterone had little inhibitory effect on the expression of MRP1 proteins (Xu et al. 2011).

22.2.2.11 Doxorubicin-resistant human myelogenousleukemia

The development of P-glycoprotein inhibitors can be effective to reverse multidrug resistance. A study was done to observe the effects of guggulsteroneon multidrug resistance in doxorubicin-resistant K562 cells (K562/DOX) and the parental K562

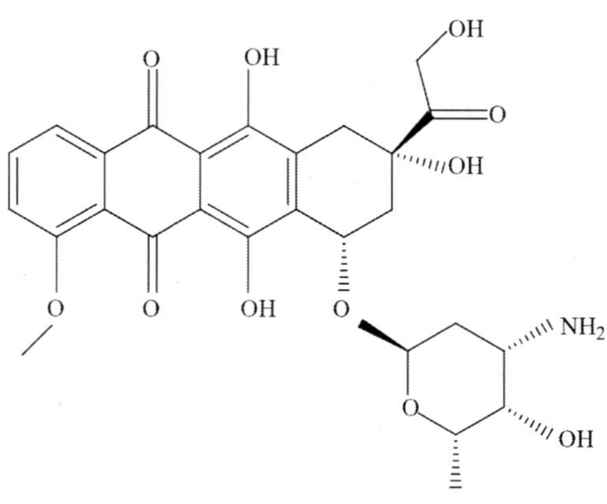

Fig. 22.8: Structure of doxorubicin.

cells. Guggulsterone up to 100 microM had little cytotoxicity against K562/DOX cells. When combined with doxorubicin, it significantly promoted the sensitivity of K562/DOX cells toward doxorubicin through increasing intracellular accumulation of doxorubicin in a dose-dependent manner (Xu et al. 2009).

22.2.2.12 Enhancement of radiosensitivity

Guggulsterone inhibited radiation induced NF-κB activation and enhanced radiosensitivity in the pancreatic cell line, PC-Sw. It reduced both cell cycle movement and cell growth. Guggulsterone reduced ERα protein in MCF7 cells and IGF1-Rβ protein in colon cancer cells and pancreatic cancer cells and inhibited DNA double strand break repair following radiation (Choudhuri et al. 2011).

22.2.2.13 Colon cancer

Guggulsterone significantly increased apoptosis in HT-29 cells by activating caspases-3 and -8. Furthermore, guggulsterone decreased cIAP-1, cIAP-2, and Bcl-2 levels and increased the levels of truncated Bid, Fas, p-JNK, and p-c-Jun. The size of HT-29 xenograft tumors in guggulsterone-treated mice was significantly smaller than of the size of tumors in control mice (An et al. 2009).

22.2.2.14 Skin tumorigenesis

Topical application of guggulsterone (1.6 micromol per mouse) 30 min prior to 2-O-tetradecanoylphorbol-13-acetate (3.2 nmol per mouse) application onto the skin of mice afforded significant inhibition against 2-O-tetradecanoylphorbol-13-acetate-mediated increase in skin edema and hyperplasia. Guggulsterone possesses anti-skin tumor-promoting effects in SENCAR mice and inhibits conventional as well as novel biomarkers of tumor promotion (Sarfaraz et al. 2008).

22.2.3 Acute pancreatitis

Acute pancreatitis was induced by intraperitoneally injecting supramaximal concentrations of the stable cholecystokinin analogcerulein (50 μg/kg) hourly for 6 h. In the guggulsterone-treated group, guggulsterone was administered intraperitoneally (10, 25, or 50 mg/kg) 1 h before the first cerulein injection. Mice were sacrificed 6 h after the final cerulein injection. Pre-treatment with guggulsterone attenuated cerulein-induced histological damage and oppressed the activation of extracellular signal-regulated protein kinase and c-Jun N-terminal kinase in the pancreas in cerulein-induced pancreatitis (Kim et al. 2015).

22.2.4 Thyroid stimulating action

Z-Guggulsterone showed a strong thyroid stimulatory action when administered to albino rats. Its administration (1 mg/100 g body weight) brought about an increase in iodine-uptake by thyroid and enhanced activities of thyroid peroxidase and protease as well as oxygen consumption by isolated slices of liver and biceps muscle (Tripathi et al. 1984; Tripathi et al. 1988).

22.2.5 Hypolipidemic

Administration of guggulsterone caused inhibition of brain dopamine ß-hydroxylase activity with marked stimulation in heart both *in vivo* and *in vitro*. The levels of catecholamines were similarily affected. The contents of serotonin and histamine were found to be enhanced in brain and decreased in heart. Alterations in biogenic amines and dopamine ß-hydroxylase activity may be one of the possible mechanisms for the antilipaemic effect of the compound (Srivastava and Kapoor 1986).

High fat diet induced diabetic rodent models resembling type II diabetic condition in human population were used to assess the anti-diabetic and hypolipidemic activity of guggulsterone. Guggulsterone demonstrated a differential effect with a significantly improved PPARgamma expression and activity in *in vivo* and *in vitro* conditions, respectively (Sharma et al. 2009).

Guggulsterone antagonized the chenodeoxycholic acid (3)-activated nuclear farnesoid X receptor, which regulates cholesterol metabolism in the liver. The cembranoids did not show a noticeable effect on FXR, but lowered the cholate (1)-activated rate of human pancreatic IB phospholipase A2 (hPLA2), which controls gastrointestinal absorption of fat and cholesterol (Yu et al. 2009).

22.2.6 Anti-dementia

Gugulipid treatment caused significant decrease in AChE activity, low level of MDA, and high concentration of GSH in brain following STZ (ic) as compared to vehicle administration in STZ (ic)-treated mice. The study demonstrated that gugulipid has significant protective affect against streptozotocin-induced memory deficits model of dementia that can be attributed to anti-oxidant and anti-AChE activity of gugulipid (Saxena et al. 2007).

22.2.7 Gastroprotective

Pretreatment with GG-52 (novel derivative of guggulsterone) suppressed TNF-α-induced activation of IκB kinase (IKK) and NF-κB signaling in MKN-45 cells. In contrast, the inactive analog GG-46 did not produce significant changes in IL-8 expression or NF-κB activation. In a model of ethanol-induced murine gastritis, administration of GG-52 significantly reduced the severity of gastritis, as assessed by macroscopic and histological evaluation of gastric mucosal damage (Kim et al. 2013).

22.2.8 Inflammatory bowel disease

Administration of guggulsterone significantly reduced the severity of DSS-induced murine colitis as assessed by the clinical disease activity score, colon length, and histology. Furthermore, tissue upregulation of IkappaB and IKK phosphorylation induced by dextran sulfate sodium was attenuated in guggulsterone-treated mice (Cheon et al. 2006).

22.2.9 Anti-inflammatory

The effect of guggulsterone on lipopolysaccharide-induced tumor necrosis factor-alpha and cyclooxygenase-2 expression was evaluated in cultured human middle ear epithelial cell by real-time reverse transcription polymerase chain reaction. Guggulsterone significantly inhibited LPS-induced upregulation of tumor necrosis factor-alpha and cyclooxygenase-2 in a dose-dependent manner. Cyclooxygenase-2 protein production by lipopolysaccharide was significantly suppressed by the guggulsterone pretreatment (Song et al. 2010).

> Endotoxin-induced uveitis was induced by subcutaneous injection of lipopolysaccharide (LPS; 150 μg) into Lewis rats treated with guggulsterone (30 mg/kg body weight, intraperitoneally) or its carrier. Compared with control, the endotoxin-induced uveitis rat eye AqH had a significantly higher number of infiltrating cells, total protein, and inflammatory markers, and the treatment of guggulsterone prevented endotoxin-induced uveitis-induced increases (Kalariya et al. 2010).

22.2.10 Anti-platelet

Inhibition of platelet aggregation has been reported by guggulsterones (Mester et al. 1979).

22.2.11 Cardioprotective

Treatment with guggulsterone and its both isomers at the dose of 50 mg/kg po, significantly protected cardiac damage as assessed by the reversal of blood and heart biochemical parameters in ischemic rats. The cardioprotective activity of guggulsterone and of both the isomers were compared with that of gemfibrozil at the same doses. Guggulsterone and both the isomers at tested concentrations (5–20 mM) inhibited oxidative degradation of lipids in human low-density lipoprotein and rat liver microsomes induced by metal ions *in vitro* (Chander et al. 2003).

22.2.12 Antifungal

Fungal transformation of (E)-guggulsterone by *Rhizopusstolonifer*, *Fusariumlini*, *Cunninghamellaelegans*, or *Gibberellafujikuroi* afforded ten hydroxylated metabolites, which exhibited significant antibacterial and radical-scavenging activities (Choudhary et al. 2005).

22.2.13 Antirheumatic

The effect of guggulsterone on interleukin (IL)-1beta-induced inflammatory responses in the fibroblast-like synoviocytes of rheumatic patients was investigated. Although the NF-kappaB binding activity and nuclear p50 and p65 subunit levels, as well as IkappaB alpha degradation in the cytoplasm was greater in cells stimulated with IL-1beta than in unstimulated cells, treatment with guggulsterone abolished all of these increases (Lee et al. 2008).

22.3 Pharmacokinetics

Serum levels of guggulsterone after intravenous administration showed a biexponential elimination phase with a mean ± s.d. terminal half-life of 10.02 ± 4.74 h and 9.24 ± 3.32 h for 1a and 1b isomers, respectively. The values of systemic clearance and AUC for both 1a and 1b were observed to be 0.71 Lh^{-1}, 4.9 μg h mL^{-1} and 1.04 Lh^{-1}, 3–65 μg h mL^{-1}, respectively. After oral administration, the concentration-time profile declined in a monoexponential fashion with the value of C_{max}, terminal half-life, clearance and AUC for 1a and 1b being 1.07 μg ML^{-1}, 4.48 h, 1.76 Lh^{-1}, 5.95 μg mL^{-1} and 0.97 μ mL^{-1}, 3.56 h, 2.24 Lh^{-1} and 4.75 μg h mL^{-1}, respectively. Absolute bioavailability of parent compound (1a) after oral administration was 42.9% (Verma et al. 1999).

22.4 Clinical study

Hyperlipidemia: A study reported ineffectiveness of guggulipid for lowering cholesterol. Despite plausible mechanisms of action, guggulipid did not appear to improve levels of serum cholesterol over the short term in this population of adults with hypercholesterolemia, and might in fact raise levels of LDL-C. Guggulipid also appeared to cause a dermatologic hypersensitivity reaction in some patients (Szapary et al. 2003).

Guggulsterone was administered orally in these patients in a daily divided dose of 75 mg for a period of eight weeks together with supportive measures like high protein diet, diuretics, and hematinics. Total serum lipid, total serum cholesterol, triglycerides, phospholipids, HDL, LDL, and VLDL were analysed at 4 and 8 weeks of therapy. Significant reduction was observed in the values of total serum lipid and total serum cholesterol (Beg et al. 1996).

Further Readings

Ahn DW, Seo JK, Lee SH, Hwang JH, Lee JK, Ryu JK, Kim YT, Yoon YB. Enhanced antitumor effect of combination therapy with gemcitabine and guggulsterone in pancreatic cancer. *Pancreas* 2012; **41**: 1048–57.

An MJ, Cheon JH, Kim SW, Kim ES, Kim TI, Kim WH. Guggulsterone induces apoptosis in colon cancer cells and inhibits tumor growth in murine colorectal cancer xenografts. *Cancer Lett* 2009; **279**: 93–100.

Beg M, Singhal KC, Afzaal S. A study of effect of guggulsterone on hyperlipidemia of secondary glomerulopathy. *Indian J Physiol Pharmacol* 1996; **40**: 237–40.

Chander R, Rizvi F, Khanna AK, Pratap R. Cardioprotective activity of synthetic guggulsterone (E and Z-isomers) in isoproterenol induced myocardial ischemia in rats: a comparative study. *Indian J Clin Biochem* 2003; **18**: 71–9.

Cheon JH, Kim JS, Kim JM, Kim N, Jung HC, Song IS. Plant sterol guggulsterone inhibits nuclear factor-kappaB signaling in intestinal epithelial cells by blocking IkappaB kinase and ameliorates acute murine colitis. *Inflamm Bowel Dis* 2006; **12**: 1152–61.

Choudhary MI, Shah SA, Sami A, Ajaz A, Shaheen F; Atta-ur-Rahman. Fungal metabolites of (E)-guggulsterone and their antibacterial and radical-scavenging activities. *Chem Biodivers* 2005; **2**: 516–24.

Choudhuri R, Degraff W, Gamson J, Mitchell JB, Cook JA. Guggulsterone-mediated enhancement of radiosensitivity in human tumor cell lines. *Front Oncol* 2011; **1**: 19.

Dixit D, Ghildiyal R, Anto NP, Ghosh S, Sharma V, Sen E. Guggulsterone sensitizes glioblastoma cells to Sonic hedgehog inhibitor SANT-1 induced apoptosis in a Ras/NFκB dependent manner. *Cancer Lett* 2013; **336**: 347–58.

Gao Y, Zeng Y, Tian J, Islam MS, Jiang G, Xiao D. Guggulsterone inhibits prostate cancer growth via inactivation of Akt regulated by ATP citrate lyase signaling. *Oncotarget* 2014. pii: 2138. Epub 2014 Jun 26.

Guan B, Hoque A, Xu X. Amiloride and guggulsterone suppression of esophageal cancer cell growth *in vitro* and in nude mouse xenografts. *Front Biol* (Beijing). 2014; **9**: 75–81.

Kalariya NM, Shoeb M, Reddy AB, Zhang M, van Kuijk FJ, Ramana KV. Prevention of endotoxin-induced uveitis in rats by plant sterol guggulsterone. *Invest Ophthalmol Vis Sci* 2010; **51**: 5105–13.

Kim DG, Bae GS, Choi SB, Jo IJ, Shin JY, Lee SK, Jeong HW, Choi CM, Seo SH, Choo GC, Seo SW, Song HJ, Park SJ. Guggulsterone attenuates cerulein-induced acute pancreatitis via inhibition of ERK and JNK activation. *Int Immunopharmacol* 2015; **26**: 194–202.

Kim JM, Kim SH, Ko SH, Jung J, Chun J, Kim N, Jung HC, Kim JS. The guggulsterone derivative GG-52 inhibits NF-κB signaling in gastric epithelial cells and ameliorates ethanol-induced gastric mucosal lesions in mice. *Am J Physiol Gastrointest Liver Physiol* 2013; **304**: G193–202.

Kong JN, He Q, Wang G, Dasgupta S, Dinkins MB, Zhu G, Kim A, Spassieva S, Bieberich E. Guggulsterone and bexarotene induce secretion of exosome-associated breast cancer resistance protein and reduce doxorubicin resistance in MDA-MB-231 cells. *Int J Cancer* 2015. doi: 10.1002/ijc.29542. [Epub ahead of print]

Koo JH, Rhee KS, Koh HW, Jang HY, Park BH, Park JW. Guggulsterone inhibits melanogenesis in B16 murine melanoma cells by downregulating tyrosinase expression. *Int J Mol Med* 2012; **30**: 974–8.

Lee YR, Lee JH, Noh EM, Kim EK, Song MY, Jung WS, Park SJ, Kim JS, Park JW, Kwon KB, Park BH. Guggulsterone blocks IL-1beta-mediated inflammatory responses by suppressing NF-kappaB activation in fibroblast-like synoviocytes. *Life Sci* 2008; **82**: 1203–9.

Leeman-Neill RJ, Wheeler SE, Singh SV, Thomas SM, Seethala RR, Neill DB, Panahandeh MC, Hahm E-R, Joyce SC, Sen M, Cai Q, Freilino ML, Li C, Johnson DE, Grandis JR. Guggulsterone enhances head and neck cancer therapies via inhibition of signal transducer and activator of transcription-3. *Carcinogenesis* 2009; **30**: 1848–1856.

Macha MA, Matta A, Chauhan SS, Siu KW, Ralhan R. Guggulsterone targets smokeless tobacco induced PI3K/Akt pathway in head and neck cancer cells. *PLoS One* 2011 24; **6**: e14728.

Mester L, Mester M, Nityanand S. Inhibition of platelet aggregation by "guggulu" steroids. *Planta Med* 1979; **37**: 367–9.

Mithila MV, Khanum F. The appetite regulatory effect of guggulsterones in rats: a repertoire of plasma hormones and neurotransmitters. *J Diet Suppl* 2014; **11**: 262–71.

Samudio I, Konopleva M, Safe S, McQueen T, Andreeff M. Guggulsterones induce apoptosis and differentiation in acute myeloid leukemia: identification of isomer-specific antileukemic activities of the pregnadienedione structure. *Mol Cancer Ther* 2005; **4**: 1982–92.

Sarfaraz S, Siddiqui IA, Syed DN, Afaq F, Mukhtar H. Guggulsterone modulates MAPK and NF-kappaB pathways and inhibits skin tumorigenesis in SENCAR mice. *Carcinogenesis* 2008; **29**: 2011–8.

Saxena G, Singh SP, Pal R, Singh S, Pratap R, Nath C. Gugulipid, an extract of *Commiphora whighitii* with lipid-lowering properties, has protective effects against streptozotocin-induced memory deficits in mice. *Pharmacol Biochem Behav* 2007; **86**: 797–805.

Sharma B, Salunke R, Srivastava S, Majumder C, Roy P. Effects of guggulsterone isolated from *Commiphora mukul* in high fat diet induced diabetic rats. *Food Chem Toxicol* 2009; **47**: 2631–9.

Song JJ, Kwon SK, Cho CG, Park SW, Chae SW. Guggulsterone suppresses LPS induced inflammation of human middle ear epithelial cells (HMEEC). *Int J Pediatr Otorhinolaryngol* 2010; **74**: 1384–7.

Srivastava M, Kapoor NK. Guggulsterone induced changes in the levels of biogenic monoamines and dopamine ß-hydroxylase activity of rat tissues. *J Biosci* 1986; **10**: 15–19.

Szapary PO, Wolfe ML, Bloedon LT, Cucchiara AJ, DerMarderosian AH, Cirigliano MD, Rader DJ. Guggulipid for the treatment of hypercholesterolemia: a randomized controlled trial. *JAMA* 13; **290**: 765–72.

Tripathi YB, Malhotra OP, Tripathi SN. Thyroid stimulating action of Z-Guggulsterone obtained from commiphora mukul. *Planta Med* 1984; **50**: 78–80.

Tripathi YB, Tripathi P, Malhotra OP, Tripathi SN. Thyroid stimulatory action of (Z)-guggulsterone: mechanism of action. *Planta Med* 1988; **54**: 271–7.

Verma N, Singh SK, Gupta RC. Pharmacokinetics of guggulsterone after intravenous and oral administration in rats. *Pharm Pharmacol Commun* 1999; **5**: 349–354.

Xu HB, Li L, Liu GQ. Reversal of P-glycoprotein-mediated multidrug resistance by guggulsterone in doxorubicin-resistant human myelogenous leukemia (K562/DOX) cells. *Pharmazie* 2009; **64**: 660–5.

Xu HB, Li L, Liu GQ. Reversal of multidrug resistance by guggulsterone in drug-resistant MCF-7 cell lines. *Chemotherapy* 2011; **57**: 62–70.

Yang MH, Lee KT, Yang S, Lee JK, Lee KH, Moon IH, Rhee JC. Guggulsterone enhances antitumor activity of gemcitabine in gallbladder cancer cells through suppression of NF-κB. *J Cancer Res Clin Oncol* 2012; **138**: 1743–51.

Yu BZ, Kaimal R, Bai S, El Sayed KA, Tatulian SA, Apitz RJ, Jain MK, Deng R, Berg OG. Effect of guggulsterone and cembranoids of *Commiphora mukul* on pancreatic phospholipase A(2): role in hypocholesterolemia. *J Nat Prod* 2009; **72**: 24–8.

Chapter 23

Phytoecdysteroids

23.1 Introduction

Phytoecdysteroids are plant-derived ecdysteroids. Phytoecdysteroids are a family of about 200 plant steroids related in structure to the invertebrate steroid hormone 20-hydroxyecdysone (Dinan 2001). Ecdysteroids are frequently detectable in leaves and flowers, but less so in stems, roots, and seeds (Dinan et al. 2001).

Chemically, phytoecdysteroids are classed as triterpenoids, the group of compounds that includes triterpenesaponins, phytosterols, and phytoecdysteroids.

23.2 Main Biological Sources

Chief sources of phytoecdysteroids are *Achyranthesbidentata* Bl (Gao et al. 2000), *Ajugaturkestanica* (Rgl.), *Leuzeacarthamoides* (Willd.) DC. (Pis et al. 1994), *Tinosporacapillipes* Gagnep (Song and Xu 1991), *Rhaponticum carthamoides* (Willd.) IIjin, *Serratulacoronata* L., and *Silenebrahuica* Boiss (Báthori et al.).

23.2.1 *Phytoecdysteroids of Diploclisia glaucescens (Bl.) Diels. (Menispermaceae)*

24-EPI-makisterone A, -20-hydroxyecdysone, makisterone A, 24(28)-dehydromakisterone A, and pterosterone have been isolated (Miller et al. 1985).

23.2.2 *Phytoecdysteroids of Axyris amaranthoides L. (Amaranthaceae)*

Bioassay/RIA-directed phytochemical examination of the seeds of *A. amaranthoides* afforded a new ecdysteroid: 1 alpha,20R-dihydroxyecdysone [1-epi-integristerone A], together with 20-hydroxyecdysone and polypodine B (Sarker et al. 1998).

23.2.3 *Phytoecdysteroids of Rhagodia baccata (Labill.) Moq. (Chenopodiaceae)*

20-hydroxy-ecdysone and polypodine B, a novel phytoecdysteroid, (20R)-22-deoxy-20,21-dihydroxyecdysone, have been reported from seeds of *R. baccata* (Dinan et al. 1999).

23.2.4 Phytoecdysteroids of *Atriplex nummularia* Lindl. (Amaranthaceae)

20-hydroxyecdysone and polypodine B have been isolated from methanol extract of the seeds of *A. nummularia* (Keckeis et al. 2000).

23.2.5 Phytoecdysteroids of *Lychnisflos-cuculi* L. (Caryophyllaceae)

Dihydrorubrosterone, 20-hydroxyecdysone 3-acetate, 20-hydroxyecdyssone, polypodine B, and rubrosterone have been isolated (Báthori et al. 2001).

23.2.6 Phytoecdysteroids of *Rhaponticum uniflorum* (L.) DC. (Asteraceae)

A new ecdysone hormone, rhaponticum, and ecdysterone have been isolated (Deng et al. 2000). Ecdysterone, ajugasterone C, ajugasterone C-20,22-monoacetonide, ajugasterone C-2,3,20,22-diacetonide and 5-deoxykaladasterone-20,22-monoacetonide have been reported (Zhang and Wang 2001).

23.2.7 Phytoecdysteroids of *Limnanthes alba* Hartw. ex Benth. (Limnanthaceae)

A new ecdysteroid glycoside, limnantheoside C, together with limnantheoside A and 20-hydroxyecdysone have been isolated by bioassay/RIA-directed HPLC analyses of a methanol extract of the seed meal of *L. alba* (Meng et al. 2001).

23.2.8 Phytoecdysteroids of *Sileneitalica* ssp. *nemoralis* (Caryophyllaceae)

A new natural ecdysteroid, 9beta,20-dihydroxyecdysone and four related compounds 5alpha-20-hydroxyecdysone, 5alpha-2-deoxy-integristerone A, integristerone A and 22-deoxy-integristerone A have been isolated (Simon et al. 2004).

23.2.9 Phytoecdysteroids of *Serratulawolffii* Andrae. (Asteraceae)

11alpha-Hydroxypoststerone and herkesterone, two new natural ecdysteroids, have been isolated (Hunyadi et al. 2004).

23.2.10 Phytoecdysteroids of *Sidarhombifolia* L. (Malvaceae)

Ecdysone, 20-hydroxyecdysone, 2-deoxy-20-hydroxyecdysone-3-O-beta-D-glucopyranoside, and 20-hydroxyecdysone-3-O-beta-D-glucopyranoside have been reported (Jadhav et al. 2007).

23.2.11 Phytoecdysteroids of *Ajuga nipponensis* Makino (Lamiaceae)

Cyasterone, ajugasterone C, cyasterone 22-acetate, and 22-dehydrocyasterone has been reported (Coll et al. 2007).

23.2.12 Phytoecdysteroids of Ajuga macrosperma var. breviflora (Lamiaceae)

Three new phytoecdysteroids, ajugacetalsterones C and D and brevifiorasterone, along with five known compounds, namely, 20-hydroxyecdysone, cyasterone, makisterone A, 20-hydroxyecdysone 3-acetate, and 20-hydroxyecdysone 2-acetate have been reported (Castro et al. 2008).

23.2.13 26-Hydroxylated ecdysteroids of Sileneviridiflora L. (Caryophyllaceae)

2-deoxy-5,20,26-trihydroxyecdysone, 5,20,26-trihydroxyecdysone 20,22-acetonide, 2-deoxy-5,20,26-trihydroxyecdysone20,22-acetonide,and20,26-dihydroxyecdysone 20,22-acetonide have been reported (Tóth et al. 2008).

23.2.14 C-29 ecdysteroids of Ajugareptans var. reptans. (Lamiaceae)

Three new ecdysteroids, named reptanslactone A, reptanslactone B, and sendreisterone, and the known 24-dehydroprecyasterone and breviflorasterone have been isolated (Ványolós et al. 2009).

23.2.15 Phytoecdysteroids of Sileneviridiflora L. (Caryophyllaceae)

A new ecdysteroid, 20-hydroxyecdysone 20,22-monoacetonide-25-acetate and a known ecdysteroid, 2-deoxypolypodine B-3-beta-D-glucoside have been reported (Mamadalieva et al. 2010).

23.2.16 Phytoecdysteroids of Braineainsignis (Hooker) J. Smith (Blechnaceae)

Phytoecdysteroid glucosides, brainesterosides A–E, along with three known phytoecdysteroids, ponasteroside A, ponasterone A, and 20-hydroxyecdysone have been isolated (Wu et al. 2010).

23.2.17 Phytoecdysteroid profiles of two Ajuga species, A. iva and A. remota

Minor ecdysteroids (abutasterone, ponasterone A, and sidisterone) have been found for the first time in the Ajuga genus (Bakrim et al. 2014).

23.3 Pharmacology

23.3.1 Erythropoiesis stimulator

Ecdysterone, sileneoside A, and, particularly turkesterone, cause a marked effect on red blood regeneration in hemotoxic phenylhydrazineanemia (Syrov et al. 1997).

23.3.2 Hypoglycemic

Alpha-ecdysone, 2-deoxy-alpha-ecdysone, and 2-deoxyecdysterone isolated from *Silenepraemixta*, integristerone A and ecdysterone isolated from *Rhaponticum carthamoides* and 22-acetylcyasterone and turkesterone isolated from *Ajuga turkestanica*, exhibit a pronounced hypoglycemic effect in experiments on intact male rats. The most active compounds–ecdysteron and turkesterone–also produce an expressed hypoglycemic effect in animals with model hyperglycemia induced by the administration of glucose, adrenalin, and alloxan (Syrov et al. 2012).

23.3.3 Anti-inflammatory

21-hydroxyshidasterone, 11β-hydroxy-20-deoxyshidasterone, and 2,3-acetonide-24-hydroxyecdysone from methanol extracts of stem bark of *Vitexdoniana* Sweet (Lamiaceae) along with known ecdysteroidsshidasterone, ajugasterone C, 24-hydroxyecdysone, and 11β,24-hydroxyecdysone showed a significant ($p \leq 0.05$) inhibitory effect at 100 mg/kg dose on rat paw oedema development due to carrageenan-induced inflammation in Sprague Dawley rats (Ochieng et al. 2013).

23.3.4 Antidepressant

Vitexdoniana Sweet (Lamiaceae) and its phytoecdysteroids constituents (11β-hydroxy-20-deoxyshidasterone, 21-hydroxyshidasterone, and 2,3-acetonide-24-hydroxyecdysone) elicited an antidepressant-like effect in a behavioral paradigm of despair. Furthermore, 21-hydroxyshidasterone produces its antidepressant-like effect through interaction with α2-adrenoceptor, 5-HT2A/2C receptor and dopamine D2-receptors but 11β-hydroxy-20-deoxyshidasterone and 24-hydroxyecdysone effects depend on interaction with α2-adrenoceptor and 5-HT2A/2C receptors while ajugasterone produces its action through interaction with post-synaptic α1-adrenoceptors (Ishola et al. 2014).

23.3.5 Antistress

Introduction of ecdysterone, turkesterone, and ñyasterone to rats subjected to prolonged immobilization stress significantly decreased involution of the thymus and spleen, contributed to normalization of increased mass of adrenal glands, and restored their content of ascorbic acid and cholesterol. With respect to the stress-protective activity, the studied phytoecdysteroids are in some cases superior to eleutherococcus extract (Syrov et al. 2014).

Further Readings

Bakrim A, Ngunjiri J, Crouzet S, Guibout L, Balducci C, Girault JP, Lafont R. Ecdysteroid profiles of two Ajuga species, A. iva and A. remota. *Nat Prod Commun* 2014; **9**: 1069–74.

Báthori M, Kalasz H, Csikkelne SA et al. Components of Serratula species; screening for ecdysteroid and inorganic constituents of some Serratula plants. *Acta Pharmacol Hungary* 1999; **69**: 72–76.

Báthori M, Lafont R, Girault JP, Máthé I. Structural diversity of ecdysteroids of Lychnis flos-cuculi. *Acta Pharm Hung* 2001; **71**: 157–67.

Castro A, Coll J, Tandrón YA, Pant AK, Mathela CS. Phytoecdysteroids from *Ajuga macrosperma* var. *breviflora* roots. *J Nat Prod* 2008; **71**: 1294–6.

Coll J, Tandrón YA, Zeng X. New phytoecdysteroids from cultured plants of Ajuga nipponensis Makino. *Steroids* 2007; **72**: 270–7.

Deng GH, Wei SL, Wei HX. A new ecdysone hormone rhaponticum from *Rhaponticum uniflorum* (L.) DC. *Zhongguo Zhong Yao Za Zhi* 2000; **25**: 417–8.

Dinan L, Sarker SD, Bourne P, Whiting P, Sik V, Rees HH. Phytoecdysteroids in seeds and plants of *Rhagodia baccata* (Labill.) Moq. (Chenopodiaceae). *Arch Insect Biochem Physiol* 1999; **41**: 18–23.

Dinan L. Phytoecdysteroids: biological aspects. *Phytochemistry* 2001; **57**: 325–339.

Dinan L, Savchenko T, Whiting P. On the distribution of phytoecdysteroids in plants. *Cellular and Molecular Life Sciences* 2001; **58**: 1121–1132.

Gao XY, Wang DW, Li FM. Determination of ecdysterone in *Achyranthes bidentata* Bl and its activity promoting proliferation of osteoblast-like cells. *Yao Xue Xue Bao* 2000; **35**: 868–870.

Hunyadi A, Tóth G, Simon A, Mák M, Kele Z, Máthé I, Báthori M. Two new ecdysteroids from *Serratula wolffii*. *J Nat Prod* 2004; **67**: 1070–2.

Ishola IO, Ochieng CO, Olayemi SO, Jimoh MO, Lawal SM. Potential of novel phytoecdysteroids isolated from *Vitex doniana* in the treatment depression: involvement of monoaminergic systems. *Pharmacol Biochem Behav* 2014; **127**: 90–100.

Jadhav AN, Pawar RS, Avula B, Khan IA. Ecdysteroid glycosides from Sida rhombifolia L. Chem Biodivers 2007 Sep; **4**(9): 2225–30.

Keckeis K, Sarker SD, Dinan LN. Phytoecdysteroids from *Atriplex nummularia*. *Fitoterapia* 2000; **71**: 456–8.

Mamadalieva NZ, Janibekov AA, Girault JP, Lafont R. Two minor phytoecdysteroids of the plant *Silene viridiflora*. *Nat Prod Commun* 2010; **5**: 1579–82.

Meng Y, Whiting P, Sik V, Rees HH, Dinan L. Limnantheoside C (20-hydroxyecdysone 3-O-beta-D-glucopyranosyl-[1-->3]-beta-D-xylopyranoside), a phytoecdysteroid from seeds of *Limnanthes alba* (Limnanthaceae). *Z Naturforsch C* 2001; **56**: 988–94.

Miller RW, Clardy J, Kozlowski J, Mikolajczak KL, Plattner RD, Powell RG, Smith CR, Weisleder D, Qi-Tai Z. Phytoecdysteroids of *Diploclisia glaucescens* Seed. *Planta Med* 1985; **51**: 40–2.

Ochieng CO, Ishola IO, Opiyo SA, Manguro LA, Owuor PO, Wong KC. Phytoecdysteroids from the stem bark of Vitex doniana and their anti-inflammatory effects. *Planta Med* 2013; **79**: 52–9.

Pis J, Budesinsky M, Vokac K, Laudová V, Harmatha J. Ecdysteroids from the roots of *Leuzea carthamoides*. *Phytochemistry* 1994; **37**: 707–711.

Sarker SD, Sik V, Rees HH, Dinan L. 1 alpha,20R-dihydroxyecdysone from *Axyris amaranthoides*. *Phytochemistry* 1998; **49**: 2305–10.

Simon A, Pongrácz Z, Tóth G, Mák M, Máthé I, Báthori M. A new ecdysteroid with unique 9beta-OH and four other ecdysteroids from Silene italica ssp. nemoralis. *Steroids* 2004; **69**: 389–94.

Song CQ, Xu RS. Phytoecdysones from the roots of *Tinospora capillipes*. *Chinese Chem Lett* 1991; **2**: 13–14.

Syrov VN, Nasyrova SS, Khushbaktova ZA. The results of experimental study of phytoecdysteroids as erythropoiesis stimulators in laboratory animals. *Eksp Klin Farmakol* 1997; **60**: 41–4.

Syrov VN, Iuldasheva NKh, Égamova FR, Ismailova GI, Abdullaev ND, Khushbaktova ZA. Estimation of the hypoglycemic effect of phytoecdysteroids. *Eksp Klin Farmakol* 2012;**75**: 28–31.

Syrov VN, Islamova ZhI, Égamova FR, Iuldasheva NKh, Khushbaktova ZA. Stress-protective properties of phytoecdysteroids. *Eksp Klin Farmakol* 2014; **77**: 35–8.

Tóth N, Simon A, Tóth G, Kele Z, Hunyadi A, Báthori M. 26-Hydroxylated ecdysteroids from *Silene viridiflora*. *J Nat Prod* 2008; **71**: 1461–3.

Ványolós A, Simon A, Tóth G, Polgár L, Kele Z, Ilku A, Mátyus P, Báthori M. C-29 ecdysteroids from *Ajuga reptans* var. *reptans*. *J Nat Prod* 2009; **72**: 929–32.

Wu P, Xie H, Tao W, Miao S, Wei X. Phytoecdysteroids from the rhizomes of Brainea insignis. *Phytochemistry* 2010; **71**: 975–81.

Zhang YH, Wang HQ. Ecdysteroids from *Rhaponticum uniflorum*. *Pharmazie* 2001; **56**: 828–9.

Chapter 24

Botany of Withanolides containing Herbs

24.1 *Acnistus arborescens* Schltdl. (Solanaceae)

Habitat: Native to Central and South America, and the Caribbean.

Botany: Large shrub or tree up to 10 meters in height. It flowers in clusters on naked branch parts below the leaves. Leaves are alternate, simple, elliptical, narrow to a long v-shape at the base, variably narrowed to a point at the tip, 15 to 30 cm long and 5 to 15 cm wide, margins entire or slightly wavy, hairless except when young. Young stems and young leaves have rusty hairs. The fragrant flowers bloom in clusters of 30 or more, with broadly funnel-shaped tubes about 1.2 cm long and recurving lobes. The protruding stamens are greenish-white to cream. The bright orange fruit is round, and about 1 cm across.

24.2 *Acnistus australis* Blue (Solanaceae)

Habitat: Native to Argentina.

Botany: A small tree bearing the true blue, bell-shaped flowers. The flowers are 2–3" long & bloom for an extended period.

24.3 *Ajuga bracteosa* Wall. ex Benth. (Lamiaceae)

Habitat: E. Asia—Himalayas from Kashmir to Nepal and China.

Botany: *A. bracteosa* is a low herb covered with soft hairs, with erect, ascending stems which arise from the rootstock, branching usually diffusely from the base and measuring 10 to 20 centimeters in length. Leaves are oblanceolate or subspatulate, 2.5 to 10 centimeters long, and 1 to 3.5 centimeters wide; the lower ones are stalked; the upper ones are stalkless and sinuate-toothed or nearly entire. Calyx is hairy, with ovate-lanceolate teeth. Corolla is pale blue or white and hairy; the tube is rarely twice as long as the calyx; the upper lip is erect and 2-fed; the side lobes or lower lobes are oblong, and the midlobe is dilated and variable in length. Stamens protrude from the upper lip. Nutlets are ellipsoid and very small.

24.4 *Ajuga parviflora* Benth (Lamiaceae)

Habitat: Afghanistan, Pakistam, and in the Himalayas from Kashmir to Nepal, at altitudes of 600–1500 m.

Botany: An annual or short-lived perennial herb. Stems are spreading or ascending, 10–25 cm, usually unbranched, sparsely to densely covered with long villous hairs. Leaves sometimes for a rosette, variable in size, up to 4.5 x 2.5 cm, obovate-spoonshaped to elliptic, narrowed into stalk, entire to irregularly toothed, with glandular hairs. Stalks on basal leaves are up to 2 cm. Stem leaves are smaller than basal and decreasing up the stem. Inflorescence is a 8–12-flowered verticillaster. Flowers could be up to 18. Sepal cup is 2.5–4 mm with a similar indumentum to stem, bell-shaped. Teeth are triangular lanceshaped, pointed, as long as tube, basally enlarged as nutlets mature. Flowers are pink, bluish white to white, 5–6 mm, hairy. Tube is slender, shortly protruding from sepals. Stamens usually remain inside the flower-tube. Nutlets are pale brown, transversely rugose with prominent ridges, about 1.5 x 1 mm.

24.5 *Aureliana fasciculata* (Vell.) Seudtn. (Solanaceae)

Habitat: Endemic to Brazil.

Botany: A shrub.

24.6 *Brachistus stramoniifolius* (Kunth) Miers (Solanaceae)

Habitat: Mexico and Central America.

Botany: Shrub, up to 6 m tall, unarmed; thin stems, ribbed, and weak pubescence of simple hairs, erect or rarely branched. Leaves mostly on rough, simple, ovate pairs, up to 18 cm long and 14 cm wide, apex acute, obtuse or truncated base and mostly somewhat oblique, entire or sinuate lobed, sparsely pubescent; petioles much shorter than leaves, slender. Inflorescences bundles with several to numerous flowers, axillary to leaves or branches dichotomies, pedicels 8–15 mm long, slender, puberulent, actinomorphic flowers, 5-mer; domed cup, 2–3 mm long, puberulent, apex shortly deltoid or elongated wolves; corolla tubular-campanulate, 7–9 mm long, lobed 2/3 of its length, glabrous outside and inside rings pubescence except on top of the tube and at the point of insertion of the stamens, yellow; filaments inserted near the middle of the corolla tube, puberulent, anthers oblong, 2–3 mm long, usually not apiculate, basifijas, with longitudinal dehiscence. Fruit a berry, ca 10 mm in diameter, mostly red, underlying the berry as a plate without separating chalice; numerous seeds, 1–1.5 mm long, compressed, with the embryo folded around the periphery of the head.

24.7 *Browallia viscosa* Saphir (Solanaceae)

Habitat: Mexico and Central America.

Botany: Compact plants have attractive, slightly sticky, oval pointed leaves, and flower continuously from May until autumn with an abundance of white-eyed, deep violet-purple flowers each an inch or so across.

24.8 *Datura ferox* L. (Solanaceae)

Habitat: *D. ferox* probably originated in southeastern China. Today it is found in all the warm parts of the earth.

Botany: *D. ferox* is an upright shrub 1½ to 3 feet high. Its thick stalks often have a red-violet color at the base. All the young shoots are noticeably hairy. The most conspicuous part of the plant is its very wide undulate, irregularly-toothed leaves, which are covered with soft, downy hairs. The yellowish white flowers are funnel-shaped and inconspicuous, and usually do not open completely.

24.9 *Datura innoxia* P Mill. (Solanaceae)

Habitat: C. and S. America, W. Indies, S. Western USA. Naturalised in the Mediterranean area, Afghanistan, Pakistan, India and Malaysia.

Botany: Plant up to 130 cm tall, branched, dense pubescent-villous. Leaves 7–18 x 4.5–12 cm, broadly ovate, repand to sinuate-dentate, base oblique to cuneate. Petiole shorter than lamina. Pedicel 10–25 mm long, reflexed in fruit, tomentose with brownish hairs. Calyx 7.5–10.5 cm long, 1/2–2/3rd the corolla length, 5-lobed, slightly inflated toward the base; basal portion persistent and strongly reflexed in fruit; lobes unequal, triangular or ovate-lanceolate, acuminate, up to 22 mm long. Corolla tube 15–17 cm long; limb 6–8.5 cm broad, 10(–12) lobed; lobes broad triangular, acute or acuminate. Anthers 9–11 mm long, white. Capsule globose, 4–5 cm broad, dense spiny, nodding; spines acicular, 5–10 mm long. Seeds 4.5–5 mm long, reniform, compressed, brown, minutely reticulate, rugose-lineate on the back.

24.10 *Datura metel* L. (Solanaceae)

Habitat: Throughout Tropical and subtropical belt.

Botany: An annual herb growing up to 3 ft. high. It is slightly furry, with dark violet shoots and oval to broad oval leaves that are often dark violet as well. The pleasantly-scented 6–8 in. flowers are immensely varied, and can be single or double. Colour range from white to cream, yellow, red, and violet. The seed capsule is covered with numerous conical humps and a few spines. It is similar to *D. innoxia*, but *D. metel* has almost glabrous leaves and fruits that are knobby, not spiny. *D. inoxia* is pilose all over and has a spiny fruit.

24.11 *Datura quercifolia* Kunth (Solanaceae)

Habitat: *D. quercifolia* grows in Mexico and the Southwestern United States.

Botany: *D. quercifolia* is an upright bush up to 3 ft tall. It produces green colored fruit with long sharp spikes or spines. The unremarkable light blue funnel-shaped flowers are less than 2 in long.

24.12 *Datura stramonium* L. (Solanaceae)

Habitat: Native to North America.

Botany: *D. stramonium* is a foul-smelling, erect, annual, freely branching herb that forms a bush up to 2 to 5 ft. The root is long, thick, fibrous and white. The stem is stout, erect, leafy, smooth, and pale yellow-green. The stem forks off repeatedly into branches, and each fork forms a leaf and a single, erect flower. The leaves are about 3 to 8 in long, smooth, toothed, soft, and irregularly undulated. The upper surface of the leaves is a darker green, and the bottom is a light green. The fragrant flowers are trumpet-shaped, white to creamy or violet, and 2 1/2 to 3 1/2 in long, and grow on short stems from either the axils of the leaves or the places where the branches fork.

24.13 *Dioscorea japonica* Thunb. (Dioscoreaceae)

Habitat: Native to Japan, Korea, China and Assam.

Botany: *D. japonica* is a climbing vine with a green stem that twines to the right. Edible bulbils, greenish-brown in colour and up to 1 cm diameter (occasionally larger), grow from the stem. Leaves grow in opposing pairs on the upper stem but alternately near the base. Leaves are triangular-lanceolate. Edible tuber is yellowish-brown in colour, long and cylindrical, up to 3 cm in diameter and grows vertically.

24.14 *Dunalia arborescens* (L.) Sleumer. (Solanaceae)

Habitat: Native to the Greater Antilles, the Lesser Antilles, Trinidad and Tobago, Mexico, Central America, and South America through Brazil and Peru.

Botany: *D. arborescens* is an evergreen shrub or small tree to 6 m in height and 15 cm in stem diameter. Multiple stems are usual for older plants. The trunk is supported by extensive lateral roots with "sinkers" and abundant fine roots. The roots are flexible, light tan, and have furrowed bark. Stem bark is light brown or gray and finely fissured. Stem wood is light brown and hard. Twigs of gallinero are stout, light brown or gray and finely hairy. The simple, alternate leaves are elliptical to lanciolate, 5 to 30 cm long by 3 to 14 cm broad, entire, and pointed at both ends.

24.15 *Hyoscyamus niger* L. (Solanaceae)

Habitat: *H. niger* is the most widely distributed and is found in Europe, Asia, Africa and the Himalayas. It has become naturalized in North America and Australia.

Botany: *H. niger* is either an annual or a biennial, depending on location. It is an upright plant that grows up to 80 cm and has undivided, very pungent leaves. The flower are in thick panicles, and this species has the largest flowers of the Hyoscyamus genus. They are generally pale yellow with violet veins, though some have lemon or bright yellow flowers without veins. The seeds are black, very small, and usually remain in the fruit.

24.16 *Hyoscyamus muticus* L. (Solanaceae)

Habitat: In the deserts of the Middle East, Africa, Europe, Asia and South America.

Botany: Herbaceous plant, 30–60 cm high, perennial, stout, green and fleshy, stem thick, richly branched from the neck, leaves succulent, alternate, long petioled below, sessile above with an acute apex and angled or toothed-lobed margin. The leaf blade is broad, the floral leaf arising on the peduncle oblong lenicular-shaped. Inflorescence in one-sided spike or raceme-like with dense flowers. Flowers are bisexual, homogeneous and slightly zygomorphic, calyx is tubular, corolla funnell-shaped, white or green or purple. Fruits are relatively small, brownish, unarmed capsules with a longitudinal opening and containing numerous seeds. It is included in the persistent calyx.

24.17 *Iochroma coccineum* Benth. (Solanaceae)

Habitat: South America.

Botany: Fast growing tropical shrubs with pendent clusters of light orange flowers. Shrubier than *Iochroma cyaneum*, which it look very similar to.

24.18 *Iochroma fuchsioides* Miers (Solanaceae)

Habitat: Found in Ecuador.

Botany: Bushy growing habit to 8–12 ft. Some sources cite growth to only 4–6 ft, but in ideal environments it can grow much taller. Foliage has a "tropical" look and the plant is evergreen is suitable climates. Flowering occurs year-round, with more intense blooms produced during spring and fall.

24.19 *Iochroma gesnerioides* (Kunth) Miers (Solanaceae)

Habitat: Found in Chile.

Botany: Bushy plant producing prolific dense pendulous clusters of orange tubular flowers Aug-Nov.

24.20 *Jaborosa caulescens* var. bipinnatifida. (Solanaceae)

Habitat: Chile.

Botany: *J. caulescens* var. bipinnatifida is a rhizomatous perennial herb.

24.21 *Jaborosa integrifolia* Lam. (Solanaceae)

Habitat: South America.

Botany: *J. integrifolia* is a persistent perennial underground rhizome and white flowers in summer. The flower is fragrant and night.

24.22 *Jaborosa laciniata*

Habitat: Argentina and Chile.

Botany: Colony-forming from a deep long main root with rhizomatous branches, the whole plant usually a few centimetres high, but the foliage sometimes ascending to 15–20 cm. Leaves rosetted, obtuse, glabrous, 22.5–26 by 1.5–3 cm, pinnatisect with the ovate or oblong lobes toothed and mucronate; petioles, often longer than the blades, sparsely hairy. Flowers clustered in the rosette centres, shallowly cup to saucer-shaped 1.5–2 cm deep and as wide or more with rather blunt, rounded lobes, almost black in the centre with a narrow white to greenish-white edging around the petal lobes.

24.23 *Jaborosa lanigera* Juss (Solanaceae)

Habitat: South America.

Botany: *J. lanigera* is a rhizomatous perennial herb.

24.24 *Lycium chinense* Mill. (Solanaceae)

Habitat: E. Asia - China, Japan. Naturalized in Britain, especially by the sea.

Botany: *L. chinense* Lycium is a deciduous perennial shrub to 5' tall. The arching branches are multiple stemmed; 2" bright green leaves are simple, and ovate to linear-lanceolate. Some plants sport thorns. Pretty purple starshaped flowers fade to a tan before falling off, yielding red fruit that is variable from plant to plant in its sweetness.

24.25 *Lycium halimifolium* Mill. (Solanaceae)

Habitat: *L. halimifolium* is grown in North China, primarily in the Ningxia Hui Autonomous Region.

Botany: *L. halimifolium* is a deciduous woody perennial plant, growing 1–3 m high. The leaves form on the shoot either in an alternating arrangement or in bundles of up to three, each having a shape that is either lanceolate (shaped like a spearhead longer than it is wide) or ovate (egg-like). Leaf dimensions are 7-cm wide by 3.5-cm broad with blunted or round tips. The flowers grow in groups of one to three in the leaf axils.

24.26 *Mandragora officinarum* L. (Solanaceae)

Habitat: *M. officinarum* originates in the eastern Mediterranean region and is distributed throughout southern Europe, the Middle East and northern Africa.

Botany: *M. officinarum* is a very variable perennial herbaceous plant with a long thick root, often branched. It has almost no stem, the leaves being borne in a basal rosette. The leaves are very variable in size and shape, with a maximum length of 45 cm (18 in). They are usually either elliptical in shape or wider towards the end (obovate), with varying degrees of hairiness. The flowers appear from autumn to spring. They are borne in the axils of the leaves. The fruit which forms in late autumn to early summer is a berry, shaped like a globe or an ellipsoid (i.e., longer than wide), with a very variable diameter of 5–40 mm (0.2–1.6 in).

24.27 *Nicandra physaloides* (L.) Gaertn. (Solanaceae)

Habitat: *N. physaloides* is native to Peru.

Botany: *N. physaloides* grow to 1 metre tall and are vigorous with spreading branches and ovate, mid-green, toothed and waved leaves. The flowers are bell-shaped and 5 centimeters or more across, pale violet with white throats. The flower becomes lantern-like towards the end of its bloom.

24.28 *Physalis alkekengi* L. (Solanaceae)

Habitat: Asia - Caucasus to China. Occasionally naturalized in Britain.

Botany: *P. alkekengi* is an herbaceous perennial plant growing to 40–60 cm tall, with spirally arranged leaves 6–12 cm long and 4–9 cm broad. The flowers are white, with a five-lobed corolla 10–15 mm across, with an inflated basal calyx which matures into the papery orange fruit covering, 4–5 cm long and broad.

24.29 *Physalis angulata* L. (Solanaceae)

Habitat: *P. angulata* is native to the Americas, but is now widely distributed and naturalized in tropical and subtropical regions worldwide.

Botany: Branched erect nearly hairless herbaceous shrub which can reach up to 2 m in height. It has a characteristic angled hollow stem. Its leaves are light green often with a toothed-edge. Its pale yellow flowers are five-sided and its edible yellow-orange flowers are surrounded by a balloon-like calyx.

24.30 *Physalis cinerascens* (Dunal) Hitchc. var. Cinerascens (Solanaceae)

Habitat: Disturbed or open habitats in the south central United States and eastern Mexico.

Botany: *P. cinerascens* is a herbaceous perennial plant. It is low-growing and the pendant blossoms are often obscured by the foliage. The flowers are yellow with dark purple or brown maculations in the center. Nectar is produced in small quantities and pollen is present in copious amounts.

24.31 *Physalis coztomatl* (Solanaceae)

Habitat: *P. coztomatl* is native to South America.

Botany: A low-growing annual to 1–2 ft. Little seems to be known or written about this plant, but growth habit is similar to other Physalis species though plant morphology is a bit different.

24.32 *Physalis divaricata* D. Don (Solanaceae)

Habitat: Afghanistan and eastward to Nepal.

Botany: A diffuse annual from 15–45 cm tall, subglabrous to pubescent. Leaves 3–8 5(–11) x 1.5–4 (–7) cm, ovate, sinuate, repand or sinuate-dentate to subentire, acute or acuminate, base cordate to oblique. Petiole up to 40 mm long, slender. Flowers solitary axillary.

24.33 *Physalis franchetii* (Solanaceae)

Habitat: It is native from southern Europe east across southern Asia to Japan.

Botany: *P. franchetti* is a highly attractive and much loved plant with a mound forming habit and mid green coloured leaves. Each year it makes a good show producing creamy-white star shaped flowers, which are followed by bright orange-red, paper-like lanterns in autumn.

24.34 *Physalis ixocarpa* Brot. (Solanaceae)

Habitat: It is native from Mexico.

Botany: *P. ixocarpa* is an annual with a much-branched, spreading form, and a rank, weedy looking appearance. It gets 3–6 ft (0.9–1.8 m) tall and falls over and sprawls on the ground if not given support. The flowers are yellow with purple markings and yield to the tomatillo fruit which is technically a berry, as is the tomato fruit. The fruit develops inside a green and purple bladder-like calyx.

24.35 *Physalis lanceifolia* Nees (Solanaceae)

Habitat: It is native from Mexico and North America.

Botany: Annual to 50 cm high, glabrescent or sparsely hairy with minute, simple hairs. Leaves 1 or 2 per node, narrow-elliptic, usually 5–7 cm long and to 20 mm wide, sometimes larger, base cuneate, margins entire or sinuate or lobed; petiole to 4 cm long, grooved above. Pedicels 15–30 mm long. Calyx 2.5–5 mm long; lobes triangular, 1–2.5 mm long. Corolla 5-angled, 5–7 mm long, yellow with darker yellow centre. Anthers 1.5–2 mm long. Style 2.5 mm long. Fruiting calyx 10-angled, 25–32 mm long, pale yellow-green. Berry 12–15 mm diam.

24.36 *Physalis longifolia* Nutt. (Solanaceae)

Habitat: Native to North America, where it is native to eastern Canada, much of the continental United States, and northern Mexico.

Botany: *P. longifolia* is a perennial herb growing 20 to 60 centimeters tall with somewhat oval-shaped leaf blades 4 to 7 centimeters long borne on petioles. Flowers occur in the leaf axils. The bell-shaped corolla is up to 2 centimeters wide and is yellow with purplish markings around the center. The husk covering the berry is up to 3.5 centimeter long with 10 veins.

24.37 *Physalis minima* L. (Solanaceae)

Habitat: Sub-Himalayas up to altitudes of 1,650 metres.

Botany: *P. minima* is a pantropical annual herb 20–50 cm high at its maturity. Leaves are soft and smooth (not furry), with entire or jagged margins, 2.5–12 cm long. Cream to yellowish flowers are followed by edible yellowish fruit encapsulated in papery cover which turns straw brown and drops to the ground when the fruit is fully ripe.

24.38 *Physalis orizabae* Dunal (Solanaceae)

Habitat: Western and Central Mexico.

Botany: An herbaceous, perennial plant.

24.39 *Physalis peruviana* L. (Solanaceae)

Habitat: Indigenous to South America.

Botany: As a perennial it develops into a diffusely branched herbaceous or somewhat woody shrub up to 2 m high, with a stout, woody, sometimes creeping rootstock, unpleasantly scented, often tinged purple or mauve and densely, softly hairy. Branches are ribbed, bearing velvety heart-shaped leaves, randomly toothed, up to 15 cm long by 10 cm wide on petioles 2–3 cm long. Flowers are formed, singly in the axils of the leaves on pedicels 1–2 cm long. The fruit is a pale yellow round berry

1.5–2 cm in diameter, enclosed in the much-inflated, lantern-like calyx 3–3.5 cm long. Seeds are numerous, pale brown, discoid up to 2 mm long, minutely reticulate.

24.40 *Physalis philadelphica* Lam (Solanaceae)

Habitat: *P. philadelphica* is cultivated in Mexico and Guatemala and originating from Mesoamerica.

Botany: *P. philadelphica* is an annual of 15 to 60 cm; it is subglabrous. sometimes with sparse hairs on the stem. The leaf lamina is 9 to 13 x 6 to 10 mm; its apices are acute to slightly acuminate, with irregularly dentate margins and two to six teeth on each side of the main tooth, of 3 to 8 mm. The pedicels are 5 to 10 mm, the calyx has ovate and hirsute lobules measuring 7–13 mm. The corolla is 8 to 32 mm in diameter, yellow and sometimes has faint greenish blue or purple spots. The anthers are blue or greenish blue. The calyx is accrescent, reaching 18 to 53 x 11 to 60 mm in the fruit, and has ten ribs. The fruit is 12 to 60 x 10 to 48 mm in size and sometimes tears the calyx.

24.41 *Physalis pruinosa* L. (Solanaceae)

Habitat: Eastern N. America.

Botany: Small shrub similar to the common tomato, can be grown as an annual or perennial. Plants are usually small, only 1–3 ft in height.

24.42 *Physalis pubescens* L. (Solanaceae)

Habitat: Native to the Americas, including the southern half of the United States, Mexico, Central and much of South America.

Botany: An annual herb producing a glandular, densely hairy stem up to about 60 centimeters in maximum height from a taproot. The oval or heart-shaped leaves are 3 to 9 centimeters long and have smooth or toothed edges. The flowers blooming from the leaf axils are bell-shaped and about a centimeter long. They are yellow with five dark spots in the throats, and have five stamens tipped with blue anthers. The five-lobed calyx of sepals at the base of the flower enlarges as the fruit develops, becoming an inflated, ribbed, lanternlike structure 2 to 4 centimeters long which contains the berry.

24.43 *Physalis virginiana* Mill. (Solanaceae)

Habitat: Eastern North America.

Botany: *P. virginiana* is a rhizomatous perennial with a deeply buried stem base. Each base typically supports one to six hairy stems that are forked with ascending branches. It leaves are palish green and lance shaped. Small greenish flowers grow on each of its stems. The flowers form a five angled bladder like structure that surrounds the plants half inch diameter sized fruit. It has been found to grow in height from 8 to 12 inches.

24.44 *Physalis viscosa* L. (Solanaceae)

Habitat: Native to South America.

Botany: A rhizomatous perennial herb producing hairy stem up to about 40 centimeters in maximum height. The oval leaves are 3 to 5 centimeters long and have smooth or toothed edges. The flowers blooming from the leaf axils are bell-shaped and about 1.5 centimeters wide. They are yellow with darker centers, and have five stamens tipped with yellow anthers. The calyx of sepals at the base of the flower enlarges as the fruit develops, becoming an inflated, ribbed, lanternlike structure 2 to 3 centimeters long which contains the berry.

24.45 *Salpichroa origanifolia* (Lam.) Baill. (Solanaceae)

Habitat: *S. origanifolia* is native to South America and is naturalised in Africa, Australasia, Europe, and North America.

Botany: *S. origanifolia* is a creeping herbaceous plant or woody vine with scrambling or trailing stems produced from a long-lived woody rootstock. The small leaves may be borne in pairs or alternately arranged along the stems. Leaves that are borne in pairs are of unequal size, with the smaller ones being about three-quarters the size of the larger ones. Its small white or cream-coloured tubular flowers (6–10 mm long) have five spreading petal lobes and are borne singly, or sometimes in pairs. The fruit is a smooth yellow or whitish-coloured berry (10–20 mm long).

24.46 *Saracha viscosa* Schrader (Solanaceae)

Habitat: Mexico and Central America.

Botany: Erect herbs or soft-wooded shrubs 1.0–2.5 m tall, viscid, glandular-pubescent; inflorescences fasciculate with 6–8 (–10) flowers per axil, corolla 5-lobed with greenish maculations in the throat, ca 4 cm in diameter, rotate; anthers bluish (drying to a yellow-green), the filaments inserted at the base; flowering calyx accrescent, exceeding the length of the corolla, deeply lobed to ca ¾ the length of the calyx, lobes acute; fruiting calyx broadly campanulate, deeply lobed, exceeding the berry but becoming reflexed at maturity, exposing the berry; berry bright red to orange-red, seeds ca 50–75, reniform, brown, ca 2 mm long, the testa rugose-reticulate.

24.47 *Solanum ciliatum* Blume ex Miq. (Solanaceae)

Habitat: Native to eastern Brazil but naturalized in other tropical regions, where it sometimes becomes an invasive weed.

Botany: Underbrush annual that grows to 30–90 cm high, with a green stem with yellowish semi-woody thorns, reaching 3 cm the diameter 5 cm soil. The leaves are alternate or in pairs, under petiole (stem or support) 3 to 7 cm length, have cartácea texture (cardboard) are broad ovate (egg-shaped), measuring 6 to 16 cm long by 4 to 8,5 cm wide. The base is cordate (like the heart) and the rest of the blade has wolves (with deep indentations) oblong (longer than wide) and apiculate. The main veins and petioles are thin spines 1.7 to 2.4 cm long and the surface of the blade is sparsely setosa (covered with wax). The flowers appear in the leaf axils and are short with 1 to4 flowers, with just one fertile. The flowers are simple, with cup (outer shell) cupuliforme (domed) of 3 to 4 mm high, divided into lobes or triangular cutouts and thorns; and the corolla (internal enclosure) consists of 5 whitish petals to 16–22 mm long, tapering to the apex. The fruit is a globular berry with thin shell of red color when ripe, with inner wall 3 to 5 mm thick with numerous seeds laminar discoidal and in the center.

24.48 *Solanum sisymbrifolium* Lam (Solanaceae)

Habitat: Native to eastern Brazil but naturalized in other tropical regions, where it sometimes becomes an invasive weed.

Botany: *S. sisymbriifolium* is an annual or perennial erect, rhizomatous herb about 1 metre in height. The stem and branches are viscid, hairy, and armed with flat, orange-yellow spines up to 15 mm in length. The ovate to lanceolate leaves are borne on petioles 1–6 cm long and are pubescent both above and below with stellate and glandular hairs. The leaves are pinnately divided into 4–6 coarse lobes and may be up to 40 cm long and 25 cm wide. Inflorescences emerge from the foliage and are internodal, unbranched racemes composed of 1–10 perfect or staminate flowers. The 5-parted flowers are white, light blue, or mauve, about 3 cm in diameter, and are subtended by a hairy calyx 5–6 mm long. Erect, converging anthers are 8–10 mm long, and ovary is puberulent with a style 1 cm long. Red, succulent, globular berries are 12–20 mm in diameter with pale yellow seeds 2.9–3.2 mm long.

24.49 *Tacca chantrieri* Andre (Taccaceae)

Habitat: Native to tropical regions of Southeast Asia including Thailand, Malaysia, and southern China: particularly Yunnan Province.

Botany: *T. chantrieri* is an unusual plant in that it has black flowers. These flowers are somewhat bat-shaped, are up to 12 inches across, and have long 'whiskers' that can grow up to 28 inches.

24.50 *Tacca integrifolia* Nivea (Taccaceae)

Habitat: Eastern India to southern China, south to Sumatra, Borneo, western Java.

Botany: *T. intergrifolia* has white bracts hovering over the nodding flowers. The bracts are beautifully veined with purple. The plants reach about 4' in height (about twice the ultimate height of *T. chantrieri*). The rhizome of this species grows vertically and the crown of large, attractive leaves emerge from the top of the rhizome.

24.51 *Tacca plantaginea* (Hance) Drenth (Taccaceae)

Habitat: South China (Type), Vietnam, Laos and Thailand.

Botany: A perennial herb with cylindrical rhizome. Leaves 3–4 together, rosulate, petiolate, 17 × 4.5 cm, lanceolate, margin entire, apex acute to acuminate, base gradually merging into the petiole, petiole up to 15 cm long, channelled on the upper surface. Inflorescence 1–2, scape simple, up to 7 cm long, up to 13-flowered. Involucral bracts 4, arranged in 2 pairs, decussate, sessile, green, outer 2 ovate-lanceolate, longer one 6.3 cm, number of bracts correlates with the number of flowers. Flowers pedicellate, light green, actinomorphic, trimerous. Tepals 3 + 3, in two series, outer 3 c. 9 × 4 mm, inner 3 c. 7 × 6 mm, broadly ovate. Stamens 6, light green, each opposite to a tepal, filaments short, c. 1 cm long, anther broad, c. 3 cm. Carpels 3, 6-ribbed, united into a compound inferior ovary, one chambered, placentation parietal, ovules many. Fruit a capsule, triangular, dehiscent, bursting longitudinally into 3 valves, valves reflexed. Seeds oblong-ovoid, grey in colour, longitudinally striped.

24.52 *Tubocapsicum anomalum* (Franch. et Sav.) Makino (Solanaceae)

Habitat: Japan.

Botany: *T. anomalum* is an erect or sprawling herb with entire leaves, few-flowered inflorescences, small campanulate flowers, and red juicy berries with discoid seeds.

24.53 *Withania adpressa* Cors (Solanaceae)

Habitat: Endemic to Morocco and Algeria.

Botany: *W. adpressa* is a shrub with very woolly leaves and a whitish green; small flowers, dioecious by abortion of one of the sexes.

24.54 *Withania aristata* (Aiton) Pauquy (Solanaceae)

Habitat: Endemic Canary Island.

Botany: *W. aristata* is a shrub that can reach 2 m in height. It differs for its flowers, yellowish green, having a cup that grows enveloping the ripe fruit. This cup has five long aristados teeth.

24.55 *Withania coagulans* (Stocks) Dunal (Solanaceae)

Habitat: Native to Afghanistan and the Indian subcontinent.

Botany: *W. coagulans* is a rigid, grey under shrub, 60–120 cm high. Leaves: Lanceolate - oblong, clothed with a persistent, greyish Memtumon on both sides, base narrowed into a stout. Flowers: Yellow in axillary cymose clusters, berries globose, red or Brownish, smooth, enclosed in leathery calyx. Seeds: Dark brown, ear shaped, glabrous, pulp brown.

24.56 *Withania frutescens* (L.) Pauquy (Solanaceae)

Habitat: South and east of the Iberian Peninsula, in the Balearic islands, in Morocco and Algeria.

Botany: *W. frutescens* is a shrub that can reach up to three meters in height, although normally usually reaches a meter and a half. It blooms from May to June and loses its leaves when summer. The berries are greenish and are between 7 and 8 mm in diameter.

24.57 *Withania somnifera* (L.) Dunal. (Solanaceae)

Habitat: *W. somnifera* is cultivated in many of the drier regions of India, such as Mandsaur District of Madhya Pradesh, Punjab, Sindh, Gujarat, and Rajasthan. It is also found in Nepal.

Botany: *W. somnifera* is a short, tender perennial shrub growing 35 to 75 centimeters tall. Tomentose branches extend radially from a central stem. The flowers are small and green. The ripe fruit is orange-red.

24.58 *Vassobia breviflora* (Sendtn.) Hunz. (Solanaceae)

Habitat: Native, not endemic to Brazil.

Botany: Shrub or small tree large, 3 m high, Shrub or small tree large, 3 m high, equipped with long thorns stem, membranous leaves, oval and elliptical, acute or obtuse, glabrous or nearly so. The inflorescence is fasciculata, pauciflora, long-campanulate calyx, corolla purple campanulate, lobed, stamens included, with whitish anthers. The fruit is fleshy, globose, and orange when ripe and edible when ripe with long thorns stem, membranous leaves, oval and elliptical, acute or obtuse, glabrous or nearly so. The inflorescence is fasciculata, pauciflora, long-campanulate calyx, corolla purple campanulate, lobed, stamens included, with whitish anthers. The fruit is fleshy, globose, and orange when ripe and edible when ripe.

Chapter 25

Anticancer Pharmacology of Withaferin A

25.1 Effect of Withaferin A on 180 tumor cells

Mouse sarcoma 180 (S-180) solid and ascites tumor cells were treated *in vivo* and *in vitro* with withaferin A. It was found to affect the spindle microtubules of cells in metaphase. An interesting finding was the double membranes surrounding the chromosomes in the treated cells; probably the nuclei were reconstructed directly from the metaphase stage in the *in vivo* withaferin A-treated cells. In addition the membranes of the cells in interphase were affected by *in vivo* or *in vitro* treatment with withaferin A (Shohat et al. 1976).

25.2 Cytotoxic effects of Withaferin A on P388 cells

P388 cells, 4-dehydrowithaferin A and withaferin A diacetate exhibited an equal inhibitory effect on thymidine, uridine, and L-valine incorporation. They stopped cell

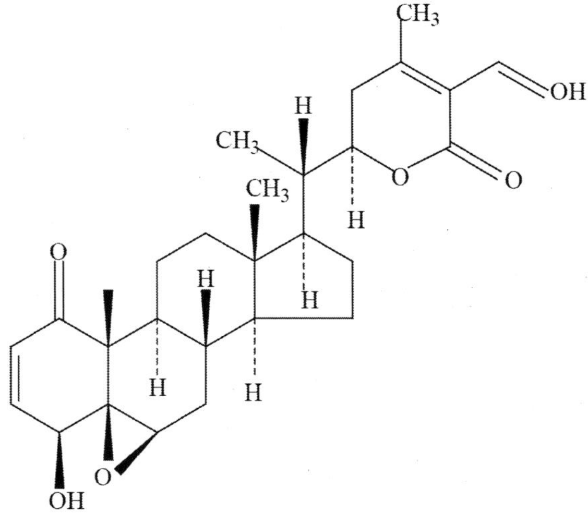

Fig. 25.1: Structure of withaferin A.

proliferation and, at the same time, killed the cells. Withaferin A promptly reacted with L-cysteine, and it was presumed that possible target sites in the cell might be the SH groups of enzymes which react with the lactone and epoxide groups of the agent (Fuskova et al. 1984).

25.3 Ehrlich ascites carcinoma and Withaferin A

Twenty-four hours after IP inoculation of 10(6) tumor cells, withaferin A was injected IP at different dose fractions (5 or 7.5 mg/kg × 8, 10 mg/kg × 5, 20 or 30 mg/kg × 2) with or without abdominal gamma irradiation (RT, 75 Gy) after the first drug dose. Increase in life span and tumor free survival was studied up to 120 days. Withaferin A inhibited tumor growth and increased survival, which was dependent on the withaferin A dose per fraction rather than the total dose. A combination of RT with all the drug schedules increased tumor cure and tumor-free survival; the best effect were seen after two fractions of 30 mg/kg each (Sharada et al. 1996).

25.4 Radiosensitizer effect in V79 cells

In a study, withaferin A reduced survival of V79 cells in a dose-dependent manner. LD50 for survival was 16 μM. One-hour treatment with a non-toxic dose of 2.1 μM before irradiation significantly enhanced cell killing, giving a sensitizer enhancement ratio of 1.5 for 37 percent survival and 1.4 for 10 percent survival. The drug induced a G2-M block, with a maximum accumulation of cells in G2-M phase at 4 hr after treatment with 10.5 μM withaferin A in 1 hr (Devi et al. 1996).

25.5 Role of Withaferin A in fibrosarcoma and melanoma

25.5.1 Fibrosarcoma and melanoma

Two mouse tumors, B16F1 melanoma and fibrosarcoma, were exposed locally to 30 or 50 Gy gamma radiation as an acute dose, or 5 fractions of 10 Gy. Withaferin A, 40 mg/kg, was injected intraperitoneally, 1 hr before acute irradiation, or 30 mg/kg before every 10 Gy fraction. Trimodality treatment synergistically increased complete response to 37 percent in melanoma and to 64 percent in fibrosarcoma. Fractionated radiotherapy (10 Gy × 5) was more effective (25 percent complete response) than acute dose of 50 Gy (0 percent complete response) on melanoma, while there was no difference between the response of fibrosarcoma in the two regimens (Devi and Kamath, 2003).

25.5.2 Melanoma

In the present investigation, the effect of withaferin A on the development and decay of thermo tolerance in B16F1 melanoma was studied in C57BL mice. Tumors of 100 ± 10 mm^3 size were subjected to repeated hyperthermia at 43°C for 30 min. Withaferin A was injected after the first hyperthermia treatment. Tumor growth delay heightened with increase in the time gap between two hyperthermia treatments and was significantly higher ($P < 0.05$ to $P < 0.001$) in withaferin A treated groups (Kalthur et al. 2009).

25.6 Inhibition of angiogenesis

It was proposed that the inhibitory action of withaferin A occurs by interference with the ubiquitin-mediated proteasome pathway as suggested by the increased levels of poly-ubiquitinated proteins. Finally, withaferin A was shown to exert potent anti-angiogenic activity *in vivo* (Mohan et al. 2004).

25.7 Mediation of action through by Annexin II

The study reported that withaferin A alters cytoskeletal architecture by covalently binding annexin II and stimulating its basal F-actin cross-linking activity. Drug-mediated disruption of F-actin organization is dependent on annexin II expression by cells and markedly limits their migratory and invasive capabilities at subcytotoxic concentrations (Falsey et al. 2006).

25.8 Withaferin A as proteasome inhibitor

Withaferin A potently inhibits the chymotrypsin-like activity of a purified rabbit 20S proteasome (IC50 = 4.5 μM) and 26S proteasome in human prostate cancer cultures (at 5–10 μM) and xenografts (4–8 mg/kg/day). Treatment of human prostate PC-3 xenografts with WA for 24 days resulted in 70 percent inhibition of tumor growth in nude mice, associated with 56 percent inhibition of the tumor tissue proteasomal chymotrypsin like activity (Yang et al. 2007).

25.9 Withaferin A as apoptosis inducer

Withaferin A induced Par-4-dependent apoptosis in androgen-refractory prostate cancer cells and regression of PC-3 xenografts in nude mice. Interestingly, restoration of wild-type AR in PC-3 (AR negative) cells abrogated both Par-4 induction and apoptosis by withaferin A. The withanolide and anti-androgen synergistically induced Par-4 and apoptosis in androgen-responsive prostate cancer cells (Srinivasan et al. 2007).

25.10 Inhibition of protein Kinase C

In *Leishmania donovani*, the inhibition of protein kinase C by withaferin A causes depolarization of Delta Psim and generates ROS inside cells. Loss of Delta Psim leads to the release of cytochrome c into the cytosol and subsequently activates caspase-like proteases and oligonucleosomal DNA cleavage (Sen et al. 2007).

25.11 Targeting of protein vimentin by Withaferin

According to one study, withaferin A binds to the intermediate filament protein, vimentin, by covalently modifying its cysteine residue. Withaferin A induces vimentin filaments to aggregate *in vitro*, an activity manifested *in vivo* as punctate cytoplasmic aggregates that colocalize vimentin and F-actin. Withaferin-A exerts potent dominant-negative effect on F-actin that requires vimentin expression and induces apoptosis (Bargagna-Mohan et al. 2007).

25.12 Withaferin A and Leukemia

25.12.1 Human myeloid leukemia and withaferin A

Withaferin A primarily induces oxidative stress in human leukemia HL-60 cells and in several other cancer cell lines. The withanolide induces early ROS generation and mitochondrial membrane potential (Deltapsi(mt)) loss, which precedes release of cytochrome c, translocation of Bax to mitochondria, and apoptosis inducing factor to cell nuclei. These events paralleled activation of caspases-9, -3 and PARP cleavage (Malik et al. 2007).

25.12.2 Leukemia

Withaferin A induces apoptosis in association with the activation of caspase-3. JNK and Akt signal pathways play crucial roles in withaferin A-induced apoptosis in U937 cells. Over expression of Bcl-2 and active Akt (myr-Akt) in U937 cells inhibited the induction of apoptosis, activation of caspase-3, and PLC-gamma 1 cleavage by withaferin A (Malik et al. 2008).

25.13 NFkappaB and Withaferin A

Leaf extract of *W. somnifera* as well as withaferin A, potently inhibits NFkB activation by preventing the tumor necrosis factor-induced activation of IkB kinase beta via a thioalkylation-sensitive redox mechanism. This prevents IkB phosphorylation and degradation, which subsequently blocks NFkB translocation, NFkB/DNA binding, and gene transcription (Kaileh et al. 2007).

25.14 Oral carcinogenesis and Withaferin A

Oral administration of withaferin A (20 mg/kg body weight) to 7,12-dimethy-lbenz[a] anthracene administered to animals for 14 weeks completely prevented tumor incidence, volume, and burden. Also, Withaferin A showed significant anti-lipid peroxidative and anti-oxidant properties and maintained the status of phase-I and phase-II detoxication agents during MBA induced oral carcinogenesis (Manoharan et al. 2009).

25.15 Effect on Breast Cancer Cells

Treatment of MDA-MB-231 (estrogen-independent) and MCF-7 (estrogen-responsive) cell lines with withaferin A resulted in a concentration and time-dependent increase in G2-M fraction, which correlated with a decrease in levels of cyclin-dependent kinase 1 (Cdk1), cell division cycle 25C (Cdc25C), and/or Cdc25B proteins, leading to accumulation of Tyrosine 15 phosphorylated (inactive) Cdk1 (Stan et al. 2008).

25.16 Withanolides and gliomas

Withaferin A, withanone, withanolide A, and the leaf extract of *W. somnifera* markedly inhibited the proliferation of glioma cells in a dose dependent manner and changed their morphology toward the astrocytic type. Molecular analysis revealed that the leaf extract of *W. somnifera* and some of its components caused enhanced expression of glial fibrillary acidic protein, change in the immunostaining pattern of mortalin from perinuclear to pancytoplasmic, delay in cell migration, and increased expression of neuronal cell adhesion molecules (Shah et al. 2009).

25.17 Inhibition of notch-1 by Withaferin A

Withaferin A inhibited Notch-1 signaling and downregulates prosurvival pathways, such as Akt/NF-kB/Bcl-2, in three colon cancer cell lines (HCT-116, SW-480, and SW-620). In addition, it downregulates the expression of mammalian target of rapamycin signaling components, pS6K and p4EBP1, and activates c-Jun-NH (2)-kinase-mediated apoptosis in colon cancer cells (Koduru et al. 2010).

25.18 Pancreatic cancer

Withaferin A exhibited potent antiproliferative activity against pancreatic cancer cells *in vitro* (with IC50 s of 1.24, 2.93 and 2.78 μM) in pancreatic cancer cell lines Panc-1, MiaPaCa2, and BxPc3, respectively. Withaferin A—biotin binds to C-terminus of Hsp90 which is competitively blocked by unlabeled withaferin A. Withaferin A—(3, 6 mg/kg), inhibited tumor growth in pancreatic Panc-1 xenografts by 30 and 58 percent, respectively (Yu et al. 2010).

25.19 Neck squamous carcinoma

A study showed that withaferin A extracted from the aerial parts of *Vassobiabreviflora* (Sendtn.) Hunz. (Solanaceae) induces apoptosis and cell death in neck squamous cell carcinoma cells as well as a cell-cycle shift from G0/G1 to G2-M. Cells treated with withaferin A exhibited inactivation of Akt and a reduction in total Akt concentration (Samadi et al. 2010).

Further Readings

Bargagna-Mohan P, Hamza A, Kim YE, Khuan Abby Ho Y, Mor-Vaknin N, Wendschlag N, Liu J, Evans RM, Markovitz DM, Zhan CG, Kim KB, Mohan R. The tumor inhibitor and antiangiogenic agent withaferin A targets the intermediate filament protein vimentin. *Chem Biol* 2007; **14**: 623–634.

Devi PU, Akagi K, Ostapenko V, Tanaka Y, Sugahara T. Withaferin A: a new radiosensitizer from the Indian medicinal plant *Withania somnifera*. *Int J Rad Biol* 1996; **69**: 193–197.

Devi PU, Kamath R. Radiosensitizing effect of withaferin A combined with hyperthermia on mouse fibrosarcoma and melanoma. *J Rad Res* 2003; **44**: 1–6.

Falsey RR, Marron MT, Gunaherath GM, Shirahatti N, Mahadevan D, Gunatilaka AA, Whitesell L. Actin microfilament aggregation induced by withaferin A is mediated by annexin II. *Nat Chem Biol* 2006; **2**: 33–38.

Fuskova A, Fuska J, Rosazza JP et al. Novel cytotoxic and antitumor agents. IV. Withaferin A: relation of its structure to the *in vitro* cytotoxic effects on P388 cells. *Neoplasma* 1984; **31**: 31–36.

Kaileh M, Vanden Berghe W, Heyerick A, Horion J, Piette J, Libert C, De Keukeleire D, Essawi T, Haegeman G. Withaferin a strongly elicits IkappaB kinase beta hyperphosphorylation concomitant with potent inhibition of its kinase activity. *J Biol Chem* 2007; **282**: 4253–4257.

Kalthur G, Mutalik S, Pathirissery UD. Development and decay of thermo tolerance in B16F1 melanoma: a preliminary study. *Integ Cancer Ther* 2009; **8**: 93–97.

Koduru S, Kumar R, Srinivasan S, Evers MB, Damodaran C. Notch-1 inhibition by Withaferin-A: a therapeutic target against colon carcinogenesis. *Mol Cancer Ther* 2010; **9**: 202–210.

Malik F, Kumar A, Bhushan S, Khan S, Bhatia A, Suri KA, Qazi GN, Singh J. Reactive oxygen species generation and mitochondrial dysfunction in the apoptotic cell death of human myeloid leukemia HL-60 cells by withaferin A with concomitant protection by N-acetyl cysteine. *Apoptosis* 2007; **12**: 2115–2133.

Manoharan S, Panjamurthy K, Menon VP, Balakrishnan S, Alias LM. Protective effect of Withaferin-A on tumour formation in 7, 12-dimethylbenz[a]anthracene induced oral carcinogenesis in hamsters. *Indian J Exper Biol* 2009; **47**: 16–23.

Mohan R, Hammers HJ, Bargagna-Mohan P, Zhan XH, Herbstritt CJ, Ruiz A, Zhang L, Hanson AD, Conner BP, Rougas J, Pribluda VS. Withaferin A is a potent inhibitor of angiogenesis. *Angiogenesis* 2004; **7**: 115–122.

Oh JH, Lee TJ, Kim SH, Choi YH, Lee SH, Lee JM, Kim YH, Park JW, Kwon TK. Induction of apoptosis by withaferin A in human leukemia U937 cells through downregulation of Akt phosphorylation. *Apoptosis* 2008; **13**: 1494–1504.

Samadi AK, Tong X, Mukerji R et al. Withaferin A, a cytotoxic steroid from *Vassobia breviflora*, induces apoptosis in human head and neck squamous cell carcinoma. *J Nat Prod* 2010; **73**: 1476–1481.

Sen N, Banerjee B, Das BB, Ganguly A, Sen T, Pramanik S, Mukhopadhyay S, Majumder HK. Apoptosis is induced in leishmanial cells by a novel protein kinase inhibitor withaferin A and is facilitated by apoptotic topoisomerase I-DNA complex. *Cell Death Differ* 2007; **14**: 358–367.

Shah N, Kataria H, Kaul SC, Ishii T, Kaur G, Wadhwa R. Effect of the alcoholic extract of Ashwagandha leaves and its components on proliferation, migration, and differentiation of glioblastoma cells: combinational approach for enhanced differentiation. *Cancer Sci* 2009; **100**: 1740–1747.

Sharada AC, Solomon FE, Devi PU, Udupa N, Srinivasan KK. Antitumor and radiosensitizing effects of withaferin-A on mouse Ehrlich ascites carcinoma *in vivo*. *Acta Oncol* 1996; **35**: 95–100.

Shohat B, Shaltiel A, Ben-Bassat M, Joshua H. The effect of withaferin A on the fine structure of S-180 tumor cells. *Cancer Lett* 1976; **2**: 71–77.

Srinivasan S, Ranga RS, Burikhanov R, Han SS, Chendil D. Par-4-dependent apoptosis by withaferin A in prostate cancer cells. *Cancer Res* 2007, **67**: 246–253.

Stan SD, Zeng Y, Singh SV. Ayurvedic medicine constituent withaferin a causes G2 and M phase cell cycle arrest in human breast cancer cells. Nutr *Cancer* 2008; **60**: 51–60.

Yang H, Shi G, Dou QP. The tumor proteasome is a primary target for the natural anticancer compound Withaferin A isolated from "Indian winter cherry". *Mol Pharmacol* 2007; **71**: 426–437.

Yu Y, Hamza A, Zhang T, Gu M, Zou P, Newman B, Li Y, Gunatilaka AA, Zhan CG, Sun D. Withaferin A targets heat shock protein 90 in pancreatic cancer cells. *Biochem Pharmacol* 2010; **79**: 542–551.

Chapter 26

Pharmacology of Withanolide A

26.1 Bio-generation and Distribution

Withanolide-A is hardly detectable in the aerial parts of field-grown *W. somnifera* (explant source). However, Withanolide-A accumulated considerably in the *in vitro* shoot cultures of *W. somnifera*. The productivity of withanolide A in the cultures varied considerably (ca. 10-fold, 0.014 to 0.14 mg per gram fresh weight) with the change in the hormone composition of the culture media as well as genotype used as source of the explant. The shoot culture of RS-Selection-1 raised at 1.00 ppm of BAP and 0.50 ppm of kinetin displayed the highest concentration of withanolide A in the green shoots of 0.238 g per 100 g dry weight tissue (Sangwan et al. 2007).

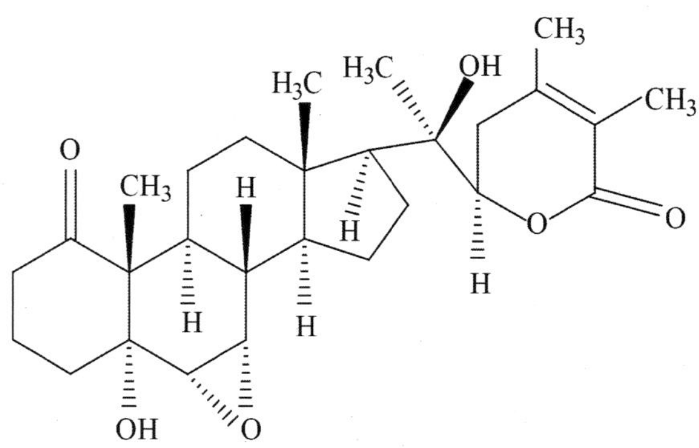

Fig. 26.1: Structure of withanolide-A.

It is thought that withanolides are synthesized in leaves and transported to roots similar to the tropane alkaloids known to be synthesized in roots and transported to leaves for storage. To study the fact, the researchers studied incorporation of (14)C from [2-(14)C]-acetate and [U-(14)C]-glucose into withanolide A in the *in vitro* cultured normal roots as well as native/orphan roots of *W. somnifera*. Analysis by thin layer chromatography (TLC) revealed that these primary metabolites were incorporated into withanolide A, demonstrating that root-contained withanolide A is *de novo* synthesized within roots from primary isoprenogenic precursors (Sangwan et al. 2008).

In vitro multiple shoots, root, callus, and cell suspension cultures of *W. somnifera* exhibited the potentiality to produce pharmacologically active withanolides. Multiple shoots cultures exhibited an increase in withanolide A accumulation compared to shoots of the mother plant. *In vitro* generated root cultures as well as callus and suspension cultures also produced withanolides *albeit* at lower levels (Sabir et al. 2008).

Distribution of withanolide A in various organs of *W. somnifera* was investigated by High Performance Liquid Chromatography (HPLC). The quantitative distribution of withanolide A was different in various organs tested; the accumulation was 386, 342, 272, 206, 102, 56, 35 and 23 $\mu g\ g^{-1}$ DW in shoot tips, leaves, nodes, whole plant, internodes, roots, flowers, and seeds respectively. The content of withanolide A gradually decreased from aerial parts that is, from young leaves to the root. In the root, the root tip accumulated higher concentration when compared to the middle and basal portion (Wang et al. 2014).

In the present study, adventitious roots from leaf explants of *W. somnifera* were induced for the production of withanolide-A by *Agrobacterium tumefaciens* strain C58C1 to obtain hair roots. Hair roots induction rate reached 30%. The withanolide A was determined by HPLC in different hair roots lines and different parts of *W. somnifera*.

The average content of withanolide A in all hair roots lines were 1.96 times as high as that in wild-plant, the concentration of withanolide A in hair roots (1.783 mg x $g^{(-1)}$ dry weight) were 1.51 times as high as the roots of wild *W. somnifera* (1.180 mg x $g^{(-1)}$ dry weight), respectively (Parveen et al. 2010).

26.1 Pharmacology

26.1.1 Axon-or dendrite-predominant outgrowth induction

Previously it has been reported that the methanol extract of *W. somnifera* induced dendrite extension in a human neuroblastoma cell line. In the present study, it was found that six compounds isolated from the methanol extract of *W. somnifera* enhanced neurite out growth in human neuroblastoma SH-SY5Y cells.

Double immunostaining was performed in rat cortical neurons using antibodies to phosphorylated NF-H as an axonal marker, and to MAP2 as a dendritic marker. In withanolide A-treated cells, the length of NF-H-positive processes was significantly increased compared with vehicle-treated cells, whereas, the length of MAP2-positive processes was increased by withanosides IV and VI (Kuboyama et al. 2002).

A study investigated whether withanolide A could regenerate neurites and reconstruct synapses in severely damaged neurons. Further, when the effect of

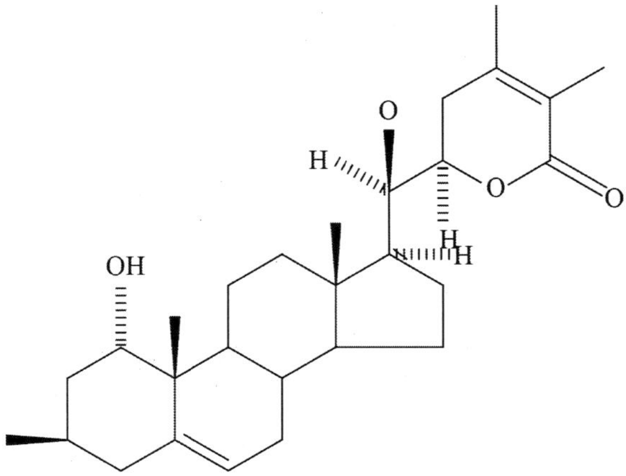

Fig. 26.2: Structure of withanoside IV.

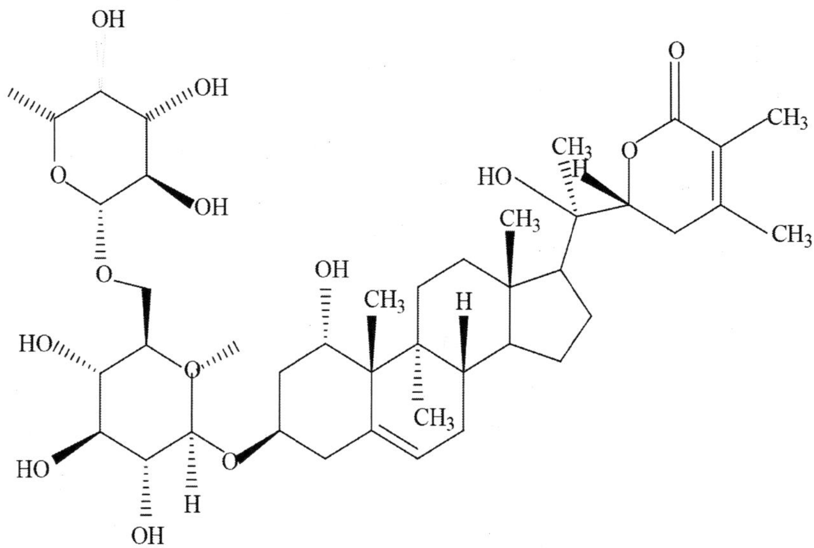

Fig. 26.3: Structure of withanoside VI.

withanolide A was investigated on memory-deficient mice, it showed neuronal atrophy and synaptic loss in the brain. Treatment with A beta(25–35) (10 microM) induced axonal and dendritic atrophy, and pre- and postsynaptic loss in cultured rat cortical neurons.

Subsequent treatment with withanolide A (1 microM) induced significant regeneration of both axons and dendrites, in addition to the reconstruction of pre- and postsynapses in the neurons. Withanolide A (10 micromol kg(–1) day (–1), for 13 days, p.o.) recovered A beta(25–35)-induced memory deficit in mice. At that time, the decline of axons, dendrites, and synapses in the cerebral cortex and hippocampus was almost recovered (Kuboyama et al. 2005).

26.1.2 Anti-alzheimer

Withanolide A and asiatic acid were investigated for their potential activities against multiple targets associated with Abeta pathways (BACE1, ADAM10, IDE, and NEP). BACE1 is a rate-limiting enzyme in the production of Abeta from amyloid-beta precursor protein (AbetaPP), while ADAM10 is involved in non-amyloidogenic processing of AbetaPP. IDE and NEP are two of the prominent enzymes involved in effectively degrading Abeta.

Withanolide A and asiatic acid significantly down-regulated BACE1 and also up-regulated ADAM10 in primary rat cortical neurons. Withanolide A significantly up-regulated IDE levels, which may help in degrading excess Abeta from the AD brain (Patil et al. 2010).

The present study investigated the cholinesterase inhibition potential of withanolide A along with the associated binding mechanism. Our docking simulation results predict the high binding affinity of the ligand to the receptor. The study provides evidence for consideration of withanolide A as a valuable small ligand molecule in treatment and prevention of Alzheimer's dementia (Grover et al. 2012).

26.1.3 Antibacterial

The growth inhibition was observed in *Pseudomonas* and *Staphylococcus aureus* at three different of Withanolide-A concentrations (100, 500, and 1000 ppm). Withanolide–A exhibited significant antibacterial activity against the *Pseudomonas* and *Staphylococcus aureus* (Mali and Singh 2013).

26.1.4 Anticancer

For the first time it was demonstrated that withanolide–A exhibits significant activity against human breast cancer cells in culture and *in vivo*. The WA treatment decreased viability of MCF-7 (estrogen-responsive) and MDA-MB-231 (estrogen-independent) human breast cancer cells in a concentration-dependent manner. The WA-mediated suppression of breast cancer cell viability correlated with apoptosis induction characterized by DNA condensation, cytoplasmic histone-associated DNA fragmentation, and cleavage of poly-(ADP-ribose)-polymerase (Stan et al. 2008).

Withanolide–A resulted in induction as well as increased S15 phosphorylation of p53 in MCF-7 cells, but RNA interference of this tumor suppressor conferred modest protection at best against WA-induced apoptosis. Withanolide–A mediated growth inhibition and apoptosis induction in MCF-7 cells were significantly attenuated in the presence of 17β-estradiol (E2) (Hahm et al. 2011).

Exposure of MDA-MB-231 and MCF-7 human breast cancer cells to pharmacological concentrations of withanolide–A resulted in cleavage (activation) of Notch 2 as well as Notch 4, which was accompanied by transcriptional activation of Notch as evidenced by RBP-Jk, HES-1A/B, and HEY-1 luciferase reporter assays.

On the other hand, WA treatment caused a decrease in levels of both transmembrane and cleaved Notch 1. Activation of Notch 2 was not observed in cells treated with withanone or withanolide A, which are structural analogs of WA (Lee et al. 2012).

26.1.5 Anti-epileptic

A study investigated the effect of *W. somnifera*, extract, withanolide A, and carbamazepine on cerebellar AMPA receptor function in pilocarpine-induced temporal lobe epilepsy.

The treatment with *W. somnifera* and withanolide A significantly reversed the motor learning deficit in rats with epilepsy when compared with control rats. There was an increase in glutamate content and IP3 content observed in rats with epilepsy which was reversed in WS- and WA-treated rats with epilepsy.

Moreover, treatment with *W. somnifera* extract, and withanolide A resulted in physiological expression of alpha-amino-3-hydroxy-5-methylisoxazole-4-propionic acid (AMPA receptors). The treatment with *W. somnifera* extract, and withanolide A reversed the GAD and GLAST expression (Soman et al. 2013).

26.1.6 Antistress

Oral administration of withanolide A once daily at the graded doses of 0.25, 0.5, 1, and 2 mg/kg p.o. caused significant recovery of stress-induced depleted T cell population causing an increase in the expression of IL-2 and IFN-gamma and a decrease in the concentration of corticosterone in stressed experimental animals. It also reversed the restraint stress-induced increase in plasma alanine aminotransferase, aspartate aminotransferase, and hepatic lipid peroxidation levels and improved the restraint stress-induced decrease in hepatic glutathione, and glycogen levels (Kour et al. 2009).

26.1.7 Immunopharmacology

Mice were administrated a chemically standardized aqueous alcoholic (1:1) root extract (AGB) of *W. somnifera* orally for 15 days. AGB stimulated cell mediated immunity, IgM and IgG titers reaching peak value with 30 mg/kg b.wt. The extract selectively, induced type 1 immunity because it guided enhanced expression of T helper cells (Th)1 cytokines interferon (IFN)-gamma and interleukin (IL)-2 while Th2 cytokine IL-4 observed a moderate decline. Confirmation of Th1 polarization was obtained from augmented levels of IgG2a over IgG1 in the blood sera of AGB treated groups.

Withanolide-A a major constituent of a chemically standardized aqueous alcoholic (1:1) root extract of *W. somnifera* appeared responsible for T helper cells Th1 skewing effect of the extract as it significantly increased the levels of Th1 cytokines, decreased moderately interleukin IL-4, and significantly restored the selective dexamethasone inhibition of Th1 cytokines in mouse splenocytes cultures *in vitro*. In addition, the extract also strongly activated macrophage functions *ex vivo* and *in vitro* indicated by enhanced secretion of nitrite, IL-12, and TNF-alpha (Malik et al. 2007).

26.1.8 Neuroprotective

Withanolide A reversed hypoxia mediated neurodegeneration; administration of buthionine sulfoximine along with withanolide A blunted its neuroprotective effects. Withanolide A seems to reduces neurodegeneration by restoring hypoxia induced glutathione depletion in the hippocampus. Withanolide A increases glutathione biosynthesis in neuronal cells by upregulating GCLC level through Nrf2 pathway in a corticosterone dependent manner (Baitharu et al. 2014).

Further Readings

Baitharu I, Jain V, Deep SN, Shroff S, Sahu JK, Naik PK, Ilavazhagan G. Withanolide A prevents neurodegeneration by modulating hippocampal glutathione biosynthesis during hypoxia. *PLoS One* 2014; **9**: e105311.

Grover A, Shandilya A, Agrawal V, Bisaria VS, Sundar D. Computational evidence to inhibition of human acetyl cholinesterase by withanolide a for Alzheimer treatment. *J Biomol Struct Dyn* 2012; **29**: 651–62.

Hahm ER, Lee J, Huang Y, Singh SV. Withaferin a suppresses estrogen receptor-α expression in human breast cancer cells. *Mol Carcinog* 2011; **50:** 614–24.

Kour K, Pandey A, Suri KA, Satti NK, Gupta KK, Bani S. Restoration of stress-induced altered T cell function and corresponding cytokines patterns by Withanolide A. *Int Immunopharmacol* 2009; **9**: 1137–44.

Kuboyama T, Tohda C, Zhao J, Nakamura N, Hattori M, Komatsu K. Axon- or dendrite-predominant outgrowth induced by constituents from Ashwagandha. *Neuroreport* 2002; **13**: 1715–1720.

Kuboyama T, Tohda C, Komatsu K. Neuritic regeneration and synaptic reconstruction induced by withanolide A. *Br J Pharmacol* 2005; **144**: 961–71.

Lee J, Sehrawat A, Singh SV. Withaferin A causes activation of Notch 2 and Notch 4 in human breast cancer cells. *Breast Cancer Res Treat* 2012; **136**: 45–56.

Mali PC, Singh AR. Isolation, characterization and evaluation of antimicrobial activity of Withanolide-A of *Withania somnifera. Int J Pharmacol Res* 2013; **3**: 48–52.

Malik F, Singh J, Khajuria A, Suri KA, Satti NK, Singh S, Kaul MK, Kumar A, Bhatia A, Qazi GN. A standardized root extract of *Withania somnifera* and its major constituent withanolide-A elicit humoral and cell-mediated immune responses by up regulation of Th1-dominant polarization in BALB/c mice. *Life Sci* 2007; **80**: 1525–38.

Parveen N, Naik PM, Manohar SH, Murthy HN. Distribution of withanolide A in various organs of *Withania somnifera* Dunal. *Int J Pharma Bio Sci* 2010; **1**: 1–5.

Patil SP, Maki S, Khedkar SA, Rigby AC, Chan C. Withanolide A and asiatic acid modulate multiple targets associated with amyloid-beta precursor protein processing and amyloid-beta protein clearance. *J Nat Prod* 2010; **73**: 1196–1202.

Sabir F, Sangwan NS, Chaurasiya ND, Misra LN, Sangwan RS. *In vitro* withanolide production by *Withania somnifera* L. cultures. *Z Naturforsch C* 2008; **63**: 409–12.

Sangwan RS, Chaurasiya ND, Lal P, Misra L, Uniyal GC, Tuli R et al. Withanolide A bio-generation *in vitro* shoot cultures of Ashwagandha (*Withania somnifera* Dunal), a main medicinal plant in Ayurveda. *Chem Pharm Bull* 2007; **55**: 1371–1375.

Sangwan RS, Das Chaurasiya N, Lal P, Misra L, Tuli R, Sangwan NS. Withanolide A is inherently de novo biosynthesized in roots of the medicinal plant Ashwagandha (*Withania somnifera*). *Physiol Plant* 2008; **133**: 278–87.

Soman S, Anju TR, Jayanarayanan S, Antony S, Paulose CS. Impaired motor learning attributed to altered AMPA receptor function in the cerebellum of rats with temporal lobe epilepsy: ameliorating effects of *Withania somnifera* and withanolide A. *Epilepsy Behav* 2013; **27**: 484–91.

Stan SD, Hahm ER, Warin R, Singh SV. Withaferin A causes FOXO3a- and Bim-dependent apoptosis and inhibits growth of human breast cancer cells *in vivo*. *Cancer Res* 2008; **68**: 7661–9.

Wang FY, Sun YM, Lv CP, Cheng MQ, Zhang L, Sun M. Hair roots induction and culture of *Withania somnifera* and its withanolide A synthesis. *Zhongguo Zhong Yao Za Zhi* 2014; **39**: 790–4.

Chapter 27

Pharmacology of Withanone

27.1 Bio-generation and Distribution

This study optimized carbon sources in half MS liquid medium for maximum biomass accumulation and withanolides production in hairy root culture of *W. somnifera*. The highest production of withaferin A and withanone was achieved when sucrose and sucrose+glucose were used individually as carbon sources.

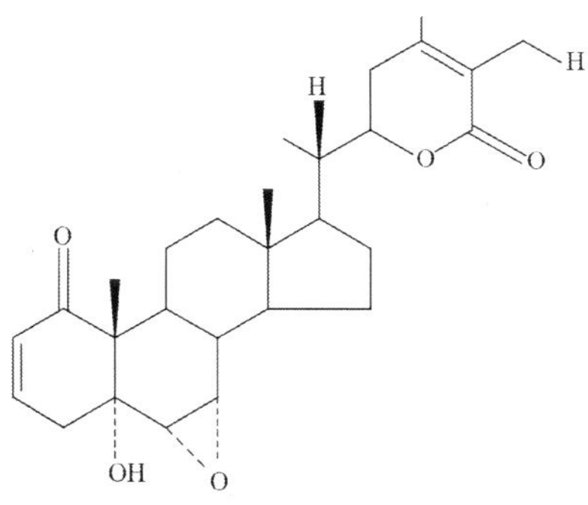

Fig. 27.1: Structure of withanone.

The hairy root suspension culture supplemented with a lower level of sucrose (2%) favored hairy root biomass accumulation (1.41 g DW) followed by sucrose+glucose (2 + 1) when compared with other carbon sources in half MS liquid medium after 40 days of culture. The hairy roots grown on sucrose (4%) enriched half MS liquid medium stimulated higher production of withaferin A (2.21 mg/g DW) and withanone (2.41 mg/g DW) on the 40th day of culture, followed by sucrose+glucose (4 + 1%) compared with glucose, fructose, maltose, and other combinations tested (Sivanandhan et al. 2012).

27.2 Pharmacology

27.2.1 Anticancer

The present study demonstrates that a major component of i-Extract and withanone (i-Factor) protected the normal human fibroblasts against the toxicity caused by withaferin A. It increased the *in vitro* division potential of normal human cells that appeared to be mediated by decreased accumulation of molecular damage, downregulation of the senescence-specific β-galactosidase activity and the senescence marker protein, p21(WAF-1), protection against oxidative damage, and induction of proteasomal activity (Widodo et al. 2009).

Randomized ribozyme library was introduced into cancer cells prior to the treatment with i-Extract. The targets were validated for their role in i-Extract induced selective killing of cancer cells by biochemical and molecular assays. Fifteen gene-targets were identified and were investigated for their role in a specific cancer cell killing activity of i-Extract and withaferin A and withanone by undertaking the shRNA-mediated gene silencing approach. The involvement of ROS-signaling components demonstrate that the selective killing of cancer cells is mediated by induction of oxidative stress (Widodo et al. 2010).

By computational approach, it found that withanone binds to TPX2-Aurora A complex. In experiment, withanone treatment to cancer cells indeed resulted in dissociation of TPX2-Aurora A complex and disruption of mitotic spindle apparatus proposing this as a mechanism of the anticancer activity of withanone. The Molecular Dynamics simulation results suggesting the thermodynamic and structural stability of TPX2-Aurora A in complex with withanone further substantiates the binding (Grover et al. 2012).

Withanone bind to mortalin in a region, earlier predicted critical for binding to p53. Cationic rhodacyanine dye, MKT-077 has also shown to bind the same region and kill cancer cells selectively. The molecular dynamic simulations reveal the thermodynamic and structural stability of the withanone-mortalin complexes. We also demonstrate the experimental evidence of abrogation of mortalin-p53 complex by withanone resulting in nuclear translocation and functional reactivation of p53 in human cancer cells (Grover et al. 2012a).

Docking studies carried out with withanone have shown strong binding affinity of—19.1088 kJ/mol with *BIR5* domain of survivin and in turn interferes with inhibitory action against caspases and may lead to apoptosis. Binding of withanone at *BIR5* domain of survivin may also interfere with chromosomal passenger complex and lead to halt the mitotic process within the cancer cell (Wadegaonkar and Wadegaonkar 2013).

Withanone and withaferin A in the i-Extract retained the selective cancer cell killing activity and found that it also has significant antimigratory, -invasive, and -angiogenic activities, in both *in vitro* and *in vivo* assays. It was demonstrated that withanone and withaferin A caused downregulation of migration-promoting proteins hnRNP-K, VEGF, and metalloproteases (Gao et al. 2014).

27.2.2 Antidote

Withanone protects cells from methoxyacetic acid-induced toxicity by suppressing the ROS levels, DNA and mitochondrial damage, and induction of cell defense signaling pathways including Nrf2 and proteasomal degradation. These findings suggest further studies of withanone as a adjuvant in consumer products where ester phthalates are cause of health concern (Priyandoko et al. 2011).

27.2.3 Neuroprotective

Scopolamine caused downregulation of the expression of BDNF and GFAP in a dose and time dependent manner, and these effects were markedly attenuated in response to i-Extract treatment. The scopolamine induced cytotoxicity in IMR32 neuronal and C6 glioma cells is associated with the downregulation of neuronal cell markers NF-H, MAP2, PSD-95, GAP-43, and glial cell marker GFAP and with upregulation of DNA damage—γH2AX and oxidative stress-ROS markers. These molecules showed recovery when cells were treated with i-Extract or its purified component, withanone (Konar et al. 2011).

Further Readings

Gao R, Shah N, Lee JS, Katiyar SP, Li L, Oh E, Sundar D, Yun CO, Wadhwa R, Kaul SC. Withanone-rich combination of Ashwagandha withanolides restricts metastasis and angiogenesis through hnRNP-K. *Mol Cancer Ther* 2014; **13**: 2930–40.

Grover A, Singh R, Shandilya A, Priyandoko D, Agrawal V, Bisaria VS, Wadhwa R, Kaul SC, Sundar D. Ashwagandha derived withanone targets TPX2-Aurora A complex: computational and experimental evidence to its anticancer activity. *PLoS One* 2012; **7**: e30890.

Grover A, Priyandoko D, Gao R, Shandilya A, Widodo N, Bisaria VS, Kaul SC, Wadhwa R, Sundar D. Withanone binds to mortalin and abrogates mortalin-p53 complex: computational and experimental evidence. *Int J Biochem Cell Biol* 2012a; **44**: 496–504.

Konar A, Shah N, Singh R, Saxena N, Kaul SC, Wadhwa R, Thakur MK. Protective role of Ashwagandha leaf extract and its component withanone on scopolamine-induced changes in the brain and brain-derived cells. *PLoS One* 2011; **6**: e27265.

Priyandoko D, Ishii T, Kaul SC, Wadhwa R. Ashwagandha leaf derived withanone protects normal human cells against the toxicity of methoxyacetic acid, a major industrial metabolite. *PLoS One* 2011; **6**: e19552.

Sivanandhan G, Rajesh M, Arun M, Jeyaraj M, Dev GK, Manickavasagam M, Selvaraj N, Ganapathi A. Optimization of carbon source for hairy root growth and withaferin A and withanone production in *Withania somnifera*. *Nat Prod Commun* 2012; **7**: 1271–2.

Wadegaonkar VP, Wadegaonkar PA. Withanone as an inhibitor of survivin: A potential drug candidate for cancer therapy. *J Biotechnol* 2013; **168**: 229–233.

Widodo N, Priyandoko D, Shah N, Wadhwa R, Kaul SC. Selective killing of cancer cells by Ashwagandha leaf extract and its component Withanone involves ROS signaling. *PLoS One* 2010; **5**: e13536.

Chapter 28

Pharmacology of Withanolide D

28.1 Introduction

Withanolide D, the major component of the leaves of *Withania somnifera* chemotype II, is a steroidal lactone of the withanolide type, isomeric with withaferin A (Lavie et al. 1968).

28.2 Pharmacology

28.2.1 Anticancer

A study demonstrated that withanolide D enhance the ceramide (Fig. 28.1) accumulation by activating N-SMase 2, modulate phosphorylation of the JNK and

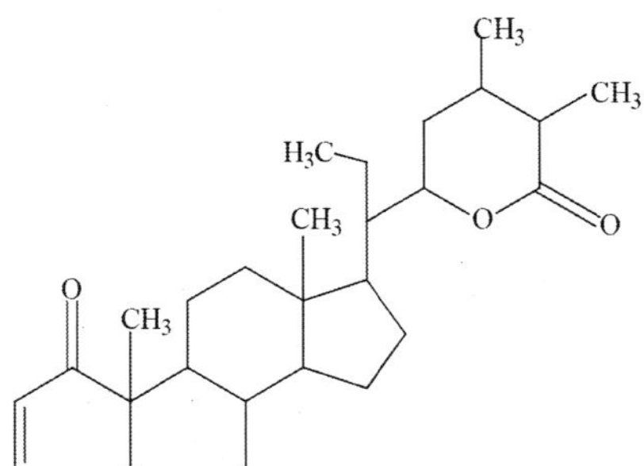

Fig. 28.1: Structure of withanolide D.

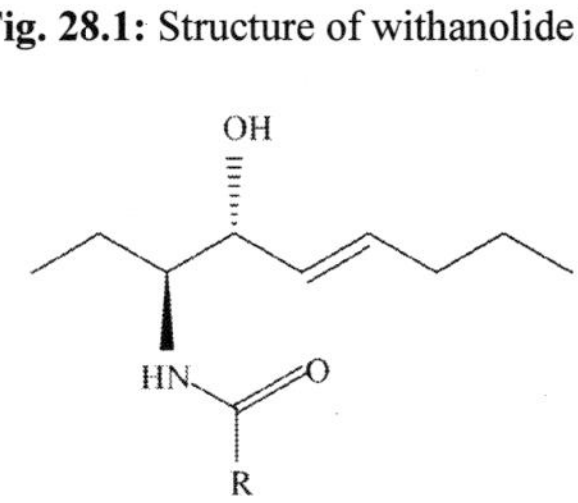

Fig. 28.2: Structure of ceramide.

p38MAPK and induced apoptosis in both myeloid and lymphoid cells along with primary cells derived from leukemia patients (Mondal et al. 2010).

28.2.2 Cytotoxicity

Cytotoxic activity of 12β-acetoxy-4-deoxy-5,6-deoxy-Δ^5–withanolide D and Withanolide D isolated from *Acnistus arborescens* were assessed against human tumor cell lines HT-29, MCF-7, MKN-45, HEp-2, HeLa, U-937 and two human normal fibroblast cultures, Fib04 and Fib05. Withanolide D presented *in vitro* cytotoxic activity against tumor cell lines at the low micromolar range (LC(50):1.0 to 1.69 microM) and showed a slightly lower activity against Fib04 (Cordero et al. 2009).

Further Readings

Chowdhury K, Neogy RK. Mode of action of Withaferin A and Withanolide D. *Biochem Pharmacol* 1975; **24**: 919–20.

Cordero CP, Morantes SJ, Páez A, Rincón J, Aristizábal FA. Cytotoxicity of withanolides isolated from *Acnistus arborescens*. *Fitoterapia* 2009; **80**: 364–8.

Lavie D, Kirson I, Glotter E. Constituents of *Withania somnifera* Dun. Part X. The Structure of Withanolide D. *Israel J Chem* 1968; **6**: 671–678.

Mondal S, Mandal C, Sangwan R, Chandra S, Mandal C. Withanolide D induces apoptosis in leukemia by targeting the activation of neutral sphingomyelinase-ceramide cascade mediated by synergistic activation of c-Jun N-terminal kinase and p38 mitogen-activated protein kinase. *Molecular Cancer* 2010; **9**: 239.

Mondal S, Roy S, Maity R, Mallick A, Sangwan R, Misra-Bhattacharya S, Mandal C. Withanolide D, carrying the baton of Indian rasayana herb as a lead candidate of antileukemic agent in modern medicine. *Adv Exp Med Biol* 2012; **749**: 295–312.

Chapter 29

Miscellaneous Withanolides

29.1 Withanolides from *Hyoscyamus niger*

Daturalactone-4 (Fig. 29.1), and Nic-3 (hyoscyamilactol) (Fig. 29.2), and a new compound, 16α-acetoxyhyoscyamilactol (Fig. 29.3), have been isolated from seeds of *Hyoscyamus niger* (Ma et al. 1999).

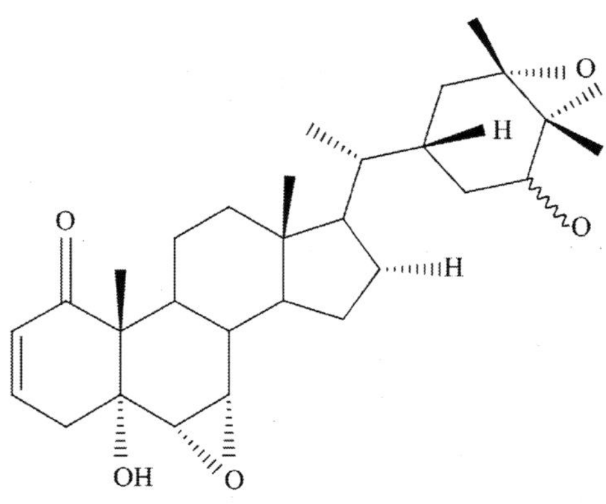

Fig. 29.1: Structure of Daturalactone-4.

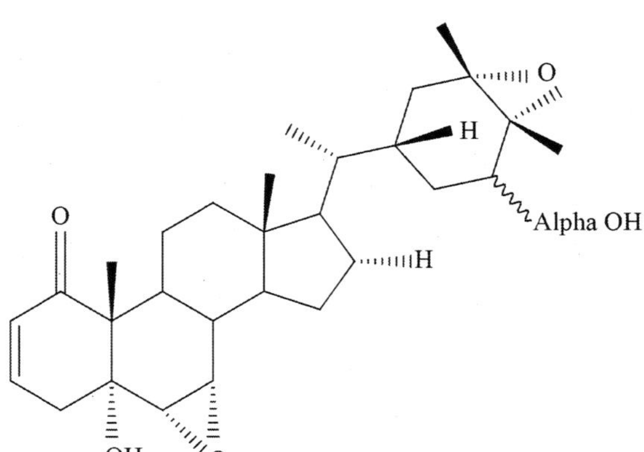

Fig. 29.2: Structure of hyoscyamilactol.

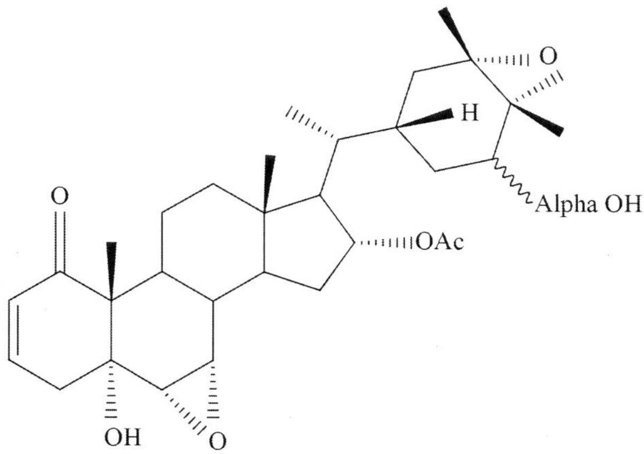

Fig. 29.3: Structure of 16α-acetoxyhyoscyamilactol.

29.2 Withanolides from *Solanum cilistum*

Six new withanolide-type steroids, designated cilistols v, t, i, j, y, and w, were obtained from the leaves of *Solanum cilistum* (Zhu et al. 2001).

29.3 Withanolides from *Ajuga bracteosa*

Bracteosin A (Fig. 29.4), bracteosin B (Fig. 29.5), and bracteosin C (Fig. 29.6) have been isolated from the whole plants of *A. bracteosa.* All three exhibited evident inhibitory potential against cholinesterase enzymes in a concentration-dependent fashion (Riaz et al. 2004).

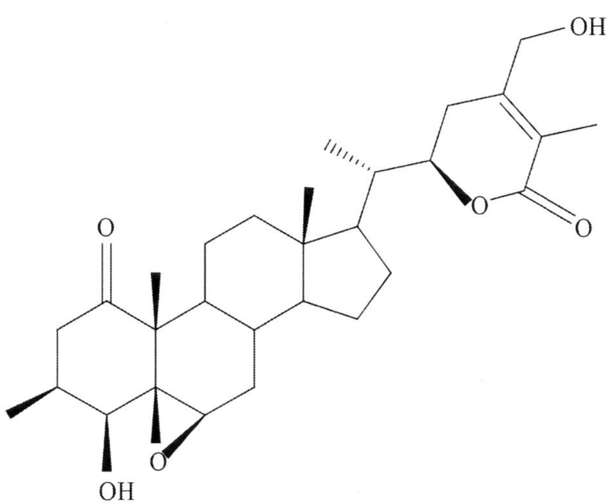

Fig. 29.4: Structure of bracteosin A.

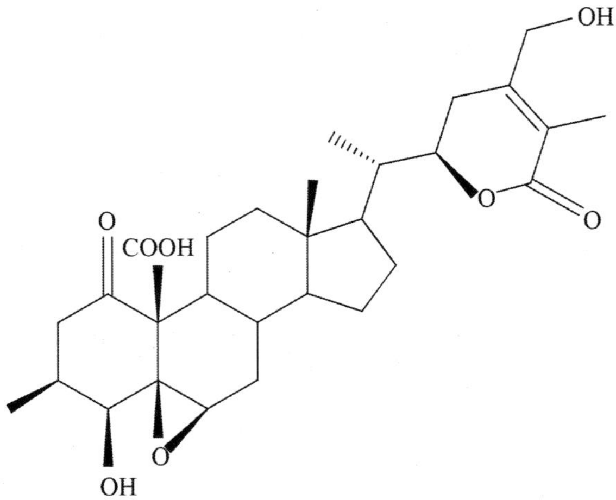

Fig. 29.5: Structure of bracteosin B.

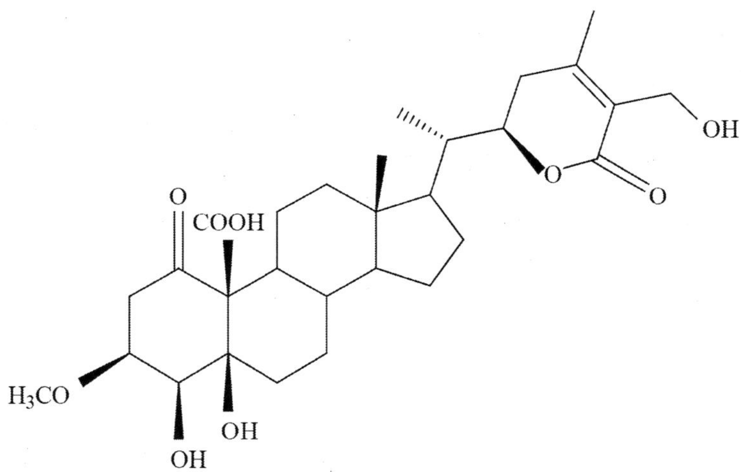

Fig. 29.6: Structure of bracteosin C.

29.4 Withanolides from *Solanum sisymbiifolium*

In addition to the known cilistol A (Fig. 29.7), two new withanolides, namely cilistepoxide (Fig. 29.8) and cilistadiol (Fig. 29.9) have been isolated from leaves and stem of *S. sisymbiifolium* (Niero et al. 2006).

29.5 Withanolides from *Jaborosa laciniata*

Six new trechonolide type withanolides, together with trechonolide A, jaborotetrol, and 12-O-methyl jaborosotetrol, has been isolated from the aerial parts of *J. laciniata* (Cirigliano et al. 2007).

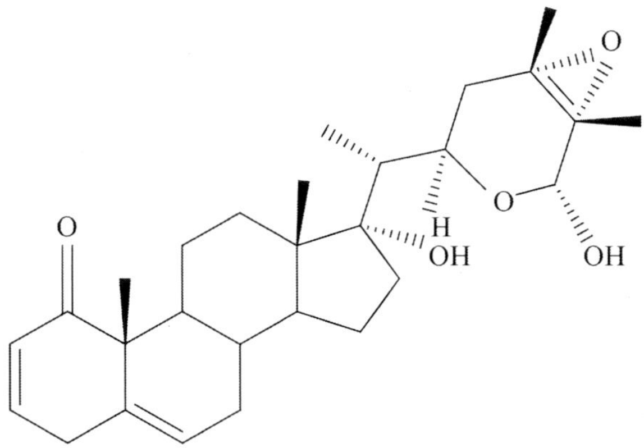

Fig. 29.7: Structure of cilistol A.

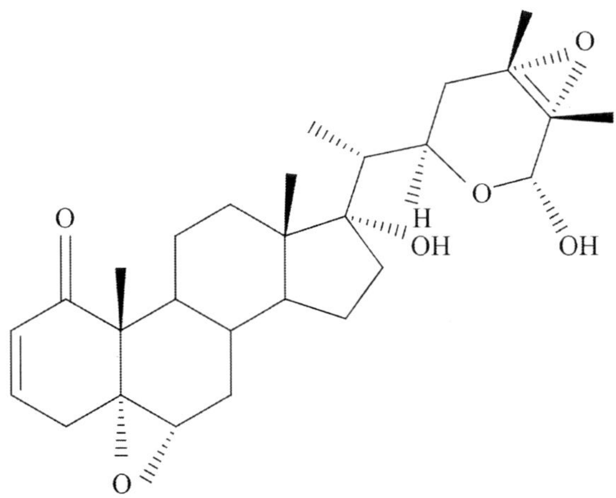

Fig. 29.8: Structure of cilistepoxide.

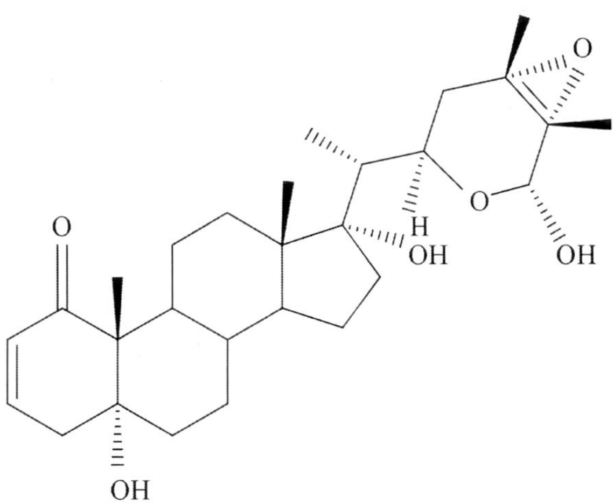

Fig. 29.9: Structure of cilistadiol.

29.6 Withanolides from *Jaborosa kurtzii*

Two new withanolides were isolated and characterized from the aerial parts of *J. kurtzii*, namely, jaborosalactone 43, with a spiranoid delta-lactone at C-22, and jaborosalactone 44 have been reported (Ramacciotti and Nicotra 2007).

29.7 Withanolide Z from *Withania somnifera*

Withanolide Z (Fig. 29. 10), along with four known withanolides, withanolide B, withanolide A, 27-hydroxywithanolide B, and withaferin A have been reported from methanolic extract of *W. somnifera* leaves (Pramanick et al. 2008).

29.8 Withanolides from *Withania adpressa* Coss. (Solanaceae)

A novel withanolide 14α,15α,17β,20β-tetrahydroxy-1-oxo-(22R)-witha-2,5,24-trienolide (Fig. 29.11), withanolides F (Fig. 29.12) and J (Fig. 29.13). The extract, semi-purified fractions, and withanolides exhibited potent cytotoxicity against cancer cell lines (Hep2, HT29, RD, Vero, and MDCK), using the MTT assay (Abdeljebbar et al. 2009).

29.9 Withanolides from *Withania aristata*

(4S,20R,22R)-27-acetoxy-4-p-bromobenzoyloxy-1-oxo-witha-2,5,16,24-tetraenolide showed cytotoxicity against all the cell lines assayed with IC(50) values ranging from 2.8 to 3.6microM, and (4S,20R,22R)-4,27-diacetoxy-4-hydroxy-1-oxo-witha-2,5,16,24-tetraenolide exhibited an IC(50) value of 5.4 microM on the MCF-7 cell line (Llanos et al. 2010).

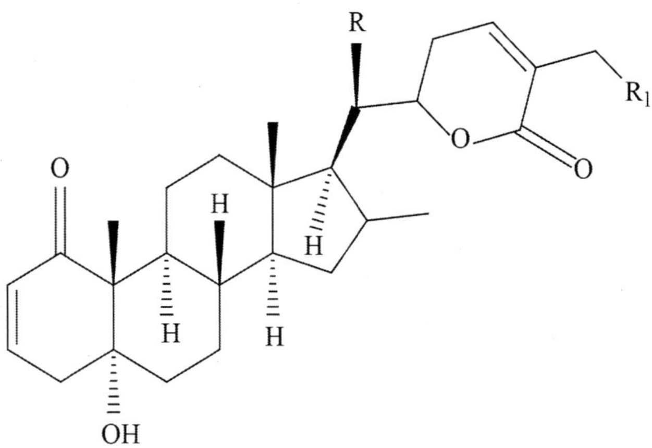

R=R1=H,6 alpha -OH,7 beta-Cl

Fig. 29.10: Structure of withanolide Z.

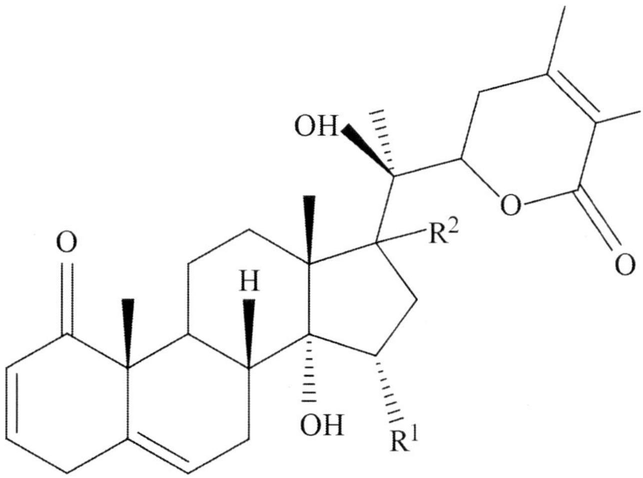

R1=OH, R2-Beta OH

Fig. 29.11: Structure of 14α,15α,17β,20β-tetrahydroxy-1-oxo-(22R)-witha-2,5,24-trienolide.

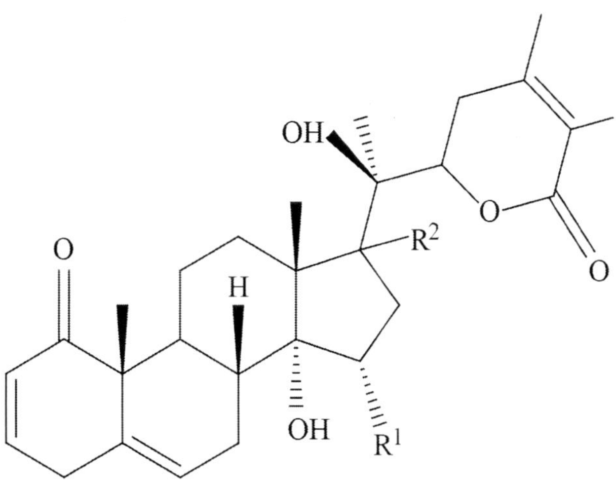

R1=H, R2-Beta OH

Fig. 29.12: Structure of withanolide F.

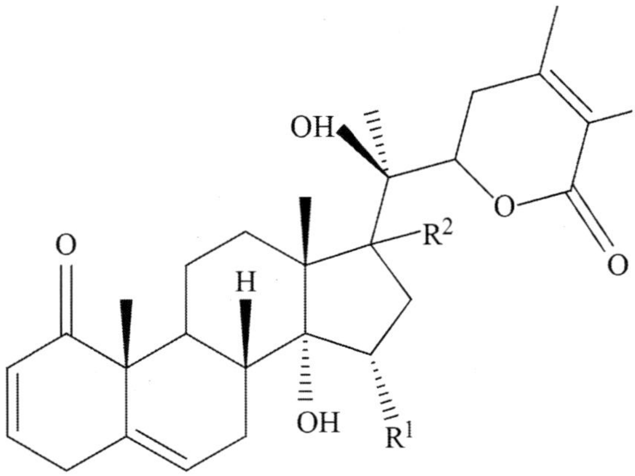

R1=OH, R2-Alpha OH

Fig. 29.13: Structure of withanolide J.

29.10 Withanolides from *Mandragora officinarum*

Two new withanolides named mandragorolide A (Fig. 29.14) and mandragorolide B (Fig. 29.15) were isolated from the MeOH extract of the whole plant of *Mandragora officinarum* of Jordanian origin, along with five known withanolides namely larnaxolide A, withanolide B, datura lactone 2, withanicandrin, and salpichrolide C (Suleiman et al. 2010).

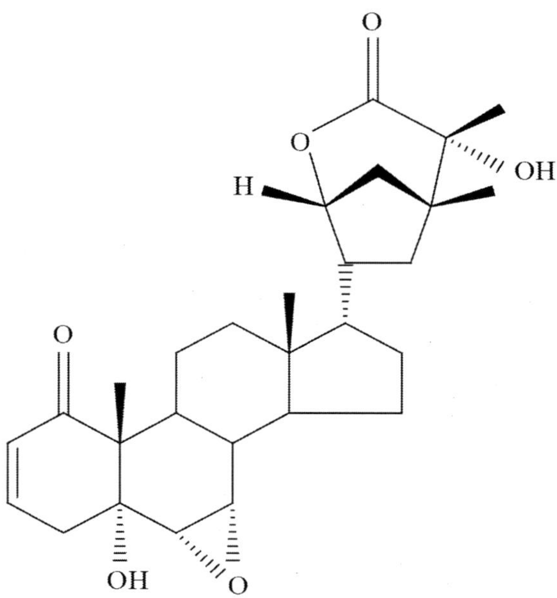

Fig. 29.14: Structure of mandragorolide A.

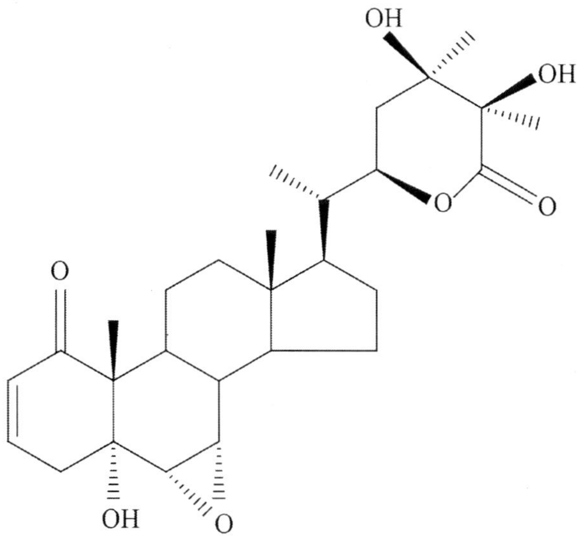

Fig. 29.15: Structure of mandragorolide B.

29.11 Withanolides from *Dioscorea japonica*

Two new withanolides, named dioscorolide A (Fig. 29.16) and dioscorolide B (Fig. 29.17) showed cytotoxicity against tumor cell lines (A549, SK-OV-3, SK-MEL-2, and HCT15) with IC(50) values ranging from 6.3 to 26.9 μM and exhibited lower activity against the normal cell line (HUVEC) with IC(50) values ranging from 27.1 to 28.8 μM, suggesting selective toxicity among tumor and normal cells (Kim et al. 2011).

29.12 Withanolides from *Withania frutescens*

The crude methanol extract of *W. frutescens* leaves was partitioned with dichloromethane, ethyl acetate, and n-butanol. MeOH extract and its fractions were tested for their cytotoxic activity against cancer cell lines (HepG2 and HT29) using the MTT assay. 2,3-dihydroxywithaferin A-3β-O-sulfate (Fig. 29.18) exhibited the strongest cytotoxic activity against HT29 cancer cell lines (IC_{50} of 1.78 ± 0.09 μM) which was comparable to that of 5-fluorouracil (El Bouzidi et al. 2013).

29.13 Withanolides from *Physalis pubescens* L.

Physapubescin B, physapubescin C, physapubescin D, and physapubescin have been isolated (Ji et al. 2013).

29.14 Withanolides from *Datura wrightii*

See wrightolide.

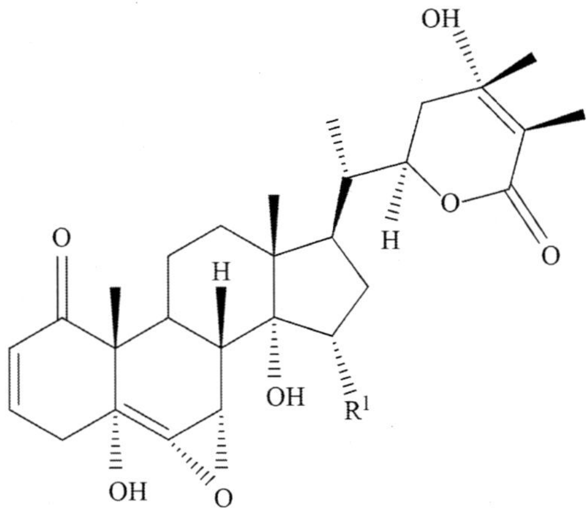

Fig. 29.16: Structure of dioscorolide A.

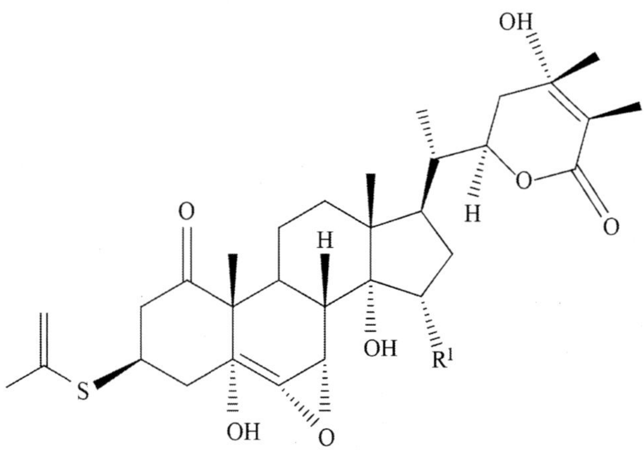

Fig. 29.17: Structure of dioscorolide B.

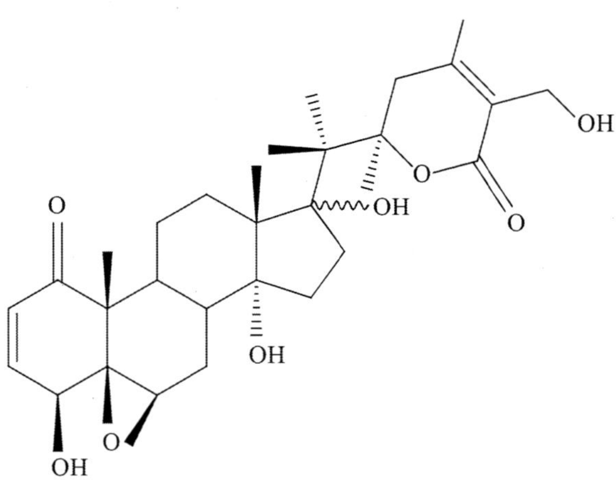

Fig. 29.18: Structure of 2,3-dihydroxywithaferin A-3β-O-sulfate.

29.15 Withanolides from *Physalis longifolia* Nutt.

Withalongolide A (Fig. 29.19), withaferin A, and withalongolide B (Fig. 29.20) have been reported from the aerial parts (Cao et al. 2014).

29.16 Withanolides from *Physalis hispida*

Withahisolides A-I, have been isolated from the aerial parts of *P. hispida* (Cao et al. 2014a).

29.17 Withanolides from *Jaborosa caulescens* var. *bipinnatifida*

Withanolides 2,3-dihydrotrechonolide A (Fig. 29.21), and 2,3-dihydro-21-hydroxytrechonolide A (Fig. 29.22) have been isolated along with two known withanolides trechonolide A and jaborosalactone 39 from *J. caulescens* var. *bipinnatifida* (Zhang et al. 2014).

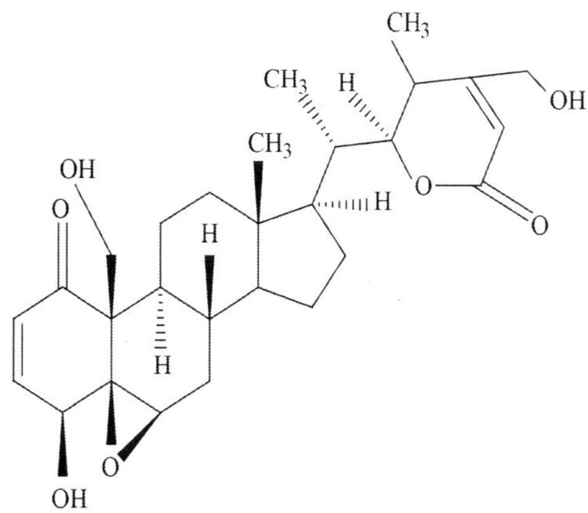

Fig. 29.19: Structure of withalongolide A.

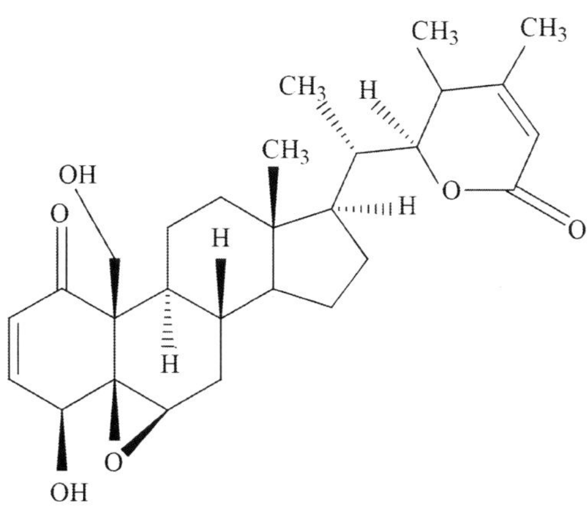

Fig. 29.20: Structure of withalongolide B.

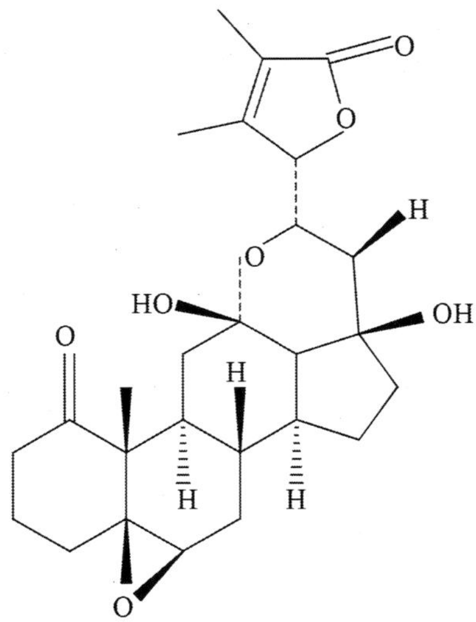

Fig. 29.21: Structure of 2,3-dihydrotrechonolide A.

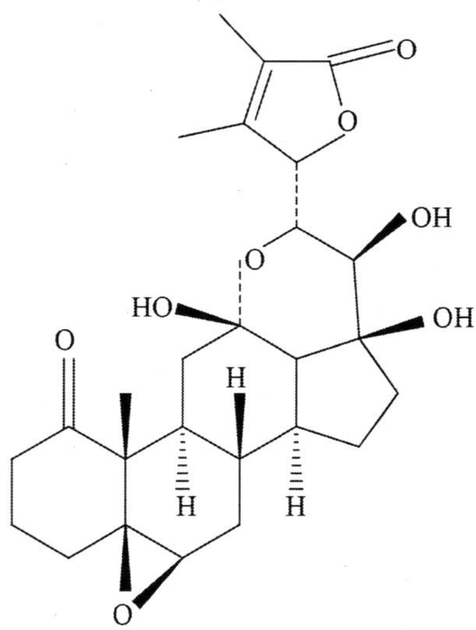

Fig. 29.22: Structure of 2,3-dihydro-21-hydroxytrechonolide A.

29.18 Coagulanolides from *Withania coagulans*

Coagulin-L (Fig. 29.23) reduces the expressions of peroxisome proliferator-activated receptor γ (PPARγ) and CCAAT/enhancer-binding protein α, the major transcription factors orchestrating adipocyte differentiation. Detailed analysis further proved that early exposure of coagulin-L is sufficient to cause significant inhibition during adipogenesis (Beg et al. 2014).

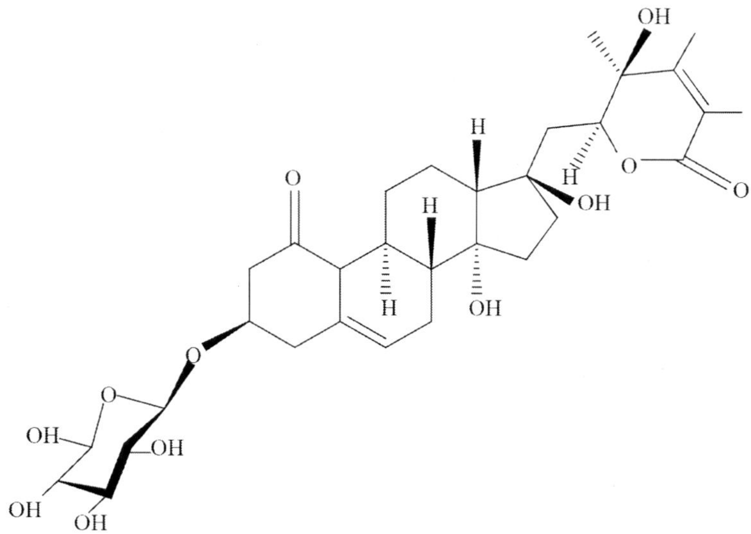

Fig. 29.23: Structure of coagulin-L.

29.19 Trichosides A and B from *Tricholepis eburnea*

Trichosides A and B, new withanolide glucosides, have been isolated from the n-butanolic fraction of the 75% methanolic extract of aerial parts of *T. eburnean* (Maher et al. 2015).

29.20 Withanolides from *Nicandra john-tyleriana*

Withanolides including two acnistins and four are withajardins, have been isolated from the aerial parts of *N. john-tyleriana* (Gutiérrez Nicolás et al. 2015).

29.21 Some lesser known withanolides

29.21.1 5,6-de-epoxy-5-en-7-one-17-hydroxy withaferin A

The cytotoxic activity of 5,6-de-epoxy-5-en-7-one-17-hydroxy withaferin A, isolated from leaves, was carried out using the MTT assay against a panel of cancer cell lines, namely MCF-7 (breast), WRL-68 (liver), PC-3 (prostate), and CACO-2 (colon). 5,6-de-epoxy-5-en-7-one-17-hydroxy withaferin A possesses strong cytotoxic activity against liver and breast cancer with an IC50 of 1.0 μg/mL and a moderate activity against colon (IC50 3.4 μg/mL) and prostate (IC50 7.4 μg/mL) cancer cells (Siddique et al. 2014).

29.21.2 Dinoxin B

A new withanolide, dinoxin B (12,21-dihydroxy-1-oxowitha-2,5,24-trienolide-27-O-β-D-glucopyranoside) (Fig. 29.24), isolated from a methanol extract of *Datura*

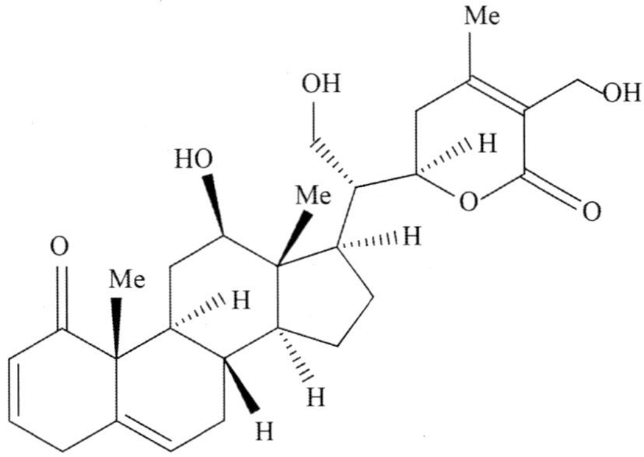

Fig. 29.24: Structure of dinoxin B.

inoxia leaves, exhibited submicromolar IC(50) values against multiple human cancer cell lines. Among the most sensitive were several breast cancer cell lines (Vermillion et al. 2011).

29.21.3 Physalins

Physalins are steroidal constituents of Physalis plants which possess an unusual 13,14-seco-16,24-cyclo-steroidal ring skeleton. Chief sources of physalins are Physalins were isolated from Physalis species, *Physalis alkekengi*, *P. angulata*, and *P. lancifolia*.

Apoptosis induced by physalin A (Fig. 29.25) in HT1080 cells is associated with up-regulation of caspase-3 and caspase-8 expression (He et al. 2013). Physalin A induces apoptosis and autophagy in A375-S2 cells (He et al. 2014).

Physalin B (Fig. 29.26) induces mito-ROS, which not only inhibits the ubiquitin-proteasome pathway but also induces incomplete autophagic response in HCT116 cells *in vitro* (Ma et al. 2015).

Physalin F (Fig. 29.27) induced cell apoptosis through the ROS-mediated mitochondrial pathway and suppressed NF-κB activation in human renal cancer A498 cells (Wu et al. 2012).

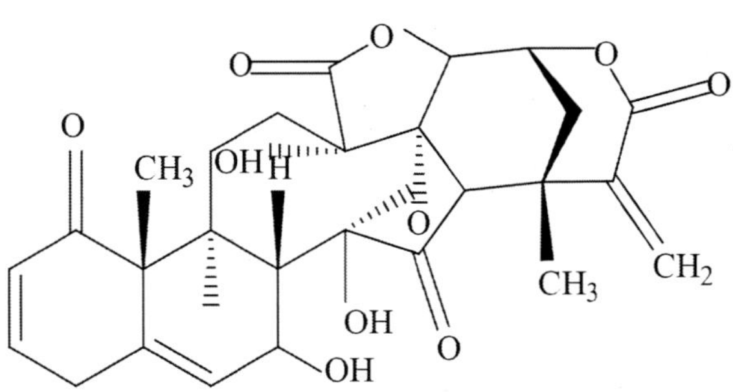

Fig. 29.25: Structure of physalin A.

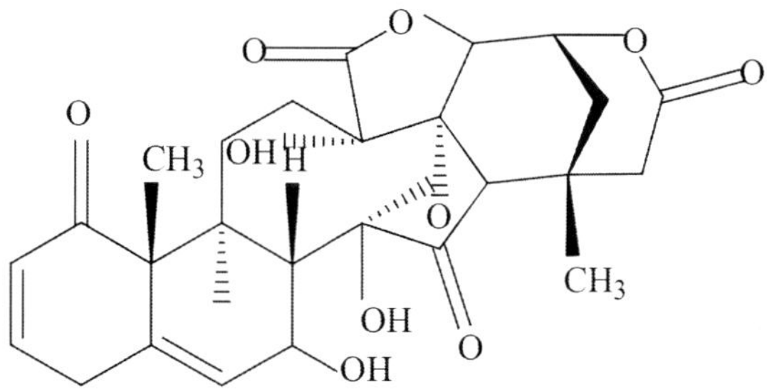

Fig. 29.26: Structure of physalin B.

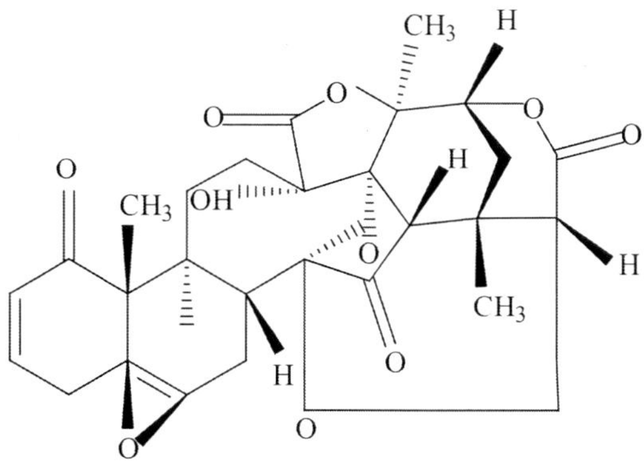

Fig. 29.27: Structure of physalin F.

29.21.4 Withaphysalins

Withaphysalins are C28-steroidal lactones are structurally based on the ergostane skeleton. *Acnistus arborescens* (Veras et al. 2004; Veras et al. 2004a; Rocha et al. 2006), *Physalis minima* Linn. var. *indica* (Sahai and Kirson 1984) and *Depreabitteriana*, *D. cuyacensis* and *D. zamorae* (Casero et al. 2015) are main sources.

The structure of withaphysalin C is closely related to withaphysalin A (Fig. 29.28) (Kirson et al. 1976). Withaphysalin D (Fig. 29.30), a new withaphysalin has been reported from *Physalis minima* Linn. var. *indica* (Sahai and Kirson 1984). A new withanolide, withaphysalin E, has also been isolated from *P. minima* var. *indica* (Oshimaa et al. 1987).

Withaphysalin M and withaphysalin O, isolated from the leaves of *Acnistus arborescens* provide potent cytotoxic activity against a panel of human cancer cell lines (Veras et al. 2004). Two new epimeric withaphysalins (17S,20R,22R)-5beta,6beta: 18,20-diepoxy-4β,18-dihydroxy-1-oxowitha-24-enolide, together with the known withaphysalin F, displayed potent cytotoxic activities against several cancer cell lines with IC50 values in the range of 0.20 to 1.46 microg/mL for (17S,20R,22R)-5β,6β: 18,20-diepoxy-4β,18-dihydroxy-1-oxowitha-24-enolide and 0.89 to 8.08 microg/mL for withaphysalin F (Veras et al. 2004a).

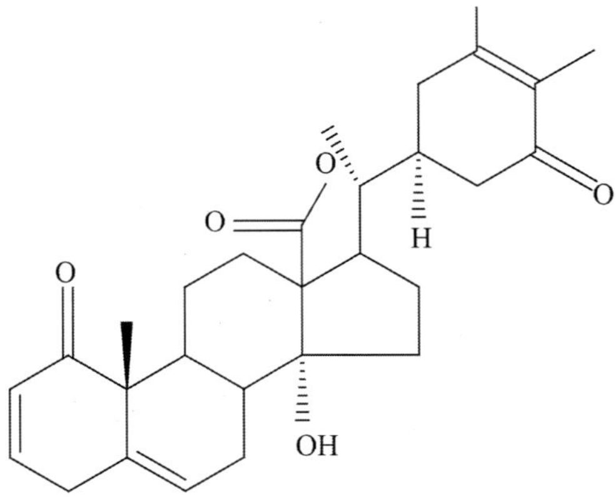

Fig. 29.28: Structure of withaphysalin A.

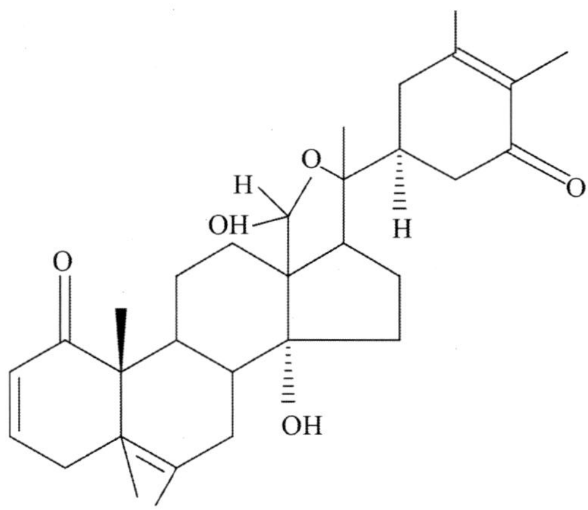

Fig. 29.29: Structure of withaphysalin B.

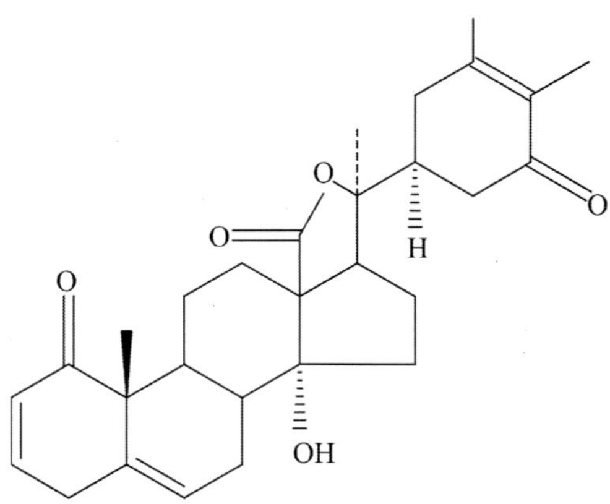

Fig. 29.30: Structure of withaphysalin D.

Withaphysalin O, M, N isolated from *A. arborescens* were evaluated for antileukemic activity against two leukemic cell lines, HL-60 and K56. Withaphysalins reduced the number of viable cells of the tumor cell lines after 24 h of exposure, except for compound 2 against the K562 cell line. The reduction was time-and concentration-dependent, and the IC(50) values ranged from 0.7 to 3.5 microM after 72 h of incubation (Rocha et al. 2006).

The antiproliferative activity of withaphysalin F and its effect in arresting cells in the G(2)/M phase of the cell cycle. These two effects are the result of the interference of withaphysalin F in the polymerization of microtubules. Withaphysalin F also induced DNA fragmentation, which can be related to an increase in mitochondrial membrane depolarization (Rocha et al. 2012).

Guinea pigs were infused with Withapysalin F (25 and 50 μg/Kg/min) and them evaluated parameters. Withaphysalin F inotropic activity appears to be dependent on the activity of RYR2 receptors regardless of increase of cAMP and activation of PKA (Nascimento et al. 2013).

29.21.5 Withametelins

Withametelins are hexacyclic withanolides isolated from the leaves of *Datura metel* (Sinha et al. 1987; Sinha et al. 1989). Withametelins I–P, 1,10-seco-withametelin B, and 12β-hydroxy-1,10-seco-withametelin B, have been isolated from methanol extract of the flowers of *D. metel.* Withametelins I, K, L, and N exhibited cytotoxic activities against A549 (lung), BGC-823 (gastric), and K562 (leukemia) cancer cell lines, with IC50 values ranging from 0.05 to 3.5 Mm (Pan et al. 2007).

29.21.6 Withawrightolide

A new withanolide, named withawrightolide (Fig. 29.31), and four known withanolides were isolated from the aerial parts of *Datura wrightii*. Withawrightolide,

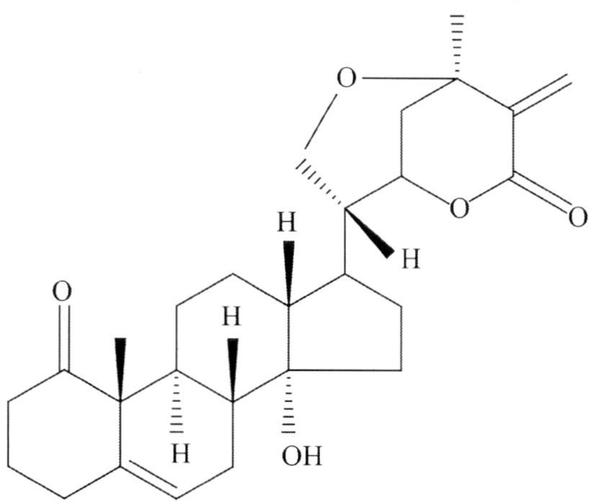

Fig. 29.31: Structure of withawrightolide.

and four withanolides showed antiproliferative activities against human glioblastoma (U251 and U87), head and neck squamous cell carcinoma (MDA-1986), and normal fetal lung fibroblast (MRC-5) cells with IC50 values in the range between 0.56 and 5.6 Mm (Zhang et al. 2013).

Further Readings

Abdeljebbar LH, Benjouad A, Morjani H, Merghoub N, El Haddar S, Humam M, Christen P, Hostettmann K, Bekkouche K, Amzazi S. Antiproliferative effects of withanolides from *Withania adpressa*. *Therapie* 2009; **64**: 121–7.

Beg M, Chauhan P, Varshney S, Shankar K, Rajan S, Saini D, Srivastava MN, Yadav PP, Gaikwad AN. A withanolide coagulin-L inhibits adipogenesis modulating Wnt/β-catenin pathway and cell cycle in mitotic clonal expansion. *Phytomedicine* 2014; **21**: 406–14.

Cao CM, Kindscher K, Gallagher RJ, Zhang H, Timmermann BN. Analysis of major Withanolides in *Physalis longifolia* Nutt. by HPLC-PDA. *J Chromatogr Sci* 2014. pii: bmu162. [Epub ahead of print]

Cao CM, Zhang H, Gallagher RJ, Day VW, Kindscher K, Grogan P, Cohen MS, Timmermann BN. Withanolides from *Physalis hispida*. *J Nat Prod* 2014a; **77**: 631–9.

Casero CN, Oberti JC, Orozco CI, Cárdenas A, Brito I, Barboza GE, Nicotra VE. Withanolides from three species of the genus Deprea (Solanaceae). Chemotaxonomical considerations. *Phytochemistry* 2015; **110**: 83–90.

Cirigliano AM, Veleiro AS, Misico RI, Tettamanzi MC, Oberti JC, Burton G. Withanolides from *Jaborosa laciniata*. *J Nat Prod* 2007; **70**: 1644–6.

El Bouzidi L, Mahiou-Leddet V, Bun SS, Larhsini M, Abbad A, Markouk M, Fathi M, Boudon M, Ollivier E, Bekkouche K. Cytotoxic withanolides from the leaves of Moroccan *Withania frutescens*. *Pharm Biol* 2013; **51**: 1040–6.

Gutiérrez Nicolás F, Reyes G, Audisio MC, Uriburu ML, Leiva González S, Barboza GE, Nicotra VE. Withanolides with antibacterial activity from *Nicandra john-tyleriana*. *J Nat Prod* 2015; **78**: 250–7.

He H, Zang LH, Feng YS, Wang J, Liu WW, Chen LX, Kang N, Tashiro S, Onodera S, Qiu F, Ikejima T. Physalin A induces apoptotic cell death and protective autophagy in HT1080 human fibrosarcoma cells. *J Nat Prod* 2013;**76**: 880–8.

He H, Feng YS2, Zang LH, Liu WW, Ding LQ, Chen LX, Kang N, Hayashi T, Tashiro S, Onodera S, Qiu F, Ikejima T. Nitric oxide induces apoptosis and autophagy; autophagy down-regulates NO synthesis in physalin A-treated A375-S2 human melanoma cells. *Food Chem Toxicol* 2014; **71**:128–35.

Ji L, Yuan Y, Ma Z, Chen Z, Gan L, Ma X, Huang D. Induction of quinone reductase (QR) by withanolides isolated from *Physalis pubescens* L. (Solanaceae). *Steroids* 2013; **78**: 860–5.

Kim KH, Choi SU, Choi SZ, Son MW, Lee KR. Withanolides from the rhizomes of *Dioscorea japonica* and their cytotoxicity. *J Agric Food Chem* 2011; **59**: 6980–4.

Kirson I, Zaretskii Z, Glotter E. Withaphysalin C, a naturally occurring 13,14-seco-steroid. *J Chem Soc Perkin Trans* 1976; **1**: 1244–1247.

Llanos GG, Araujo LM, Jiménez IA, Moujir LM, Vázquez JT, Bazzocchi IL. Withanolides from *Withania aristata* and their cytotoxic activity. *Steroids* 2010; **75**: 974–81.

Ma CY, Williams ID, Che CT. Withanolides from hyoscyamus niger seeds. *J Nat Prod* 1999; **62**: 1445–7.

Ma YM, Han W, Li J, Hu LH, Zhou YB. Physalin B not only inhibits the ubiquitin-proteasome pathway but also induces incomplete autophagic response in human colon cancer cells *in vitro*. *Acta Pharmacol Sin* 2015; **36**: 517–27.

Maher S, Rasool S, Mehmood R, Perveen S, Tareen RB. Trichosides A and B, new withanolide glucosides from *Tricholepis eburnea*. *Nat Prod Res* 2015; **13**: 1–6.

Nascimento NR, Amorim LS, Gomes VM, Santos CF, Fonteles MC. Withapysalin induced cardiac positive inotropism: *in vitro* and *in vivo* studies. *The Faseb Jl* 2013; **27**: 879.

Niero R, Da Silva IT, Tonial GC, Santos Camacho BD, Gacs-Baitz E, Monache GD, Monache FD. Cilistepoxide and cilistadiol, two new withanolides from *Solanum sisymbiifolium*. *Nat Prod Res* 2006; **20**: 1164–8.

Oshimaa Y, Bagchia A, Hikinoa H, Sinha SC, Ray AB. Withaphysalin E, a withanolide of *Physalis minima* var. *indica*. *Phytochemistry* 1987; **26**: 2115–2117.

Pan Y, Wang X, Hu X. Cytotoxic Withanolides from the flowers of *Datura metel*. *J Nat Prod* 2007; **70**: 1127–1132.

Pramanick S, Roy A, Ghosh S, Majumder HK, Mukhopadhyay S. Withanolide Z, a new chlorinated withanolide from *Withania somnifera*. *Planta Med* 2008; **74**: 1745–8.

Ramacciotti NS, Nicotra VE. Withanolides from *Jaborosa kurtzii*. *J Nat Prod* 2007; **70**: 1513–5.

Riaz N, Malik A, Aziz-ur-Rehman, Nawaz SA, Muhammad P, Choudhary MI. Cholinesterase-inhibiting withanolides from *Ajuga bracteosa*. *Chem Biodivers* 2004; **1**: 1289–95.

Rocha DD, Militão GC, Veras ML, Pessoa OD, Silveira ER, Alves AP, de Moraes MO, Pessoa C, Costa-Lotufo LV. Selective cytotoxicity of withaphysalins in myeloid leukemia cell lines versus peripheral blood mononuclear cells. *Life Sci* 2006 27; **79**: 1692–701.

Rocha DD, Balgi A, Maia AI, Pessoa OD, Silveira ER, Costa-Lotufo LV, Roberge M, Pessoa C. Cell cycle arrest through inhibition of tubulin polymerization by withaphysalin F, a bioactive compound isolated from *Acnistus arborescens*. *Invest New Drugs* 2012; **30**: 959–66.

Sahai M, Kirson I. Withaphysalin D, a new withaphysalin from *Physalis minima* Linn. var. *indica*. *J Nat Prod* 1984; **47**: 527–529.

Siddique AA, Joshi P, Misra L, Sangwan NS, Darokar MP. 5,6-de-epoxy-5-en-7-one-17-hydroxy withaferin A, a new cytotoxic steroid from *Withania somnifera* L. Dunal leaves. *Nat Prod Res* 2014; **28**: 392–8.

Sinha SC, Kundu S, Maurya R, Ray AB, Oshima Y, Bagchi A, Hikino H. Withametelin, a hexacyclic withanolide of *Datura metel*. *Tetrahedron Letters* 1987; **28**: 2025–2027.

Sinha SC, Kundu S, Maurya R, Ray AB, Oshima Y, Bagchi A, Hikino H. Structures of withametelin and isowithametelin, withanolides of leaves of *Datura metel*. *Tetrahedron* 1989; **45**: 2165–2176.

Suleiman RK, Zarga MA, Sabri SS. New withanolides from *Mandragora officinarum*: first report of withanolides from the Genus *Mandragora*. *Fitoterapia* 2010; **81**: 864–8.

Veras ML, Bezerra MZ, Lemos TL, Uchoa DE, Braz-Filho R, Chai HB, Cordell GA, Pessoa OD. Cytotoxic withaphysalins from the leaves of *Acnistus arborescens*. *J Nat Prod* 2004; **67**: 710–3.

Veras ML, Bezerra MZ, Braz-Filho R, Pessoa OD, Montenegro RC, do O Pessoa C, de Moraes MO, Costa-Lutufo LV. Cytotoxic epimeric withaphysalins from leaves of *Acnistus arborescens*. *Planta Med* 2004a; **70**: 551–5.

Vermillion K, Holguin FO, Berhow MA, Richins RD, Redhouse T, O'Connell MA, Posakony J, Mahajan SS, Kelly SM, Simon JA. Dinoxin B, a withanolide from Datura inoxia leaves with specific cytotoxic activities. *J Nat Prod* 2011; **74**: 267–71.

Wu SY, Leu YL, Chang YL, Wu TS, Kuo PC, Liao YR, Teng CM, Pan SL. Physalin F induces cell apoptosis in human renal carcinoma cells by targeting NF-kappaB and generating reactive oxygen species. *PLoS One* 2012; **7**: e40727.

Zhang H, Bazzill J, Gallagher RJ, Subramanian C, Grogan PT, Day VW, Kindscher K, Cohen MS, Timmermann BN. Antiproliferative withanolides from *Datura wrightii*. *J Nat Prod* 2013; **76**: 445–9.

Zhang H, Cao CM, Gallagher RJ, Day VW, Montenegro G, Timmermann BN. Withanolides from *Jaborosa caulescens* var. *bipinnatifida*. *Phytochemistry* 2014; **98**: 232–5.

Zhu XH, Ando J, Takagi M, Ikeda T, Nohara T. Six new withanolide-type steroids from the leaves of *Solanum cilistum*. *Chem Pharm Bull* (Tokyo) 2001; **49**: 161–4.

Chapter 30

Sitoindosides

30.1 Introduction

Sitoindosides are glycowithanolides obtained from *Withania somnifera* Dunal (Ghosal et al. 1988). A sitoindoside is a withanolide containing a glucose molecule at carbon 27. The sitoindosides IX (withaferin-A-C_{27}*O*-β-D-glucoside) and X(6'-O-palmitoyl-withaferin-A-C_{27}*O*-β-D-glucoside), represent C-27-glycowithanolides, the sitoindosides VII (3β-Hydroxyergost-5,24-diene 3-O-[6′-O-palmitoyl-β-D-glucopyranoside) (Fig. 30.1) and VIII (Fig. 30.2), acyl-esterylglucosides. The sitoindosides VII, VIII, IX, and X represent the adaptogenic active substances of *W. somnifera*, in spite of diverse steroidal structures (Bhattacharya et al. 1988; Ghosal et al. 1989).

30.2 Pharmacology

30.2.1 Antistress

The total MeOH-H2O (1:1) extractives of the roots of *W. somnifera* and equimolecular combination of sitoindosides VII, VIII, and withaferin-A exhibited significant anti-

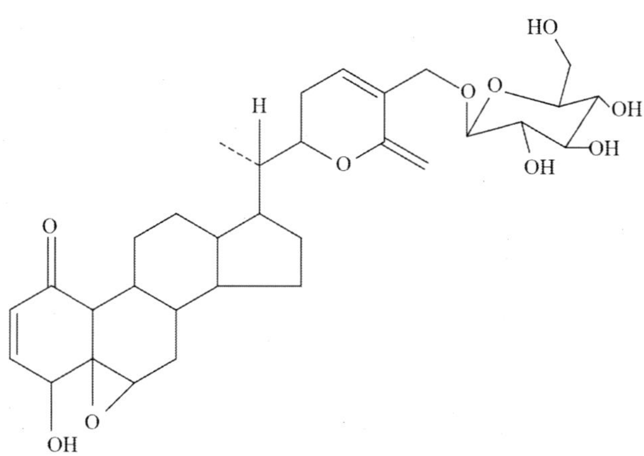

Fig. 30.1: Structure of sitoindoside IX (withaferin-A-C_{27}*O*-β-D-glucoside).

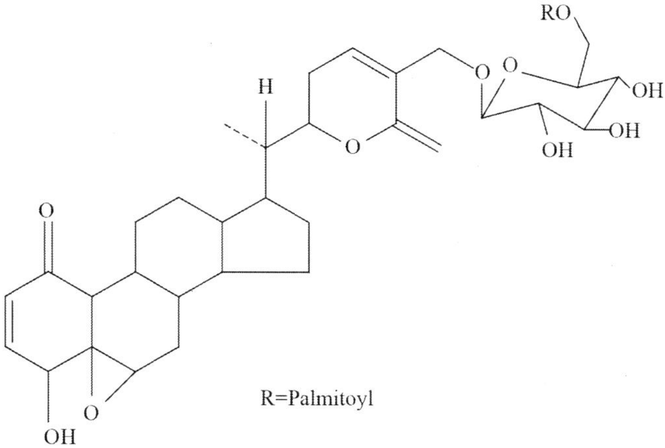

Fig. 30.2: Structure of sitoindoside X (6'-O-palmitoyl-withaferin-A-C_{27}*O*-β-D-glucoside).

stress activity in all the test parameters used. Sitoindosides VII and VIII displayed *per se* anti-stress activity which was potentiated by withaferin-A. A preliminary acute toxicity study indicated that sitoindosides VII, VIII, and withaferin-A have a low order of acute toxicity (Bhattacharya et al. 1988).

30.2.2 Immunomodulatory and CNS effects

Sitoindoside IX (Fig. 30.1) and sitoindoside X (Fig. 30.2), isolated from *W. somnifera* in doses of 100–400 μg/mouse, produced statistically significant mobilization and activation of peritoneal macrophages, phagocytosis, and increased activity of the lysosomal enzymes secreted by the activated macrophages. Sitoindoside IX and sitoindoside X (50–200 mg/kg p.o.) produced significant anti-stress activity in albino mice and rats and augmented learning acquisition and memory retention in both young and old rats (Ghosal et al. 1989).

30.2.3 Antioxidant

Equimolar concentrations of sitoindosides VII–X and withaferin A (10 and 20 mg/kg, i.p.), administered once daily for 21 days, induced a dose-related increase in SOD, CAT, and GPX activity in frontal cortex and striatum, which was statistically significant on days 14 and 21, except with the lower dose of WSG on GPX activity, where the effect was evident only on day 21 (Bhattacharya et al. 1997).

30.2.4 Hepatoprotective

In the present study, the effect of 10 days of oral administration of sitoindosides VII–X and withaferin A, in graded doses (10, 20 and 50 mg/kg), was noted on iron overload (FeSo(4), 30 mg/kg, i.p.) induced hepatotoxicity in rats. Silymarin (20 mg/kg, p.o.) was used for comparison. Iron overload induced marked increase in hepatic LPO and serum levels of the enzymes, which was attenuated by sitoindosides

VII–X and withaferin A in a dose-related manner, and by silymarin (Bhattacharya et al. 2000).

Further Readings

Bhattacharya A, Ramanathan M, Ghosal S, Bhattacharya SK. Effect of *Withania somnifera* glycowithanolides on iron-induced hepatotoxicity in rats. *Phytother Res* 2000; **14**: 568–70.

Bhattacharya SK, Goel RK, Kaur R, Ghosal S. Anti-stress activity of sitoindosides VII and VIII, new acylsterylglucosides from *Withania somnifera*. *Phytother Res* 1987; **1**: 32–37.

Bhattacharya SK, Satyan KS, Ghosal S. Antioxidant activity of glycowithanolides from *Withania somnifera*. *Indian J Exp Biol* 1997; **35**: 236–9.

Ghosal S, Kaur R, Bhattacharya SK. Chemistry and bioactivity of sitoindosides IX and X. *Planta Med* 1988; **54**: 561.

Ghosal S, Lal J, Srivastava R, Bhattacharya SK, Upadhyay SN, Jaiswal AK, Chattopadhyay U. Immunomodulatory and CNS effects of sitoindosides IX and X, two new glycowithanolides from *Withania somnifera*. *Phytother Res* 1989; **3**: 201–206.

Chapter 31

Pyrrolizidine Alkaloids

31.1 What are Pyrrolizidine Alkaloids?

Pyrrolizidine alkaloids, abbreviated as PAs, are sometimes called as necine bases. Pyrrolizidine alkaloids are naturally occurring alkaloids, based on the structure of pyrrolizidine (Fig. 31.1). Chemically, pyrrolizidine alkaloids are esters composed of 1-hydroxymethylpyrrolizidin (necine base) and aliphatic mono or dicarbon acids-necine acids (Bull et al. 1968; Mattocks 1986).

More than 500 different PAs are found in over 6,000 plant species (Stegelmeier 2011). They occur in some range plants that animals eat (Wiedenfeld et al. 2002). PA may enter the human food supply if cereal crops are contaminated with weeds containing the PA, or in small amount in meat and milk of animals ingesting the PA-containing plants. PA are also found in some herbal teas and herbal medicine preparations (Smith and Culvenor 1981).

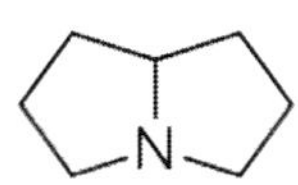

Fig. 31.1: Chemical structure of pyrrolizidine.

31.2 Physical and Chemical Properties

The pure alkaloids are mostly crystalline solids; some are gums or amorphous solids. Some are only slightly soluble in water, but all dissolve when neutralized with acid. They occur in the plants partly as *N*-oxides, which are water soluble (Mattocks 1986). The alkaloids are fairly stable, but are subject to hydrolysis in alkaline solution and to enzymatic decomposition. The latter occurs in some plant species during wilting and drying. The stability of the alkaloids when the plants are cooked is not known (Rizk 1990).

31.3 Botany of Medicinal Plants containing Pyrrolizidine Alkaloids

Adenostyles alliariae **(Gouan) A. Kern.**

Common name: Hedge-leaved Adenostyle.

Family: Asteraceae.

Distribution: The mountainous southern Europe. Altitude: 1,300–2,400 metres (4,300–7,900 ft) above sea level.

Botany: *A. alliariae* can reach a height of 40–70 centimetres. The inflorescence consists of dense corymbs hold by hairy peduncles. The small heads are usually composed of 3 to 4 flowers. The receptacle is naked or hairless. The flowers are of a tubular type and hermaphroditic. The corolla is cylindrical and pink violet. The length of the flower is of 7–8 mm. The period of flowering is from June until August. Basal leaves are large, kidney-shaped or heart-shaped, leaf margin is toothed. Size of leaves at the base: width 12–14 centimetres, length 6–9 centimetres. Cauline leaves are arranged in alternating fashion with successively smaller size and are petiolated. At the base of the petiole are present two large leaflets enveloping the stem.

Phytochemistry: Pyrrolizidine alkaloids: seneciphylline, spartioidine, acetyl-senciphylline, and senecionine. Seneciphylline is the main pyrrolizidine alkaloid (it accounts for more than 90% of the alkaloid fraction in all above ground plant parts). Inflorescences have the highest alkaloid contents with 21.1 mg/g. Stems and leaves had 2–3 times lower contents (Chizzola 2015).

Adenostyles glabra (Miller) DC

Common name: Alpine plantain.

Family: Asteraceae.

Distribution: The mountainous southern Europe. Altitude: 1,300–2,400 metres (4,300–7,900 ft) above sea level.

Botany: *A. glabra* grows to a height of about 60 centimetres. The inflorescence consists of dense corymbs hold by hairy peduncles. The small heads are usually composed of 3 to 4 flowers. The receptacle is naked or hairless. The flowers are of a tubular type and hermaphroditic. The corolla is cylindrical and pink violet. The length of the flower is of 7–8 mm. The period of flowering is from June until August. Basal leaves are large, kidney-shaped, leaf margin is toothed. The leaves are glabrous on both sides. Size of leaves at the base: width 12–14 centimetres, length 10–11 centimetres. Cauline leaves are arranged in alternating fashion with successively smaller size and are petiolated. Size of lower cauline leaves: width 10–14 centimetres, length 6–8.

Phytochemistry: Pyrrolizidine alkaloids: seneciphylline, spartioidine, acetyl-senciphylline, and senecionine. Seneciphylline is the main pyrrolizidine alkaloid (it accounts for more than 90% of the alkaloid fraction in all above ground plant parts). Inflorescences have the highest alkaloid contents with 13.4 mg/g. Stems and leaves had 2–3 times lower contents (Chizzola 2015).

Amsinckia intermedia Fisch. & Mey.

Common name: Tarweed.

Family: Boraginaceae.

Distribution: Native to California.

Botany: An annual herb with bristly leaves which are lanceolate to linear in shape and which sometimes are finely toothed. The yellow, tabular flowers are produced in terminal spikes in mid spring.

Phytochemistry: Pyrrolizidine alkaloids: echumine, intermedine, lycopsamine, and sincamidine.

Parts used: Leaves.

Note: *A. intermedia* is toxic to livestock because of the high nitrate and alkaloid levels in the leaves. The bristles can cause skin irritation.

Ageratum houstonianum Mill.

Common name: Blue Ageratum, Bluemink, Blueweed, Flossflower, Mexican paintbrush, Pussy foot.

Family: Asteraceae.

Distribution: Native to Central America.

Botany: *A. glabra* grows to a height of about 60 centimetres. The inflorescence consists of dense corymbs held by hairy peduncles. The small heads are usually composed of 3 to 4 flowers. The receptacle is naked or hairless. The flowers are of a tubular type and hermaphroditic. The corolla is cylindrical and pink violet. The length of the flower is of 7–8 mm. The period of flowering is from June until August. Basal leaves are large, kidney-shaped, leaf margin is toothed. The leaves are glabrous on both sides. Size of leaves at the base: width 12–14 centimetres, length 10–11 centimetres. Cauline leaves are arranged in alternating fashion with successively smaller size and are petiolated. Size of lower cauline leaves: width 10–14 centimetres, length 6–8.

Phytochemistry: Essential oil (Kurade et al. 2001), and pyrrolizidine alkaloids including lycopsamine (Wiedenfeld and Andrade-Cetto 2001).

Pharmacology: Antibacterial (Kurade et al. 2001).

Arnebia euchroma (Royle) Johnston

Family: Boraginaceae.

Distribution: Open slopes and rocks, only in the drier areas of the Himalayas, 3300–4500 metres.

Botany: An erect perennial, hairy herb; leaves linear with white bristles; flowers pink/purple in clusters.

Phytochemistry: Pyrrolizidine alkaloids: 0^9-angeloylretronecine-95% and 0^7-angeloyl-retronecine-5% (Röder and Rengel-Mayer 1993), naphthoquinone:

shikonin, naphthazarins (Shen et al. 2002), a new naphthoquinone dimer: arnebiabinone, a new phenolic compound: ethyl 9-(2',5'-dihydroxyphenyl) nonanoate), and a new natural product, octyl ferulate (Liu et al. 2010).

Actions: Antipyretic, anti-cancer, contraceptive, emollient, and vulnerary.

Therapeutics: Measles, mild constipation, burns, frostbite, eczema, and dermatitis. In cold desert Ladakh, *A. euchroma* is used in the treatment of cold, cough, and fever (Ballabh and Chaurasia 2007).

Parts used: Roots.

Pharmacology: Anti-inflammatory (Kaith et al. 1996) and wound-healing (Ashkani-Esfahani et al. 2012; Nasiri et al. 2015; Nasiri et al. 2015a).

Borago officinalis L.

Common name: Borage.

Family: Boraginaceae.

Distribution: Most parts of Europe.

Botany: The whole plant is rough with white, stiff, prickly hairs. The round stems, about 1 1/2 feet high, are branched, hollow, and succulent; the leaves alternate, large, wrinkled, deep green, oval, and pointed, 3 inches long or more, and about 1 1/2 inch broad, the lower ones stalked, with stiff, one celled hairs on the upper surfaces and on the veins below, the margins entire, but wavy. The flowers, which terminate the cells, are bright blue and star-shaped, distinguished from those of every plant in this order by their prominent black anthers, which form a cone in the centre and have been described as their beauty spot. The fruit consists of four brownish-black nutlets.

Phytochemistry: Pyrrolizidine alkaloids: amabiline and thesinine (Dodson and Stermitz 1986). PA are found in leaves and seeds. However they are absent in the borage oil (Parvais et al. 1994). The seed yields 30% oil, 20% of which is gamma-linolenic acid.

Actions: Diuretic, demulcent, and emollient.

Therapeutics: Fever and ling implications.

Parts used: Leaves.

Pharmacology: Amoebicidal (Leos-Rivas et al. 2011), antioxidant (Singh et al. 2013), antinociceptive (Shahraki et al. 2015).

Brachyglottis adamsii (Cheeseman) B. Nord.

Family: Asteraceae.

Distribution: Native to New Zealand.

Botany: A flowering plant.

Phytochemistry: Pyrrolizidine alkaloids: senecionine and retrorsine (Benn and Gul 2007).

Brachyglottis huntii (Muell.) B. Nord.

Common name: Chatham Island Christmas tree.

Family: Asteraceae.

Distribution: The Chatham Islands in New Zealand.

Botany: A flowering plant.

Phytochemistry: Pyrrolizidine alkaloids: senkirkine and retrorsine (Benn and Gul 2007).

Brachyglottis perdicioides (Hook. f.) B. Nordenstam

Family: Asteraceae.

Distribution: Native to New Zealand.

Botany: A small tree or shrub.

Phytochemistry: Pyrrolizidine alkaloids: 7-*O*-angelylheliotridine (Benn and Gul 2007).

Brachyglottis repens B. Nord.

Family: Asteraceae.

Distribution: Native to New Zealand.

Botany: A flowering plant.

Phytochemistry: Pyrrolizidine alkaloids (Benn and Gul 2007).

Therapeutics: It is used in Homeopathic System of Medicine.

Parts used: Leaves.

Note: *B. repens* is a forage crop for livestock, toxic to horses, causing paralysis of the limbs. It is used in Homeopathic System of Medicine.

Cacalia hastata L.

Syn: *Parasenecio hastatus*

Family: Asteraceae.

Distribution: E. Asia - Japan.

Botany: A perennial growing to 6 ft 7 in. It is in flower from Sep to October, and the seeds ripen from Sep to October. The flowers are hermaphrodite (have both male and female organs) and are pollinated by insects.

Phytochemistry: Coumarins, triterpenes, and pyrrolizidine alkaloids.

Therapeutics: Used as a flavouring agent.

Parts used: Leaves.

Chromolaena odorata (L.) King & H.E. Robins.

Syn: *Eupatorium odoratum* Linn.

Common name: Boneset.

Family: Asteraceae.

Distribution: Native to North America.

Botany: *C. odorata* is a woody herbaceous perennial growing as a climbing shrub to 3 meters in height, typically shorter. The leaves are arranged oppositely, to 15 cm in length, triangular to ovate with an acuminate leaf apex and dentate leaf margin with large teeth. The vegetative structures are covered with articulate hairs throughout. The actinomorphic flowers are arranged in corymbs of heads subtended an involucre made of four series of phyllaries. The calyx is modified as hairs forming a pappus. The corolla has five fused white to lavender petals. There are five stamens fused to the base of the corolla. The ovary is inferior with a single locule. The fruit is an achene at maturity that retains the modified calyx (pappus).

Phytochemistry: *N*-oxides of five pyrrolizidine alkaloids: 7- and 9-angeloylretronecine, intermedine, rinderine, and 3′-acetylrinderine. Highest concentrations occur in roots and mature flower heads, while leaves and stems are almost devoid of alkaloids, and no PAs are present in nectar (Biller et al. 1994) and essential oil (Baruah and Leclercq 1993).

Therapeutics: Wounds, rashes, and diabetes.

Parts used: Leaves.

Pharmacology: Antimicrobial (Iwu and Chiori, 1984; Ravishankar et al. 2010), anti-inflammatory (Owoyele et al. 2005), and wound healing (Odoh and Ezugwu 2008; Ravishankar et al. 2010).

Crassocephalum crepidioides (Benth.) S. Moore

Syn: *Gynura crepidioides* Benth.

Common name: Fireweed.

Family: Asteraceae.

Distribution: Africa and Madagascar.

Botany: An erect or straggling annual herb. Leaves are alternately arranged along the stem. Leaf shape is variable, smaller upper leaves are generally elliptical, larger lower leaves have two lobes at the base. All leaves have toothed margins. Flower heads droop while forming and are orange-pink. Base of plant, lower stems, some veins, and some leaves have a purple colouration. Seed has hairs (pappus) which are white and tinged reddish or mauve. Stem is round in cross-section and exudes watery liquid when cut. This plant is easily removed by pulling.

Phytochemistry: Pyrrolizidine alkaloids: jacobine.

Therapeutics: A lotion of the leaves is used as a mild medicine that strengthens the stomach and excites its action.

Parts used: Leaves and stem.

Pharmacology: Free radical scavenging and hepatoprotective (Aniya et al. 2005) and antitumor (Tomimori et al. 2012).

Crotalaria albida Heyne ex Roth

Syn: *C. deflexa* Benth.

Family: Fabaceae.

Distribution: Native in India, China, and many countries in South East Asia from Pakistan to the Philippines and extending south to Papua New Guinea and north to Nepal.

Botany: A much-branched under shrub, 30–60 cm high, branches slender, terete. Leaves small, 2.5–5 cm long, linear-oblong or oblanceolate, more or less silky-pubescent. Flowers in terminal 6–20 flowered racemes, 5–12.5 cm long; corolla 4 mm long, pale yellow. Pods sessile, oblong-cylindric, 1.2–1.6 cm long.

Phytochemistry: Pyrrolizidine alkaloids: croalbidine, neocroalbidine and neocroalbidinone (Suna et al. 2013) and isoflavonoids.

Therapeutics: It is used in treatment for warts, especially on the sole of the foot, and a juice obtained from the roots is given for indigestion.

Parts used: Leaves and flowers.

Pharmacology: Antibacterial.

Crotalaria assamica Benth.

Syn: *C. burmannii* DC.

Common name: Indian rattle box.

Family: Fabaceae.

Distribution: Native to China, India, Myanmar, the Philippines, Thailand, and Viet Nam.

Botany: Herbs, erect, to 1.5 m tall. Branches terete, sericeous. Stipules linear, minute. Leaves simple; petiole 2–3 mm; leaf blade oblanceolate to narrowly elliptic, 5–15 × 2–4 cm, thin, abaxially sericeous, adaxially glabrous, base cuneate, apex obtuse, and mucronate. Racemes terminal or leaf-opposed, to 30 cm, 20–30-flowered; bracts linear, 1–2 mm. Bracteoles similar to bracts but shorter. Calyx 2-lipped, 1–1.5 cm, pubescent; lobes lanceolate-triangular, ± as long as tube. Corolla deep golden yellow; standard suborbicular to elliptic, 1.5–2 cm, base with 2 appendages, apex retuse; wings 1.5–1.8 cm; keel rounded through 90°, narrowed apically from middle and extended into a long twisted beak exserted beyond calyx. Ovary glabrous. Legume oblong, 4–6 × ca. 1.5 cm, 20–30-seeded; stipe ca. 5 mm. Fl. May–Sep, fr. Aug–Dec.

Phytochemistry: Pyrrolizidine alkaloids: assamicadine and monocrotaline (Cui and Cheng 1989; Edgar et al. 1992), flavonoids and pterocarpanoid.

Therapeutics: A root extract is also used to treat bladder stones.

Parts used: Leaves and flowers.

Crotalaria crispata F. Muell. ex Benth.

Common name: Kimberley horse poison.

Family: Fabaceae.

Distribution: North Australia.

Botany: A low much-branched herb, under a foot high, covered with hairs. The leaves are yellow, oblong, and unifoliate.

Phytochemistry: Pyrrolizidine alkaloids: monocrotaline, fulvine, and crispatine (Culvenor and Smith 1963).

Parts used: Leaves and flowers.

Note: Walk about disease, which is also known as Kimberley horse disease, affects horses that graze *C. crispata* (Gardiner et al. 1965).

Crotalaria dura Wood & Evans

Family: Fabaceae.

Distribution: South Africa.

Botany: Virgate perennial herb or subshrub with a number of ascending stems to 60 cm tall, branched above, appressed pubescent to mentellous. Leaves shortly petiolate, 3-foliolate; leaflets mostly 8–25 × 3–10 mm, narrowly oblong-elliptic to elliptic or oblanceolate to obovate, appressed pubescent to sericeous at least beneath; stipules narrow, variously developed. Racemes mostly well developed, with a number of flowers laxly to rather closely arranged towards the top; bracts 2–6 mm long, linear to linear-lanceolate; bracteoles on upper part of the pedicel, small, filiform.

Calyx 6–7 mm long, appressed pubescent; lobes narrowly attenuate-triangular, longer than the tube. Standard elliptic or elliptic-obovate, yellow, glabrous or with small scattered hairs near the apex outside; wings a little shorter than the keel; keel 7.5–8.5 mm long, strongly rounded, and crested behind the small beak. Pod 8–11 mm long, shortly stipitate, obliquely obovoid to oblong-ellipsoid, sericeous, 2–4-seeded. Seeds c. 3.5 mm long, oblique-cordiform, smooth, brown.

Phytochemistry: Pyrrolizidine alkaloids.

Parts used: Leaves and flowers.

Note: *C. dura* causes pulmonary and liver diseases in horses and cirrhosis of the liver in cattle.

Crotalaria globifera E. Mey.

Family: Fabaceae.

Distribution: South Africa.

Botany: An herb with yellow flowers.

Phytochemistry: Pyrrolizidine alkaloids: globiferine, trichodesmine, grantaline, and grantianine (Brown et al. 1984).

Parts used: Leaves and flowers.

Crotalaria mucronata Desv.

Syn: *C. striata* DC., *C. pallida* Ait., *C. saltiana.*

Family: Fabaceae.

Distribution: Australia.

Botany: A perennial herb.

Phytochemistry: Pyrrolizidine alkaloids and flavanoids (Subramanian and Nagarajan 1969).

Parts used: Leaves and flowers.

Note: *C. mucronata* is toxic to sheep.

Crotalaria retusa L.

Syn: *C. cuneifolia* (Forssk.) Schrank.

Common name: Devil bean.

Family: Fabaceae.

Distribution: Tropical Asia, Africa and Australia.

Botany: The stems are erect, slightly ridged, and pubescent. The leaves are alternate, simple, oblanceolate, up to 9 cm long, and 1–4 cm in width. The flowers are borne in a terminal raceme; they are typical pea flowers, yellow with fine purple lines near the base. The seed pods are inflated, green, maturing to dark brown or black, 3–4 cm long, with the 20-or-so tan to black seeds in each pod.

Phytochemistry: *P*yrrolizidine alkaloid: monocrotaline (Adams and Rogers 1939) and flavonoids (Subramanian and Nagarajan 1969).

Therapeutics: In Africa, roots are used against coughing up blood. Leaves are used to treat fever, scabies, lung diseases, and impetigo.

Parts used: Leaves and flowers.

Note: *C. retusa* causes acute intoxication in sheep (Nobre et al. 2005).

Crotalaria pallida Aiton

Common name: Smooth rattlebox.

Family: Fabaceae.

Distribution: Tropical Africa.

Botany: *C. pallida* is an erect, well-branched, sometimes robust perennial herb with stems that become more or less woody. It can grow up to 2 metres tall.

Phytochemistry: Pyrrolizidine alkaloid: monocrotaline.

Actions: Febrifuge.

Therapeutics: Fever, urinary diseases, and thrush.

Parts used: Whole plant.

Crotalaria sesseliflora L.

Family: Fabaceae.

Distribution: Central and southern Japan.

Botany: An annual herb.

Phytochemistry: Pyrrolizidine alkaloids: monocrotaline, integerrimine, and trichodesmine (Räder et al. 1992) and flavonoids (Subramanian and Nagarajan 1969).

Therapeutics: A paste of the plant is applied as a poultice to treat headaches.

Parts used: Seeds.

Crotalaria spectabilis Roth

Common name: Showy rattlebox.

Family: Fabaceae.

Distribution: Native to the Indo-Malaysian area.

Botany: An erect, summer annual.

Phytochemistry: Pyrrolizidine alkaloids: monocrotaline (Tinker and Lauter 1956).

Parts used: Flowers and leaves.

Note: *C. spectabilis* is toxic to horses and cattle.

Crotalaria tetragona Roxb. ex Andr.

Common name: Eastern Rattlepod.

Family: Fabaceae.

Distribution: The Himalayas, from Kumaun to Bhutan, Assam, SE Asia, and China, at altitudes of 200–1700 m.

Botany: An erect herb, up to 2 m tall. Branches are 4-angled, silky velvety. Stipules are linear to linear-lance shaped, 4–5 mm. Leaves are simple, carried on about 4 mm long stalks. Leaves are oblong-elliptic to linear-lance shaped, 10–20 cm long, 1–2.5 cm wide, both surfaces finely velvety, mid vein pale, and prominent on the underside. Flowers are borne in racemes at branch ends or leaf-opposed, 6–10-flowered. Flowers are yellow, pea-shaped. Standard petal is circular to oblong, about 2.5 cm, base with 2 appendages. Seed-pod is oblong, 4–5 cm, 10–20-seeded, densely brownish yellow pubescent.

Phytochemistry: Pyrrolizidine alkaloids.

Parts used: Leaves.

Cryptantha crassipes I.M. Johnst.

Common name: Terlingua Creek cat's-eye.

Family: Boraginaceae.

Distribution: Endemic to Brewster County, Texas.

Botany: A perennial herb producing several erect stems reaching a maximum height around 25 centimeters. There is a clump of basal leaves around the stem bases. The herbage is covered in silvery soft and bristly hairs. The inflorescence is a head of yellow-throated white flowers.

Phytochemistry: N-oxides of lycopsamine and intermedine (El-Shazly et al. 1956).

Parts used: Leaves.

Cynoglossum amabile Stapf & J.R. Drumm.

Family: Boraginaceae.

Distribution: Native to Asia.

Botany: An annual reaching around 12" in height, with dark greyish-green pointed leaves and rounded bright blue flowers in summer.

Phytochemistry: Pyrrolizidine alkaloids: supinine, amabiline, rinderine, echinatine, and 3′-*O*-acetylechinatine (El-Shazly et al. 1956).

Therapeutics: The root has been used in the treatment of stomach aches and venereal diseases.

Parts used: Roots.

Cynoglossum furcatum Wall. ex Roxb.

Common name: Ceylon hound's tongue.

Family: Boraginaceae.

Distribution: Afghanistan, India, Japan, Malaysia, Pakistan, Philippines, Thailand, Vietnam.

Botany: Herbs erect, 40–60 cm tall. Stems single or several and cespitose, densely yellowish brown strigose. Basal and lower stem leaves long petiolate, oblong to oblong-lanceolate, 15–20 × 3–5 cm, densely appressed pubescent, base attenuate, apex obtuse; upper stem leaves sessile, smaller. Inflorescences terminal and axillary, subdichotomously branching; branches spreading at obtuse angle, ebracteate. Nutlets ovoid-globose, 2–3 × 1.5–2.5 mm, abaxially concave, with dense glochids, margin wingless, or winged below middle.

Phytochemistry: Pyrrolizidine alkaloids: isoechinatine, lactodine, and viridinatine (Ravi et al. 2008).

Therapeutics: The root has been used in the treatment of stomach aches and venereal diseases.

Parts used: Roots.

Cynoglossum lanceolatum Forssk.

Common name: Chinese forget-me-not.

Family: Boraginaceae.

Distribution: Native to Asia.

Botany: Annual or biennial herb, the taproot 1–8 mm in diam.; stems erect, to c. 1 m tall, with sparse to moderate, appressed to spreading pubescence. Basal leaves in an evident rosette or smaller plants apparently immediately erect and lacking a basal rosette. Inflorescences terminal, once to several times dichotomously branched cymes, the branches strigillose; flowers on pedicels 1–7 mm long, bisexual; sepals narrowly ovate. Fruits 4.5–5.5 mm broad; nutlets ovoid, 2–3 mm broad.

Phytochemistry: Pyrrolizidine alkaloids: cynaustraline and cynaustine, supinine, amabiline, rinderine, echinatine, and 3′-*O*-acetylechinatine (El-Shazly et al. 2012).

Therapeutics: Acute nephritis, periodontitis, acute submandibular lymphadenitis, and snake bite.

Parts used: Whole plant.

Pharmacology: Diuretic, anti-inflammatory, and analgesic (Yua et al. 2012).

Cynoglossum officinale L.

Common name: Hounds tongue.

Family: Boraginaceae.

Distribution: Europe and North America.

Botany: A biennial herb. Entire plant covered with long, soft hairs. Leaves alternate, long and narrow, smaller higher up the stem, 1 to 3 inches wide, rough, hairy, lacking teeth and lobes, with distinctive veins. Lower leaves up to a foot long and resemble a hound's tongue, broader at the tips and tapering to a petiole (leaf stalk) at the base; upper leaves are reduced, narrower, and lack petioles (stalks). Flowers dull reddish-purple, drooping slightly along slender stalks, about 1/3 inch wide. Each flower produces four nutlets (seeds), about 1/3 inch long.

Phytochemistry: Pyrrolizidine alkaloids: heliosupine, heliosupine N-oxide, 3′-acetylheliosupine, and viridiflorine (El-Shazly et al. 2012).

Actions: Sedative, calming, and slightly narcotic.

Therapeutics: Inflammations.

Parts used: Whole plant.

Note: *C. officinale* is toxic to cows and is especially dangerous to pasture owners.

Cynoglossum zeylanicum (Vahl Ex Hornem) Thunb. Ex. Lehm.

Family: Boraginaceae.

Distribution: Native to Asia.

Botany: Biennial, perennial, or annual herbs. Cymes elongating in fruit, usually terminal, ebracteate, Calyx 5-partite, lobes in fruit stellately spreading. Corolla rotate or funnel-form with a short tube. Throat scales present. Stamens included; anthers small. Nutlets 4, ovoid to ± globose. Dorsal surface depressed forming a raised margin or not. Margin glochidiate. Gynobase narrow conical to ± pyramidal, with or without awn coherent to the persistent style.

Phytochemistry: Pyrrolizidine alkaloids.

Therapeutics: Decoction prepared from the whole plant is used to arrest vomiting by the Badaga community in the Nilgiri Biosphere Reserve, Tamil Nadu.

Parts used: Whole plant.

Pharmacology: Antihyperglycemic (Anitha et al. 2011), antihyperlipidameic and antioxidant (Anitha et al. 2012), hepatoprotective (Anitha et al. 2012a), antitumor (Anitha et al. 2012b) and anti-Inflammatory (Anitha et al. 2013).

Echium plantagineum L.

Common name: Purple viper's-bugloss.

Family: Boraginaceae.

Distribution: Native to western and southern Europe, northern Africa, and southwestern Asia.

Botany: *E. plantagineum* is a winter annual plant growing to 20–60 cm tall, with rough, hairy, lanceolate leaves up to 14 cm long. The flowers are purple, 15–20 mm long, with all the stamens protruding, and borne on a branched spike.

Phytochemistry: Pyrrolizidine alkaloids occurred mainly as their N-oxides (Colegate et al. 2005).

Therapeutics: An infusion of the plant is taken internally as a diuretic and in the treatment of fevers, headaches, and chest conditions.

Parts used: Whole plant.

Echium vulgare L.

Common name: Viper's-bugloss.

Family: Boraginaceae.

Distribution: Native to Europe, and western and central Asia.

Botany: *E. vulgare* is a biennial or monocarpic perennial plant growing to 30–80 cm tall, with rough, hairy, lanceolate leaves. The flowers start pink and turn vivid blue and are 15–20 mm in a branched spike, with all the stamens protruding. The pollen is blue but the filaments of the stamens remain red, contrasting against the blue flowers.

Phytochemistry: Pyrrolizidine alkaloids: echimidine, acetylechimidine, uplandicine, 9-O-angelylretronecine, echiuplatine, leptanthine, and echimiplatine, one unidentified (echivulgarine) (Boppré et al. 2005) and allantion in roots.

Therapeutics: An infusion of the plant is taken internally as a diuretic and in the treatment of fevers, headaches, and chest conditions.

Parts used: Leaves and roots.

Emilia sonchifolia (L.) DC. ex Wight

Common name: Tassel flower.

Family: Asteraceae.

Distribution: Native to Asia and naturalized in Africa, Australia, and the Americas.

Botany: *E. sonchifolia* is a branching, perennial herb up to 40 cm tall. Leaves are lyrate-pinnatilobed, up to 10 cm long, sometimes becoming purplish as they get old. One plant can produce several pink or purplish flower heads.

Phytochemistry: Pyrrolizidine alkaloids: senecionine, seneciphylline integerrimine, senkirkine, otosenine, neosenkirkine, petasitenine, acetylsenkirkine, desacetyldoronine, acetylpetasitenine, and doronine (Hsieha et al. 2015) and flavonoids (Cibin et al. 2006).

Therapeutics: Dysenetry, treating eye inflammations, night blindness, sore ears, infantile tympanites, bowl complaints, diarrhoea.

Parts used: Whole plant.

Pharmacology: Antioxidant and anti-inflammatory (Shylesh and Padikkala 1999), anti-tumour (Shylesh and Padikkala, 2000; Shylesh et al. 2005), antioxidant and antiproliferative (Cibin et al. 2006).

Eupatorium cannabinum L.

Common name: Hemp-agrimony, Fever wort.

Family: Asteraceae.

Distribution: Native to Europe.

Botany: A perennial herb up to 1.5 meters tall or more and 1.2 meters wide. It is dioecious, with racemes of mauve flower heads which are pollinated by insects from July to early September. The flowers are visited by many types of insects, and can be characterized by a generalized pollination syndrome. The flower heads are tiny, fluffy, and can be pale dusty pink or whitish. The fruit is an achene about 2 or 3 mm long, borne by a pappus with hairs 3 to 5 mm long, which is distributed by the wind.

Phytochemistry: Pyrrolizidine alkaloids: viridiflorine, cynaustraline, amabiline, supinine, echinatine, and rinderine (Edgar et al. 1992) and sesquiterpene lactones: eupafolin, euperfolitin, eufoliatin, eufoliatorin, euperfolide, eucannabinolide, and helenalin (Woerdenbag 1986).

Actions: Diaphoretic, bitter, laxative, tonic, anti-spasmodic, carminative, and astringent.

Therapeutics: Constipation, influenza, and rheumatism.

Parts used: Whole plant.

Pharmacology: Choleretic and hepatoprotective (Lexa et al. 1969).

Eupatorium chinense L.

Family: Asteraceae.

Distribution: Indigenous to China.

Botany: Herbs, perennial, or small shrubs or subshrubs. Leaves opposite, sessile or subsessile with petiole to 2–4 mm; median stem leaves simple or 3-lobed, ovate or broadly ovate, 4.5–10(–20) × (2–)3–5(–6.5) cm, both surfaces scabrid, white puberulent and glandular. Synflorescences terminal, of large laxly compound corymbs, 20–30 cm in diam. Capitula numerous, 5-flowered; involucre campanulate, ca. 5 mm; phyllaries 3-seriate, imbricate; outer phyllaries short, ovate or lanceolate-ovate, outside puberulent and sparsely glandular, 1–2 mm; median and inner longer, elliptic or elliptic-lanceolate, 5–6 mm, apically and marginally white, membranous, glabrous, but with yellow glands; corollas white, pink, red, or reddish purple, ca. 5 mm, with yellow glands. Achenes pale black-brown, elliptic, ca. 3 mm, 5-ribbed, yellow glandular; pappus setae white, ca. 5 mm.

Phytochemistry: Pyrrolizidine alkaloids and sesquiterpenoids: eupachifolins and eupachinilides.

Actions: Anodyne, carminative, diuretic, nervine, and vermifuge.

Therapeutics: Hot water extract is traditionally used for the treatments of cold, snakebite, and inflammation. An infusion is used in the treatment of colds, diphtheria, and rheumatoid arthritis.

Parts used: Whole plant.

Eupatorium fortunei Turcz.

Family: Asteraceae.

Distribution: Native to Asia.

Botany: *E. fortunei* is herbaceous perennial that grows 40 to 100 centimeters tall, growing from procumbent rhizomes. Plants are upright growing with green stems that are often tinted with reddish or purple dots. The stems have few branches and the inflorescence is apically branched. The flowers are in capitula, which are numerous and arranged in apical compound corymbs; inflorescence 3 to 10 cm across. The phyllaries are purple-red, they also lack hairs and glands. The florets are white to reddish in color, have 5 cm wide corolla that also lack glands. The elliptically shaped fruits are 5 angled achenes which are black-brown in color and 3–4 mm long. Pappus is white and about 5 mm long.

Phytochemistry: Pyrrolizidine alkaloids and sesquiterpenoids: eupachifolins and eupachinilides.

Actions: Prescribed as a diuretic and detoxifying drug in Chinese medicine.

Therapeutics: Nausea and poor appetite.

Parts used: Whole plant.

***Eupatorium japonicum* Thunb.**

Common name: Japanese bog orchid.

Family: Asteraceae.

Distribution: Native to China, Japan, and Korea.

Botany: *E. japonicum* is a herbaceous perennial growing 50–200 cm tall from short rhizomes with many fibrous roots. The stems are upright and marked with purplish red, ending with simple or corymbose, (flat) inflorescence that branch near their ends. The leaves are oppositely arranged on the stems and have short but rather thick petioles that are 1–2 cm long. The flowers are collected together into capitula (heads) that are densely corymbose, the inflorescence are usually 3 to 6 cm wide and rarely form large compound corymbose inflorescence that can be up to 20 cm wide. The fruits are black-brown, 5-angled, hairless achenes, that are elliptic in shape and about 3.5 mm long, and are covered with yellow glands. The 5 mm long pappus are white.

Phytochemistry: Pyrrolizidine alkaloids: viridiflorine, cynaustraline, amabiline, supinine, echinatine, and rinderine (Edgar et al. 1992) and essential oil containing thymol.

Actions: Anodyne, antibacterial, antiviral, carminative, diaphoretic, diuretic, nervine, and vermifuge.

Therapeutics: Indigestion, nausea, vomiting, diarrhoea, and colds.

Parts used: Whole plant.

***Eupatorium purpureum* L.**

Common name: Gravel root.

Family: Asteraceae.

Distribution: Native to eastern and central North America.

Botany: *E. purpureum* is a clump forming herb that grows to 1.5–2.4 meters (4.9–7.9 ft) tall and about 1.2 meters (3.9 feet) wide. Plants are found in full sun to part shade in moisture retentive to wet soils. Stems are upright, thick, round, and purple, with whorls of leaves at each node. As the plant begins to bloom the stems often bend downward under the weight of the flowers. The leaves grow to 30 cm (12 in) long and have a somewhat wrinkled texture. The purplish flowers are produced in large loose, convex shaped compound corymbiform arrays.

Phytochemistry: Pyrrolizidine alkaloids.

Actions: Diaphoretic, astringent, diuretic, nervine, and tonic.

Therapeutics: Urinary incontinence in children, cystitis, urethritis, and impotence.

Parts used: Whole plant.

Farfugium jaonicum **(L.) Kitam.**

Syn: *L. tussilaginea.*

Common name: Leopard plant.

Family: Asteraceae.

Distribution: Native to Japan.

Botany: *F. japonicum* grows in a loose clump about 60 cm tall and wide, spreading by rhizomes. Daisy-like yellow flowers, 2.5–5 cm across, are borne in loose clusters.

Phytochemistry: Pyrrolizidine alkaloids: farfugine and petasitenine.

Therapeutics: Colds and flus.

Parts used: Whole plant.

Gynura bicolor **(Roxb. ex Willd.) DC.**

Common name: Edible gynura.

Family: Asteraceae.

Distribution: Native to Japan.

Botany: A perennial plant.

Phytochemistry: Pyrrolizidine alkaloids and flavonoids (Lu et al. 2010).

Therapeutics: Post-labor recovery, blood circulation improvement, treatment of dysmenorrhea, haemoptysis, and diabetes.

Parts used: The stems and roots.

Pharmacology: Hypoglycemic (Zheng et al. 2007) and anti-inflammatory (Wu et al. 2013).

Gynura divaricata **(L.) DC**

Family: Asteraceae.

Distribution: Asia.

Botany: A perennial herb which is fleshy in its upper parts but woody and procumbent at the base, having ascending scapose or leafy flowering shoots, and ribbed stems which are usually purple tinged when dried.

Phytochemistry: Pyrrolizidine alkaloids (Roeder et al. 1996).

Therapeutics: In the treatment of opium addiction, diabetes, hypertension, herpes, inflammation, and cancer.

Parts used: The stems and roots.

Pharmacology: Antiproliferative (Chen et al. 2003).

Gynura segetum (L.) Fourr

Family: Asteraceae.

Distribution: Indonesia and Malaysia.

Botany: A perennial plant.

Phytochemistry: Pyrrolizidine alkaloids and phenolics (Lu et al. 2010).

Therapeutics: Cancer, diabetes, and hypertension.

Parts used: Leaves.

Pharmacology: Anti-inflammatory and antioxidant (Seow et al. 2014).

Heliotropium amplexicaule Vahl

Common name: Blue heliotrope.

Family: Boraginaceae.

Distribution: Native to South America.

Botany: a clumpy perennial herb growing branching, hairy stems to about half a meter in maximum height. It has abundant foliage of oblong wavy-edged green leaves four to nine centimeters long. The curving terminal spike inflorescences hold several tiny bright purple flowers with rounded lobes and tubular yellow throats. The fruits are paired rough-surfaced nutlets.

Phytochemistry: Pyrrolizidine alkaloids.

Parts used: Leaves.

Heliotropium crassifolium Boiss.

Family: Boraginaceae.

Distribution: Western region of Iran.

Botany: An annual herb.

Phytochemistry: Pyrrolizidine alkaloids: europine, europine N-oxide, ilamine, and its N-oxide (Farsam et al. 2000).

Parts used: Leaves.

Heliotropium europaeum L.

Common name: European heliotrope.

Family: Boraginaceae.

Distribution: Native to Europe, Asia, and North Africa.

Botany: An annual herb growing from a taproot and reaching maximum heights near 40 centimeters. The stem and oval-shaped leaves are covered in soft hairs. The inflorescences are coiled spikes of white flowers with fuzzy or bristly sepals. Each flower is just a few millimeters wide. The fruit is a bumpy nutlet.

Phytochemistry: Pyrrolizidine alkaloids.

Parts used: Leaves.

Heliotropium rotundifolium Lehm.

Common name: Turnsole.

Family: Boraginaceae.

Distribution: Native to Isreal.

Botany: A softly tomentose perennial plant with a woody base. Stems are erect or ascending, and terete. Leaves are orbicular, revolute margins, apiculate apex, and cuneate base. Flowers are ebracteate, sessile or subsessile with white, tubular corolla silky outside and glabrous inside and are arranged in terminal, spike-like, simple or branched inflorescence. The fruit is composed of four ovoid, glabrous, minutely tuberculate nutlets.

Phytochemistry: Pyrrolizidine alkaloids: uropine, heliotrine, lasiocarpine, and 5'-acetyleuropine (Asibal et al. 1989).

Parts used: Leaves.

Lappula intermedia (Ledebour) Popov in Komarov

Family: Boraginaceae.

Distribution: China.

Botany: Herbs annual. Taproots stout, conical, ca. 7 cm. Stems erect, usually single, much branched above middle, to 60 cm tall, gray hispid; branches ascending. Inflorescences terminal on stem and branches, to 5–20 cm in fruit; bracts leaf like, slightly longer than fruit, apex gradually reduced. Fruit broadly ovoid to subglobose, ca. 3 mm; nutlets broadly ovoid, 2.5–3 mm, granulose, adaxially wrinkled.

Phytochemistry: Sesquiterpene lactones, phenol derivatives and Pyrrolizidine alkaloids.

Parts used: Leaves.

Lappula squarrosa (Retz.) Dumort.

Common name: European stick seed.

Family: Boraginaceae.

Distribution: Native to Europe and Asia.

Botany: An annual herb producing an erect stem often with sprays of many long, bending branches, its form varying in different regions and climates. The plant may approach a meter in height. The stems are lined with linear to oval leaves up to 5 centimeters long and coated in whitish hairs, and the herbage emits a scent generally considered unpleasant. The inflorescence is a long, leafy raceme of tiny flowers near the ends of the branches. Each flower is 2 to 4 millimeters wide with five light blue corolla lobes. White-flowered plants are occasionally seen. The fruit is a cluster of four nutlets which are coated in hooked prickles.

Phytochemistry: Pyrrolizidine alkaloids: supinine, amabiline, intermedine, lycopsamine, and 3'-acetylintermedine (Letsyo et al. 2016).

Parts used: Leaves.

Ligularia cymbulifera (W. W. Smith) Handel-Mazzetti.

Family: Asteraceae.

Distribution: China.

Botany: Stem erect, 80–120 cm tall, to 2.5 cm in diam. at base, shortly pilose and white arachnoid-puberulent. Distal most stem leaves sheath like. Compound corymb much branched, to 40 cm, white arachnoid-puberulent and shortly pilose; leaf like and supplementary bracts smaller, linear; peduncles 2–15(–22) mm. Capitula numerous. Involucre campanulate, 8–10 mm, mouth to 1 cm in diam., outside white arachnoid-puberulent or glabrous; phyllaries 7–10, in 2 rows, ovate-lanceolate or lanceolate, margin brown membranous, apex acute. Ray florets yellow; lamina linear, 1–1.4 cm × 1.5–2 mm; tube 4–5 mm. Tubular florets numerous, deep yellow, 6–7 mm; tube ca. 2 mm. Achenes blackish gray, narrowly cylindric, 3–6 mm. Pappus white or yellowish, as long as tubular corolla.

Phytochemistry: Pyrrolizidine alkaloids and bisabolane sesquiterpenes (Liu et al. 2008).

Parts used: Roots.

Ligularia dentata (A. Gray) H. Hara

Common name: Summer ragwort.

Family: Asteraceae.

Distribution: China.

Botany: A robust herbaceous perennial growing to 1–1.5 m tall by 1 m wide. The dark green leaves are large, long-stalked, leathery, cordate-based, and very rounded, with serrated edges. Orange-yellow daisy-like composite flowers bloom on thick red, mostly leafless stalks, rising above the foliage in early summer.

Phytochemistry: Pyrrolizidine alkaloids: clivorine and ligularidine (Hikichi et al. 1979; Kuhara et al. 1980).

Parts used: Roots.

Note: Clivorine is carcinogenic (Kuhara et al. 1980).

Ligularia hodgsonii Hook.

Common name: Golden torch.

Family: Asteraceae.

Distribution: China and North Japan.

Botany: A perennial herb.

Phytochemistry: Pyrrolizidine alkaloids: clivorine and ligularine (Lin et al. 2000).

Actions: Antitussive, diuretic, and expectorant.

Therapeutics: Cancer.

Parts used: Roots.

Ligularia intermedia Nakai

Common name: Golden torch.

Family: Asteraceae.

Distribution: N. China, Japan, Korea.

Botany: A perennial herb with bright yellow flowers blooming from July to September. Interesting red stems. Grows up to 36" tall.

Phytochemistry: Sesquiterpene lactones, phenol derivatives and Pyrrolizidine alkaloids.

Parts used: Leaves.

Ligularia lapathifolia (Franchet) Handel-Mazzetti

Family: Asteraceae.

Distribution: N. China.

Botany: Stem erect, to 120 cm tall, 8–15 mm in diam. at base. Basal leaves petiolate; petiole 7–25 cm, white arachnoid-puberulent, base sheathed; leaf blade ovate or ovate-oblong, 19–40.5 × 8.5–23 cm. Compound corymb branched; branches to 23 cm, spreading or fascinated. Involucre hemispheric or broadly campanulate, 1–1.2 × ca. 2 cm, outside white arachnoid-puberulent. Achenes brown, cylindric, 3–6 mm. Pappus reddish brown or yellowish, as long as tubular corolla.

Phytochemistry: Eremophilenolides (Feia et al. 2006), sesquiterpenes (Lia et al. 2004) and pyrrolizidine alkaloids.

Parts used: Roots.

Liparis nervosa (Thunb.) Lindl.

Common name: Pantropical Widelip orchid.

Family: Orchidaceae.

Distribution: Tropical Asia, America, and Africa.

Botany: Herbs, terrestrial. Stem cylindric, 2–8(–10) cm, 5–7(–10) mm in diam., thick, fleshy, with many nodes, usually ± enclosed by sheaths, upper part sometimes naked. Leaves 3–6; petiole sheath like, 2–3(–5) cm, amplexicaul, long, not articulate. Flowers purple; pedicel and ovary 8–16 mm. Capsule obovate-oblong or narrowly elliptic, ca. 1.5 cm × 6 mm; fruiting pedicel 4–7 mm.

Phytochemistry: Pyrrolizidine alkaloids: nervosine VII, nervosine VIII, and nervosine IX (Huang et al. 2016).

Parts used: Whole plant.

Pharmacology: Inhibitory activities against LPS-induced NO production (Huang et al. 2013).

Lithospermum erythrorhizon Siebold & Zucc.

Common name: The purple gromwell.

Family: Boraginaceae.

Distribution: China, Japan, Korea, and East Russia.

Botany: Herbs perennial. Roots dark red, with a copious purple dye. Stems usually 1–3, erect, appressed or spreading, branching distally, 40–90 cm tall. Leaves sessile, ovate-lanceolate to broadly lanceolate, 3–8 × 0.7–1.7 cm, short strigose, base attenuate, apex acuminate; veins prominent abaxially, more densely strigose. Inflorescences terminal, 2–6 cm, elongated in fruit; bracts similar to leaves but smaller. Nutlets white or pale yellowish brown, ovoid, ca. 3.5 mm, smooth, shiny, concave adaxially with center line forming a longitudinal groove.

Phytochemistry: Pyrrolizidine alkaloids: myoscorpine, intermedine and third, an isomer of echimidine and glycanes: lithospermans A, B, and C and napthoquinone: shikonin.

Actions: Antitumor, cardiotonic, contraceptive, depurative, and febrifuge.

Therapeutics: Irritant skin conditions, measles, chicken pox, boils, carbuncles, hepatitis, and skin cancer. Externally it is used to treat nappy rash, burns, cuts, wounds, abscesses, eczema, and haemorrhoids.

Parts used: The dried root.

Madhuca pasquieri (Dubard). H. J. Lam

Common name: Jackass bitters.

Family: Sapotaceae.

Distribution: Yunnan and Northern provinces of Viet Nam.

Botany: Trees, to 30 m tall. Trunk to 60 cm d.b.h., bark blackish. Branchlets densely lenticellate, rust colored tomentose, glabrescent. Leaves scattered or more often closely clustered at end of branchlets. Flowers several, axillary, fascicled. Fruit ellipsoid to globose, with lengthened style, 2–3 x 1.5–2 cm, rust colored tomentose but glabrescent; pericarp fleshy; 1–5-seeded. Seeds ellipsoid.

Phytochemistry: Pyrrolizidine alkaloids: madhumidine A, lindelofidine benzoic acid ester and minalobine B (Hoang et al. 2015).

Parts used: Leaves.

Pharmacology: Anti-inflammatory (Hoang et al. 2015).

Myosotis scor Pioides L.

Common name: Water forget-me-not.

Family: Boraginaceae.

Distribution: Native to Europe and Asia.

Botany: An erect plant which ranges in height from 6 in to two feet, bearing small blue flowers with yellow centers. It blooms from mid-spring to first frost in temperate climates.

Phytochemistry: Pyrrolizidine alkaloids: myoscorpine, scropioidine, and symphytine.

Actions: Sedative and tonic.

Therapeutics: Wash for sore eyes.

Parts used: Leaves.

Neurolaena lobata L.

Common name: Jackass bitters.

Family: Asteraceae.

Distribution: Native to Belize but also found between Mexico and Peru.

Botany: *N. lobata* grows as a many-branching shrub to 3 meters in height. The vegetative growth is covered with a fine glandular pubescence particularly on the abaxial leaf surfaces. The leaves are arranged alternately, to 25 cm in length, oblong to obovate to elliptical with an entire or serrulate margin and acuminate leaf apex. The actinomorphic flowers are arranged in corymbs of heads.

Phytochemistry: Pyrrolizidine alkaloids: tussilagine, isotussilagine (Passreiter 1998).

Therapeutics: In Belize *N. lobata* is used to treat and prevent malaria, fungus, ringworm, amoebas, and intestinal parasites. *N. lobata* is used medicinally in the Bahamas to treat dermatological problems, and colds and fevers. It is used elsewhere in the Caribbean to treat sprains and dislocations.

Parts used: Leaves.

Pharmacology: Antiprotozoal (Berger et al. 2001), and wound healing (Bijoor et al. 2014).

Note: The leaves of *N. lobata* have been used as a tobacco substitute.

Packera candidissima (Greene) Weber & Löve

Syn: *Senecio candidissimus* Greene.

Common name: Golden ragwort.

Family: Asteraceae.

Distribution: Native to Mexico.

Botany: A perennial herb.

Phytochemistry: Pyrrolizidine alkaloids: senecionine, integerrimine, retrorsine, and usaramine (Bah et al. 1994).

Actions: Antispetic.

Therapeutics: Nephrolithasis.

Parts used: Aerial parts.

Petasites jaPonicus (Siebold & Zucc.) Maxim.

Syn: *Senecio candidissimus* Greene.

Common name: Giant butterbur.

Family: Asteraceae.

Distribution: Native to Mexico.

Botany: An herbaceous perennial plant.

Phytochemistry: Pyrrolizidine alkaloid: petasitenine (Maxim et al. 1977).

Actions: Antispetic.

Therapeutics: Nephrolithasis.

Parts used: Aerial parts.

Pharmacology: Antioxidant (Park et al. 2010) and anti-inflammatory (Lee et al. 2011).

Pulmonaria obscura **Dumort.**

Common name: Suffolk Lungwort.

Family: Asteraceae.

Distribution: Central Sweden and southern Finland to Central Europe.

Botany: A perennial herb reaches a height of about 10 to 20, sometimes up to 30 cm. The stem grows upright and is covered in the upper part with strong bristles, stem glands, and soft hairs. The leaves have softer hairs. They are 4–6 cm long, 1–2 cm wide, and very rarely with pale green patches along the veins. The basal leaves are heart-shaped and oblong, have an approximately 5 to 10 cm long stem, and a length of 4 to 12 cm and are about twice as long as wide. The flowers are about 10 to 15 mm long, initially pink and later red-violet. The corolla tube is bald under the hair ring. The seeds have a length of 3.5 to 4 mm and are brown to black.

Phytochemistry: Pyrrolizidine alkaloids: intermedine, lycopsamine, and their O(7)-derivatives (Haberer et al. 2002).

Parts used: Leaves.

Pulmonaria officinalis **L.**

Common name: Lungwort.

Family: Asteraceae.

Distribution: Native to Europe.

Botany: A herbaceous rhizomatous evergreen perennial plant. The basal leaves are green, cordate, more or less elongated and pointed and always with rounded and often sharply defined white or pale green patches. The upper surface of the leaves has tiny bumps and it is quite hairy. In spring, the plant produces small bunches of flowers. The 5-petals flowers are red or pink at first, later turn to blue-purple.

Phytochemistry: Pyrrolizidine alkaloids (Lüthy et al. 1984).

Actions: Pulmonary.

Therapeutics: Chronic bronchitis.

Parts used: Leaves.

Rindera umbellata **(Waldst. et Kit.) Bunge**

Syn: *Cynoglossum umbellatum* Waldst. et Kit.

Family: Boraginaceae.

Distribution: Serbia.

Botany: A biennial to perennial herbaceous plant. Numerous hairy leaves are set in a rosette. During flowering, ribbed and hairy stems may grow up to 60 cm in height, to be crowned with beautiful inflorescences with bronze, amber, or apricot flowers.

Phytochemistry: Pyrrolizidine alkaloids: 7-angeloyl heliotridane, 7-angeloyl heliotridine, lindelofine, 7-angeloyl rinderine, punctanecine and heliosupine (Mandić et al. 2015).

Parts used: The aerial parts, roots, and seeds.

Senecio alpines Koch

Family: Asteraceae.

Distribution: Native to Australia.

Botany: A perennial plant.

Phytochemistry: Pyrrolizidine alkaloids: senecionine, integerrimine, seneciphylline, senkirkine, and andretrorsine.

Parts used: Aerial parts.

Senecio ambraceus Turcz.

Family: Asteraceae.

Distribution: Native to Mangolia.

Botany: A perennial plant.

Phytochemistry: Pyrrolizidine alkaloids: seneciphylline and small amounts of cristalline jacozine.

Parts used: Aerial parts.

Senecio arcticus Rupr.

Syn: *Tephroseris palustris* (L.) Rchb.

Family: Asteraceae.

Distribution: China, Japan, Korea, Mongolia, Russia (Far East, SE Siberia).

Botany: A perennial herb.

Phytochemistry: Pyrrolizidine alkaloids: senecionine, platyphylline, neoplatyphylline, and acetylplatyphylline.

Parts used: Aerial parts.

Senecio argunensis Turcz.

Family: Asteraceae.

Distribution: Native to Mangolia.

Botany: A perennial plant.

Phytochemistry: Pyrrolizidine alkaloids: senecionine; integerrimine, seneciphylline, otosenine, erucifoline, and 21-hydroxyintegerrimine (Liua and Röder 1991).

Parts used: Aerial parts.

Senecio brasiliensis (Spreng.) Less.

Common name: Flower of souls.

Family: Asteraceae.

Distribution: Central South America.

Botany: *S. brasiliensis* is a densely leafy perennial herb, 1 metre to 2 metres, with yellow flowers. It stands very upright with a branched hairless and grooved stem. The leaves are alternate, pinnate, and deeply lobed dark green on the top, whitish green on the underside. The lower part of the plant is smooth, while the upper part is hairy and the leaves cluster at the highest point with the flower stalks (corymbs). Yellow flowers dense on corymbs; two types of flowers, disc florets with both male and female flowers and ray flowers which are simply female. Seeds: Small seed with white hairs that use the wind to get around with.

Phytochemistry: Pyrrolizidine alkaloids: senecionine, integerrimine, retrorsine, usaramine, and seneciphylline (Adams and Gianturco 1956).

Parts used: Aerial parts.

Senecio chrysanthemoides DC.

Common name: Flower of souls.

Family: Asteraceae.

Distribution: Bhutan, NE and NW India, Nepal, NW Pakistan.

Botany: A robust branched perennial, leaves deeply lobed, flower heads yellow, Ray florets more than 10.

Phytochemistry: Pyrrolizidine alkaloids.

Parts used: Aerial parts.

2.70 *Senecio cineraria* DC.

Syn: *Jacobaea maritime* (L.) Pelser & Meijden.

Common name: Silver ragwort.

Family: Asteraceae.

Distribution: Native to the western and central Mediterranean region.

Botany: *S. cineraria* is a very white-wooly, heat and drought tolerant evergreen subshrub growing to 0.5–1 m tall. The stems are stiff and woody at the base, densely branched, and covered in long, matted grey-white to white hairs. The leaves are pinnate or pinnatifid, 5–15 centimetres long and 3–7 centimetres broad, stiff, with oblong and obtuse segments, and like the stems, covered with long, thinly to thickly matted with grey-white to white hairs; the lower leaves are petiolate and more deeply lobed, the upper leaves sessile and less lobed. The flowers are yellow, daisy-like in dense capitula 12–15 millimetres diameter, with central disc florets surrounded by a ring of 10–13 ray florets, and enclosed in a common whorl of bracts at the base of the capitulum. The seeds are cylindrical achene.

Phytochemistry: Pyrrolizidine alkaloids: senecionine, seneciphylline, integerrimine, jacobine, jacozine, jaconine, otosenine, florosenine, floridanine, and doronine (Tundis et al. 2007).

Actions: Ophthalmic.

Therapeutics: The fresh juice of the leaves is ophthalmic. Applied to the eyes it has a mildly irritating effect that increases blood flow to the area, helping to strengthen resistance and clear away infections. One or two drops put into the eyes is said to be of use in removing cataracts and also in the treatment of conjunctivitis. This remedy should only be used under the supervision of a trained practitioner.

Parts used: Leaves.

Senecio glabellus Pior.

Common name: Butterweed.

Family: Asteraceae.

Distribution: Native to central and southeastern North America.

Botany: *S. glabellus* is a winter annual or biennial that initially forms a low rosette of basal leaves. The hollow central stem is stout, light green or reddish green, and glabrous; it has conspicuous longitudinal veins. The alternate leaves are up to 10" long and 2½" across, becoming smaller as they ascend the central stem and any lateral stems. These leaves are either simple-pinnate or deeply pinnatifid, usually with a larger terminal leaflet and smaller lateral leaflets. Each flower head is about ½" across, consisting of 5–15 ray florets that surround numerous disk florets in the center. After the blooming period, the flower heads are replaced by achenes with small tufts of white hair.

Phytochemistry: Pyrrolizidine alkaloids: senecionine and integerrimine (Ray et al. 1987).

Parts used: Aerial parts.

Senecio interggerrimus Nutt.

Common name: Lambs tongue ragwort.

Family: Asteraceae.

Distribution: Western and central North America.

Botany: A biennial or perennial herb producing a single erect stem 20 to 70 centimeters tall from a caudex with a fleshy root. The linear to lance-shaped or triangular leaves have blades up to 25 centimeters long. The herbage is slightly hairy to woolly or cobwebby. The inflorescence bears several flower heads in a cluster, the middle, terminal head often largest and held on a shorter peduncle, making the cluster look flat. The heads contain many disc florets and usually 8 or 13 ray florets which may be yellow to cream to white in color. Some heads lack ray florets.

Phytochemistry: Pyrrolizidine alkaloids: senecionine and integerrimine in traces (Manske 1939).

Parts used: Aerial parts.

Senecio integrifolius (L.) Clairv. subsp. karsianus

Common name: Common fire weed.

Family: Asteraceae.

Distribution: Coastal areas of Australia.

Botany: A perennial herb.

Phytochemistry: Pyrrolizidine alkaloids: O7-angeloylturneforcidine, 1,2-dihydrosenkirkine, O7-angeloylheliotridine, and its N-oxide (Roeder and Liu 1991).

Parts used: Aerial parts.

Senecio jacobaea L.

Common name: Common ragwort.

Family: Asteraceae.

Distribution: Northern Eurasia.

Botany: The plant is generally considered to be biennial. The stems are erect, straight, have no or few hairs, and reach a height of 0.3–2.0 metres. The leaves are pinnately lobed and the end lobe is blunt. The hermaphrodite flower heads are 1.5–

2.5 centimetres in diameter, and are borne in dense, flat-topped clusters; the florets are bright yellow.

Phytochemistry: Pyrrolizidine alkaloids: jacobine (Macel et al. 2004).

Parts used: Aerial parts.

Senecio lautus

Common name: Common fire weed.

Family: Asteraceae.

Distribution: Coastal areas of Australia.

Botany: A much-branched prostrate herb up to 30 cm high. A plant that is extremely variable in size, leaf shape, and the number of flower heads. Yellow daisy-like flowers 1–2 cm diameter with an orange central discs. Reddish-brown or green achene 2–3 mm long with a ring of hairs at their apex.

Phytochemistry: Pyrrolizidine alkaloids.

Parts used: Aerial parts.

Senecio longilobus Benth.

Common name: Thread leaf ground seal.

Family: Asteraceae.

Distribution: Coastal areas of Australia.

Botany: A shrubby, erect, branched, leafy plant, 30–60 cm tall. It has narrowly linear leaves which are thick, white, and occasionally pinnately lobed, up to 10 cm long. The composite yellow flower heads contain numerous clusters.

Phytochemistry: Pyrrolizidine alkaloid: longilobine (Manske 1939).

Parts used: Aerial parts.

Senecio madagascariensis L.

Common name: Fire weed, Madagascar Ragwort.

Family: Asteraceae.

Distribution: Australia.

Botany: An erect, invasive, glabrous or sparsely hairy herb that grows up to 20–60 cm high and is a perennial exotic weed. The stem is glabrous. The leaves are alternate, narrow-lanceolate to elliptic or oblanceolate. The flower heads (capitula) are small, yellow and daisy-like and are from 1–2 cm in diameter and can number from 2–200 per plant.

Phytochemistry: Pyrrolizidine alkaloids: enecivernine, senecionine, integerrimine, senkirkine, mucronatinine, retrorsine, usaramine, otosenine, acetylsenkirkine, desacetyldoronine, florosenine, and doronine (Gardner et al. 2006).

Parts used: Aerial parts.

Senecio nemorensis L.

Common name: Fire weed, Madagascar Ragwort.

Family: Asteraceae.

Distribution: Endemic to Turkey.

Botany: A perennial plant that can reach 2 metres in height.

Phytochemistry: Pyrrolizidine alkaloids:7-Senecioyl-9-sarracinoyl-retronecine, retroisosenine, doriasenine, and bulgarsenine (Wiedenfeld et al. 2000).

Actions: Hypoglycaemic.

Therapeutics: Urinary tract infections, gout, rheumatism, and stone formation in the urinary tract.

Parts used: Aerial parts.

Senecio riddelli Torr. & A. Gray

Common name: Riddell's ragwort.

Family: Asteraceae.

Distribution: North America.

Botany: A grey-white half-shrub, 30–90 cm tall with pinnatifid and relatively hairless leaves, revealing its bright green leaf colour. It has bright yellow flowers on the stems at about the same height above the ground.

Phytochemistry: Pyrrolizidine alkaloid: riddelline (Manske 1939).

Parts used: Aerial parts.

Senecio pterophorus DC.

Common name: African daisy.

Family: Asteraceae.

Distribution: South-eastern parts of South Australia.

Botany: An upright herbaceous plant or small bushy shrub usually growing 1–1.5 m tall. Its greyish-green stems that are ribbed lengthwise and the lower parts of the stems develop distinctive toothed 'wings'. These 'wings' extend down from the bases of the lance-shaped leaves. These leaves have coarsely toothed or entire margins, green upper surfaces, and whitish woolly undersides. The small flower-heads have several yellow 'petals' and a darker yellow centre. The reddish-brown or dark brown ribbed 'seeds' are topped with a ring of silky hairs.

Phytochemistry: Pyrrolizidine alkaloid: retronecine (Castells et al. 2014).

Parts used: Aerial parts.

Senecio scandens Buch-Ham ex D. Don

Common name: Climbing Senecio.

Family: Asteraceae.

Distribution: East Asia.

Botany: A climber with zig-zag, grooved branches, with arrow-shaped long-pointed leaves, and with many lax, domed branched clusters of bright yellow flower-heads at branch ends and on branch sides. Lower-heads are about 8 mm long, with few ray florets, about 5 mm. Involucral bracts are linear-oblong, pointed, nearly hairless. Leaf margins are entire or coarsely toothed. Leaves are stalked. 7.5–10 cm long. The climber is up to 4 m long.

Phytochemistry: Pyrrolizidine alkaloids: *seneciobipyrrolidine* and *seneciopiperidine* (Daopeng et al. 2014).

Actions: Depurative, febrifuge, and ophthalmic.

Therapeutics: A decoction is used in the treatment of epidemic influenza, malaria, boils and abscesses, acute conjunctivitis, dysentery and enteritis.

Parts used: Aerial parts.

Senecio vulgaris L.

Common name: Groundsel.

Family: Asteraceae.

Distribution: Europe, North Africa, and temperate Asia.

Botany: An erect herbaceous annual growing up to 45 cm tall. Upper leaves lack petioles and are sessile, lacking their own stem, alternating in direction along the length of the plant, two rounded lobes at the base of the stem and sub-clasping above. Leaves are pinnately lobed and + 2.4 inches long and 1 inch wide, smaller towards the top of the plant. Leaves are sparsely covered with soft, smooth, fine hairs. Lobes typically sharp to rounded saw-toothed. Open clusters of 10 to 22 small cylinder shaped rayless yellow flower heads ¼ to ½ inch with a highly conspicuous ring of black tipped bracts at the base of the inflorescence.

Phytochemistry: Pyrrolizidine alkaloids: senecionine and vulgarine (Xie et al. 2010).

Actions: Diaphoretic, an antiscorbutic, a purgative.

Therapeutics: Kidney stones.

Parts used: Aerial parts.

Solanecio gigas (Vatke) C. Jeffrey

Family: Asteraceae.

Distribution: Native to Ethiopia.

Botany: A giant rosette herb or shrub with soft woody stems growing to 4 m high. The leaves are large and shallowly lobed with dull yellow flower heads which are clustered in large terminal panicles. All parts of the plant, but especially the flowers, have an unpleasant or mushy odour.

Phytochemistry: Pyrrolizidine alkaloids:integerrimine, senecionine, and usaramine (Asres et al. 2007).

Actions: Antiabortifacient.

Therapeutics: Colic, diarrhea, gout, otitis media, and typhoid.

Parts used: Stems, leaves, and flowers.

Solenanthus lanatus DC.

Family: Boraginaceae.

Distribution: Native to Turkey.

Botany: A perennial herb.

Phytochemistry: Pyrrolizidine alkaloids: 7-O-angeloylechinatine N-oxide, 3'-O-acetylheliosupine N-oxide, heliosupine N-oxide, and heliosupine (Benamar et al. 2016).

Parts used: Whole plant.

Pharmacology: Acetylcholinesterase inhibitory (Benamar et al. 2016).

Symphytum asperum Lepech.

Common name: Prickly comfrey.

Family: Boraginaceae.

Distribution: Native to Asia and it is known in Europe and North America.

Botany: *S. asperum* is a coarse, hairy, rhizomatous perennial that is typically grown in shaded wildflower areas or naturalized areas for its attractive foliage and spring flowers. It grows to 3–4' tall. Ovate to elliptic leaves are dark green and prickly hairy. Mature stems are not winged (leaf bases are not decurrent as is the case with *Symphytum officinale*). Small tubular flowers in scorpioid cymes open rose-pink in spring but mature to blue or purple. Flowers bloom May to August.

Phytochemistry: Pyrrolizidine alkaloids: echimidine, 7-acetyllycopsamine, 3'-acetyllycopsamine, triangularine, and heliosupine.

Actions: Anodyne, mildly astringent, demulcent, emollient, expectorant, haemostatic, refrigerant, and vulnerary.

Therapeutics: They are used as an external poultice in the treatment of cuts, bruises, and sprains. Internally, they are used as a tea in the treatment of chest complaints.

Parts used: Leaves.

Note: *S. asperum* is a good source of allantoin.

Symphytum officinale L.

Common name: Common comfrey.

Family: Boraginaceae.

Distribution: Native to Europe and some parts of Asia.

Botany: *S. officinale* is a perennial shrub. Fond in moist soils, comfrey has a thick, hairy stem, and grows 2–5 feet tall. Its flowers are dull purple, blue or whitish, and densely arranged in clusters. The leaves are oblong, and often look different depending on where they are on the stem: Lower leaves are broad at the base and tapered at the ends while upper leaves are broad throughout and narrow only at the ends. The root has a black outside and fleshy whitish inside filled with juice.

Phytochemistry: Pyrrolizidine alkaloids: 7-acetylintermedine, 7-acetyllycopsamine, echimidine, intermedine, lasiocarpine, lycopsamine, myoscorpine, symlandine, symphytine, and symviridine (Oberlies et al. 2014; Mei et al. 2010).

Actions: Vulnerary, demulcent, antihaemorrhagic, antirheumatic, and anti-inflammatory.

Therapeutics: Gastric and duodenal ulcer, rheumatic pain, arthritis. Topically as a poultice or fomentation in bruises, sprains, athlete's foot, crural ulcers, and mastitis.

Parts used: Leaves.

Clinical studies: Acute unilateral ankle sprains (Oberlies et al. 2004), myalgia (Kucera et al. 2005), osteoarthritis (Grube et al. 2007), and acute upper or lower back pain (Giannetti et al. 2010).

Syneilesis aconitifolia Max.

Common name: Umbrella plant.

Family: Asteraceae.

Distribution: China.

Botany: An unusual hardy deciduous foliage plant. Each delicate drooping radial leaflet is forked, 65 cm.

Phytochemistry: Pyrrolizidine alkaloids: syneilesine and acetylsyneilesine (Boeder et al. 1995).

Actions: Analgesic and antirheumatic.

Therapeutics: Arthritis.

Parts used: Whole plant.

Tussilago farfara L.

Common name: Coltsfoot.

Family: Asteraceae.

Distribution: Native to Europe and parts of western and central Asia.

Botany: A perennial herbaceous plant that spreads by seeds and rhizomes. The flowers, which superficially resemble dandelions, appear in early spring before dandelions appear. The leaves, which resemble a colt's foot in cross section, do not appear usually until after the seeds are set. Thus, the flowers appear on stems with no apparent leaves, and the later appearing leaves then wither and die during the season without seeming to set flowers. The plant is typically 10–30 cm in height. The leaves have angular teeth on their margins.

Phytochemistry: Pyrrolizidine alkaloids: senecionine and senkirkine (Jiang et al. 2009).

Actions: Antitussive, astringent, bitter, demulcent, diaphoretic, emollient, expectorant, and tonic.

Therapeutics: Respiratory tract, skin, locomotor system, viral infections, flu, colds, fever, rheumatism, and gout.

Parts used: Leaves.

Pharmacology: Carcinogenic (Hirono et al. 1976), and anti-inflammatory (Hwangbo et al. 2009).

References

Adams R, Rogers EF. The structure of monocrotaline, the alkaloid in *Crotalaria spectabilis* and *Crotalaria retusa*. I. *J Am Chem Soc* 1939; **61**: 2815–2819.

Adams R, Gianturco M. Senecio Alkaloids: The Alkaloids of *Senecio brasiliensis*, *fremonti* and *ambrosioides*. *J American Chem Soc* 1956; 78.

Aliasl J, Barikbin B, Khoshzaban F, Naseri M, Sedaghat R, Kamalinejad M, Talei D, Emadi F, Akbari Z, Aliasl F, Jalaly NY, Mohseni-Moghaddam P. Effect of *Arnebia euchroma* ointment on post-laser wound healing in rats. *J Cosmet Laser Ther* 2015a; **17**: 41–5.

Anitha M, Daffodil ED, Muthukumarasamy S, Mohan VR. Effect of *Cynoglossum zeylanicum* (Vehl ex Hornem) Thunb. Ex Lehm on oral glucose tolerance in rats. *J Appl Pharm Sci* 2011; **2**: 75–78.

Anitha M, Rajalakshmi K, Muthukumarasamy S, Mohan VR. Antihyperglycemic, antihyperlipidameic and antioxidant activity of *Cynoglossum zeylanicum* (Vahl Ex hornem) Thunb. Ex Lehm. in alloxan induced diabetic rats. *Int J Pharm Pharmaceut Sci* 2012; **4**: 490–495.

Anitha M, Daffodil ED, Muthukumarasamy S, Mohan VR. Hepatoprotective and antioxidant activity of ethanol extract of *Cynoglossum zeylanicum* (Vahl ex Hornem) Thunb. ex Lehm. in CCl4-treated rats. *J Appl Pharmaceut Sci* 2012a; **2**: 99–103.

Anitha M, Daffodil ED, Muthukumarasamy S, Mohan VR. Antitumor activity of *Cynoglossum zeylanicum* (Vahl Ex Hornem) Thunb. Ex. Lehm. against Dalton Ascites Lymphoma in Swiss Albino mice. *Int J Appl Biol Pharmaceut Technol* 2012b; **3**: 457–462.

Anitha M, Daffodil ED, Muthukumarasamy S, Mohan VR. Anti-inflammatory activity of whole plant of *Cynoglossum zeylanicum. J Adv Pharm Edu & Res* 2013; **3**: 25–27.

Aniya Y, Koyama T, Miyagi C, Miyahira M, Inomata C, Kinoshita S, Ichiba T. Free radical scavenging and hepatoprotective actions of the medicinal herb, *Crassocephalum crepidioides* from the Okinawa Islands. *Biol Pharm Bull* 2005; **28**: 19–23.

Anon. Pyrrolizidine Alkaloids, Environmental Health Criteria No. 80. WHO Geneva, 1998; pp. 275–337.

Ashkani-Esfahani S, Imanieh MH, Khoshneviszadeh M, Meshksar A, Noorafshan A, Geramizadeh B, Ebrahimi S, Handjani F, Tanideh N. The healing effect of *Arnebia euchroma* in second degree burn wounds in rat as an animal model. *Iran Red Crescent Med J* 2012; **14**: 70–4.

Asibal CF, Gelbaum LT, Zalkow LH. Pyrrolizidine alkaloids from *Heliotropium rotundifolium. J Nat Prod* 1989 Jul-Aug; **52**(4): 726–31.

Asres K, Sporer F, Wink M. Identification and quantification of hepatotoxic pyrrolizidine alkaloids in the Ethiopian medicinal plant *Solanecio gigas* (Asteraceae). *Pharmazie* 2007; **62**: 709–13.

Ballabh B, Chaurasia OP. Traditional medicinal plants of cold desert Ladakh-used in treatment of cold, cough and fever. *J Ethopharmacol* 2007; **112**: 341–349.

Baruah RN, Leclercq PA. Constituents of the essential oil from the flowers of *Chromolaena odorata. Planta Med* 1993; **59**: 283.

Benamar H, Tomassini L, Venditti A, Marouf A, Bennaceur M, Nicoletti M. Pyrrolizidine alkaloids from *Solenanthus lanatus* DC. with acetylcholinesterase inhibitory activity. *Nat Prod Res* 2016; 1–8. [Epub ahead of print]

Benn M, Gul W. Pyrrolizidine alkaloids in the antipodean genus Brachyglottis (Asteraceae). *Biochem System Ecol* 2007; **35**: 676–681.

Bah M, Bye R, Miranda RP. Hepatotoxic pyrrolizidine alkaloids in the Mexican medicinal plant *Packera candidissima* (Asteraceae: Senecioneae). *J Ethnopharmacol* 1994; **43**: 19–30.

Berger I, Passreiter CM, Cáceres A, Kubelka W. Antiprotozoal activity of *Neurolaena lobata. Phytother Res* 2001; **15**: 327–30.

Bijoor SN, Ramlogan S, Chalapathi Rao AV, Maharaj S. *Neurolaena lobata* L. promotes wound healing in Sprague Dawley rats. *Int J Appl Med Res* 2014; **4**: 106–110.

Biller A, Boppréa M, Witte L, Hartmann T. Pyrrolizidine alkaloids in *Chromolaena odorata.* Chemical and chemoecological aspects. *Phytochemistry* 1994; **35**: 615–619.

Boeder E, Wiedenfeld H, Liu K, Kroge R. Pyrrolizidine Alkaloids from *Syneilesis aconitifolia. Planta Med* 1995; **61**: 97–8.

Boppré M, Colegate SM, Edgar JA. Pyrrolizidine alkaloids of *Echium vulgare* honey found in pure pollen. *J Agric Food Chem* 2005; **53**: 594–600.

Brown K, Devlin JA, Robins DJ. Globiferine a pyrrolizidine alkaloid from *Crotalaria globifera. Phytochemistry* 1984; **23**: 457–459.

Bull LB, Culvenor CCJ, Dick AT. The Pyrrolizidine Alkaloids. Amsterdam: North Holland Publishing Company, 1968; pp. 115–132.

Castells E, Mulder PPJ, Pérez-Trujillo M. Diversity of pyrrolizidine alkaloids in native and invasive Senecio pterophorus (Asteraceae): implications for toxicity. *Phytochemistry* 2014; **108**: 137–146.

Chen SC, Hong LL, Chang CY, Chen CJ, Hsu MH, Huang YC. Antiproliferative constituents from *Gynura divaricata* subsp. formosana. *Chin Pharm J* 2003; **55**: 109–19.

Chizzola R. Pyrrolizidine alkaloids in *Adenostyles alliaria*e and *A. glabra* from the Austrian Alps. *Nat Prod Commun* 2015; **10**: 1179–80.

Cibin TR, Srinivas G, Gayathri Devi D, Srinivas P, Lija Y, Abraham A. Antioxidant and antiproliferative effects of flavonoids from *Emilia sonchifolia* Linn. on human cancer cells. *Int J Pharmacol* 2006; **2**: 520–524.

Colegate SM, Edgar JA, Knill AM, Lee ST. Solid-phase extraction and LCMS profiling of pyrrolizidine alkaloids and their N-oxides: a case study of *Echium plantagineum. Phytochem Anal* 2005; **16**: 108–119.

Cui Y, Cheng J. Assamicadine, a new *pyrrolizidine* alkaloid from *Crotalaria assamica. J Nat Prod* 1989; **52**: 1153–1155.

Culvenor CCJ, Smith LW. The alkaloids of Amsinckia species: *A. intermedia* Fisch. & Mey., *A. hispida* (Ruiz. & Pav.) Johnst. and *A. lycopsoides* Lehm. *Aust J Chem* 1966; **19**: 1955–1964.

Culvenor CCJ, Smith LW. Alkaloids of *Crotalaria crispata* F. Muell. ex Benth., The Structures of Crispatine and Fulvine. *Australian J Chem* 1963; **16**: 239–245.

D'Anchise R, Bulitta M, Giannetti B. Comfrey extract ointment in comparison to diclofenac gel in the treatment of acute unilateral ankle sprains (distortions). *Arzneimittelforschung* 2007; **57**: 712–6.

Daopeng T, Chou G, Wang Z. Three new Alkaloids from *Senecio scandens. Chem Nat Comp* 2014; **50**: 329–332.

Dodson CD, Stermitz FR. Pyrrolizidine alkaloids from borage (*Borago officinalis*) seeds and flowers. *J Nat Prod* 1986; **49**: 727–728.

Edgar JA, Lin HJ, Kumana CR, Ng MMT. Pyrrolizidine alkaloid composition of three Chinese medicinal herbs, *Eupatorium cannabinum, E.japonicum* and *Crotalaria assamica. American J Chin Med* 1992; **20**: 281–288.

Edgar JA, Kumana CR, Ng MMT. Pyrrolizidine alkaloid composition of three Chinese medicinal herbs, *Eupatorium cannabinum, E. japonicum* and *Crotalaria assamica. Am J Chin Med* 1992; **20**: 281.

El-Shazly A, Sarg T, Ateya A, Abdel Aziz E, Witte L, Wink M. Pyrrolizidine alkaloids of *Cynoglossum officinale* and *Cynoglossum amabile* (family boraginaceae). *Biochem System Ecol* 1996; **24**: 415–421.

Farsam H, Yassa N, Sarkhail P, Shafiee A. New pyrrolizidine alkaloids from *Heliotropium crassifolium. Planta Med* 2000; **66**: 389–91.

Feia DQ, Hana YF, Wua G, Gaoa K. Two new eremophilenolides from *Ligularia lapathifolia. J Asian Nat Prod Res* 2006; **8**: 99–103.

Gardiner MR, Royce R, Bokor A. Studies on *Crotalaria crispata*, a newly recognised cause of Kimberley horse disease. *J Pathol Bacteriol* 1965; **89**: 43–55.

Gardnera DR, Thorneb MS, Molyneuxc RJ, Pfistera JA, Seawrightd AA. Pyrrolizidine alkaloids in *Senecio madagascariensis* from Australia and Hawaii and assessment of possible livestock poisoning. *Biochem System Ecol* 2006; **34**: 736–744.

Giannetti BM, Staiger C, Bulitta M, Predel HG. Efficacy and safety of comfrey root extract ointment in the treatment of acute upper or lower back pain: results of a double-blind, randomised, placebo controlled, multicentre trial. *Br J Sports Med* 2010; **44**: 637–41.

Grube B, Grünwald J, Krug L, Staiger C. Efficacy of a comfrey root (Symphyti offic radix) extract ointment in the treatment of patients with painful osteoarthritis of the knee: results of a double-blind, randomised, bicenter, placebo-controlled trial. *Phytomedicine* 2007; **14**: 2–10.

Habib AA. Senecionine, seneciphylline, Jacobine and Otosenine from *Senecio cineraria. Planta Med* 1974; **26**: 279–82.

Haberer W, Witte L, Hartmann T, Dobler S. Pyrrolizidine alkaloids in *Pulmonaria obscura. Planta Med* 2002; **68**: 480–2.

Hikichi M, Asada Y, Furuya T. Ligularidine, a new pyrrolizidine alkaloid from Ligularia dentata. *Tetrahedron Lett* 1979; **20**: 1233–1236.

Hirono I, Mori H, Culvenor CC. Carcinogenic activity of coltsfoot, *Tussilago farfara* L. *Gan* 1976; **67**: 125–9.

Hoang le S, Tran MH, Lee JS, To DC, Nguyen VT, Kim JA, Lee JH, Woo MH, Min BS. Anti-inflammatory activity of pyrrolizidine alkaloids from the leaves of *Madhuca pasquieri* (Dubard). *Chem Pharm Bull* (Tokyo) 2015; **63**: 481–4.

Hsieha CH, Chenb HW, Leea CC, Hea BJ, Yang YC. Hepatotoxic pyrrolizidine alkaloids in *Emilia sonchifolia* from Taiwan. *J Food Comp Anal* 2015; **42**: 1–7.

Huang S, Zhong DX, Shan LH, Zheng YZ, Zhang ZK, Bu YH, Ma HW, Zhou XL. Three new pyrrolizidine alkaloids derivatives from *Liparis nervosa. Chin Chem Lett* 2016. In Press.

Huanga S, Zhoua X, Wanga CJ, Wanga YS, Xiaoa F, Shana LH, Guoa ZU, Weng J. Pyrrolizidine alkaloids from *Liparis nervosa* with inhibitory activities against LPS-induced NO production in RAW264.7 macrophages. *Phytochemistry* 2013; **93**: 154–161.

Huiqing X, Xuemei M, Xuelei X, Hanqing W. Eremophilane sesquiterpene lactones from *Ligularia intermedia* Nakai of Shanxi. *Chem Nat Comp* 2007; **43**: 746–748.

Hwangbo C, Lee HS, Park J, Choe J, Lee JH. The anti-inflammatory effect of tussilagone, from *Tussilago farfara*, is mediated by the induction of heme oxygenase-1 in murine macrophages. *Int Immunopharmacol* 2009; **9**: 1578–84.

Iwu MM, Chiori CO. Antimicrobial activity of *Eupatorium odoratum* extracts. *Fitoterapia.* 1984; **55**: 354–356.

Jiang Z, Liu F, Goh JJ, Yu L, Li SF, Ong ES, Ong CN. Determination of senkirkine and senecionine in *Tussilago farfara* using microwave-assisted extraction and pressurized hot water extraction with liquid chromatography tandem mass spectrometry. *Talanta* 2009; **79**: 539–546.

Kaith BS, Kaith NS, Chauhan NS. Anti-inflammatory effect of *Arnebia euchroma* root extracts in rats. *J Ethnopharmacol* 1996; **55**: 77–80.

Kucera M, Barna M, Horàcek O, Kàlal J, Kucera A, Hladìkova M. Topical symphytum herb concentrate cream against myalgia: a randomized controlled double-blind clinical study. *Adv Ther* 2005; **22**: 681–92.

Kuhara K, Takanashia H, Hironoa I, Furuyab T, Asadab Y. Carcinogenic activity of clivorine, a pyrrolizidine alkaloid isolated from *Ligularia dentata. Cancer Lett* 1980; **10**: 117–122.

Kurade NP, Jaitak V, Kaul VK, Sharma OP. Chemical composition and antibacterial activity of essential oils of *Lantana camara*, *Ageratum houstonianum* and *Eupatorium adenophorum. Pharm Biol* 2001; **48**: 539–544.

Lee JS, Yang EJ, Yun CY, Kim DH, Kim IS. Suppressive effect of Petasites japonicus extract on ovalbumin-induced airway inflammation in an asthmatic mouse model. *J Ethnopharmacol* 2011; **133**: 551–557.

Leos-Rivas C, Verde-Star MJ, Torres LO, Oranday-Cardenas A, Rivas-Morales C, Barron-Gonzalez MP, Morales-Vallarta MR, Cruz-Vega DE. *In vitro* amoebicidal activity of borage (*Borago officinalis*) extract on *Entamoeba histolytica. J Med Food* 2011; **14**: 866–9.

Letsyo E, Jerz G, Winterhalter P, Horn G, Beuerle T. Survey of pyrrolizidine alkaloids in seven varieties of *Lappula squarrosa*: An alternative source of heart-healthy vegetable oil. *Phytochem Anal* 2016; **27**: 133–9.

Lexa A, Fleurentin J, Lehr PR, Mortier F, Pruvost M, Pelt JM. Choleretic and hepatoprotective properties of *Eupatorium cannabinum* in the rat. *Tetrahedron* 1969; **25**: 1603–15.

Lia YS, Wanga ZT, Zhanga M, Chenb JJ, Luo SD. Two new norsesquiterpenes from *Ligularia lapathifolia. Nat Prod Res* 2004; **18**: 99–104.

Lin G, Rose P, Chatson KB, Hawes EM, Zhao XG, Wang ZT. Characterization of two structural forms of otonecine-type pyrrolizidine alkaloids from *Ligularia hodgsonii* by NMR spectroscopy. *J Nat Prod* 2000; **63**: 857–60.

Liu CM, Wang HX, Wei SL, Gao K. Pyrrolizidine alkaloids and bisabolane sesquiterpenes from the roots of *Ligularia cymbulifera. Helvetica Chimica Acta* 2008; **91**: 308–316.

Liu H, Jin YS, Song Y, Yang XN, Yang XW, Geng DS, Chen HS. Three new compounds from *Arnebia euchroma. J Asian Nat Prod Res* 2010; **12**: 286–92.

Liua K, Röder E. Pyrrolizidine alkaloids from *Senecio argunensis. Phytochemistry* 1991; **30**: 1303–1305.

Lu H, Pei Y, Li W. Studies on flavonoids from *Gynura bicolor* DC. *Zhongguo Xian Dai Ying Yong Yao Xue* 2010; **27**: 613–614.

Lüthy J, Brauchli J, Zweifel U, Schmid P, Schlatter C. Pyrrolizidine alkaloids in medicinal plants of Boraginaceae: *Borago officinalis* L. and *Pulmonaria officinalis* L. *Pharm Acta Helv* 1984; **59**: 242–6.

Macel M, Vrieling K, Klinkhamer PGL. Variation in pyrrolizidine alkaloid patterns of *Senecio jacobaea. Phytochemistry* 2004; **65**: 865–873.

Mandić BM, Vlajić MD, Trifunović SS, Simić MR, Vujisić LV, VuČković IM, Novaković MM, Nikolić-Mandić SD, Tešević VV, Vajs VV, Milosavljević SM. Optimisation of isolation procedure for pyrrolizidine alkaloids from *Rindera umbellata* Bunge. *Nat Prod Res* 2015; **29**: 887–90.

Manske RHF. The alkaloids of Senecio species: III. *Senecio interggerrimus*, *S. longiglobus*, *S. spartioides* and *S. ridellii. Canadian J Res* 1939; **1**: 7–9.

Mattocks AR. Chemistry and Toxicology of Pyrrolizidine Alkaloids. London. U.K.: Academic Press, 1986.

Mei N, Guo L, Fu PP, Fuscoe JC, Luan Y, Chen T. Metabolism, genotoxicity, and carcinogenicity of comfrey. *J Toxicol Environ Health B Crit Rev* 2010; **13**(7-8).

Nasiri E, Hosseinimehr SJ, Azadbakht M, Akbari J, Enayati-Fard R, Azizi S, Azadbakht M. The healing effect of *Arnebia euchroma* ointment versus silver sulfadiazine on burn wounds in rat. *World J Plast Surg* 2015; **4**: 134–44.

Nobre VMT, Dantas AFM, Riet-Correa F, Barbosa Filho JM, Tabosa IM, Vasconcelos JS. Acute intoxication by *Crotalaria retusa* in sheep. *Toxicon* 2005; **45**: 347–352.

Oberlies NH, Kim NC, Brine DR, Collins BJ, Handy RW, Sparacino CM, Wani MC, Wall ME. Analysis of herbal teas made from the leaves of comfrey (*Symphytum officinale*): reduction of N-oxides results in order of magnitude increases in the measurable concentration of pyrrolizidine alkaloids. *Public Health Nutr* 2004; **7**: 919–24.

Odoh UE, Ezugwu CO. Wound healing activity of *Chromolaena odorata* leaves. *J Trop Med Plants* 2008; **8**: 9–13.

Owoyele VB, Adediji JO, Soladoye AO. Anti-inflammatory activity of aqueous leaf extract of *Chromolaena odorata*. *Inflammopharmacology* 2005; **13**: 479–484.

Park CH, Kim MY, Sok DE, Kim JH, Lee JH, Kim MR. Butterbur (*Petasites japonicus* Max.) extract improves lipid profiles and antioxidant activities in monosodium L-glutamate-challenged mice. *J Med Food* 2010; **13**: 1216–1223.

Parvais O, Stricht BV, Vanhaeln-Fastre R, Vanhaelen M. TLC detection of pyrrolizidine alkaloids in oil extracted from the seeds of *Borago officinalis*. *J Planar Chromatography* 1994; **7**: 80–82.

Passreiter CM. Pyrrolizidine alkaloids from *Neurolaena lobata*. *Biochem System Ecol* 1998; **26**: 839–843.

Räder H, Hang XT, Kabus KJ. Pyrrolizidine alkaloids from the seeds of *Crotalaria sessiliflora*. *Planta Med* 1992; **58**: 283.

Ravi S, Ravikumar R, Lakshmanan AJ. Pyrrolizidine alkaloids from *Cynoglossum furcatum*. *J Asian Nat Prod Res* 2008; **10**: 349–54.

Ravishankar K, Suresh M, Kiranmayi GVK. Evaluation of *in-vitro* antibacterial and wound healing activity of ethanolic extract of *Eupatorium odoratum*. *Adv Pharmacol Toxicol* 2010; **11**: 15–24.

Ray AC, Williams HJ, Reagor JC. Pyrrolizidine alkaloids from *Senecio longilobus* and *Senecio glabellus*. *Phytochemistry* 1987; **26**: 2431–2433.

Rizk AM. Naturally Occurring Pyrrolizidine Alkaloids. Boca Raton, FL, USA: CRC Press, 1990.

Röder ET, Rengel-Mayer B. Pyrrolizidine alkaloids from *Arnebia euchroma*. *Planta Med* 1993; **59**: 192.

Roeder E, Rengerl B. Pyrrolizidine alkaloids from *Lithospermum erythrorhizon*. *Phytochemistry* 1990; **29**: 690–693.

Roeder E, Liu K. Pyrrolizidine alkaloids of *Senecio integrifolius*. *Phytochemistry* 1991; **30**: 1734–1737.

Roeder E, Eckert A, Wiedenfeld H. Pyrrolizidine alkaloids from *Gynura divaricata*. *Planta Med* 1996; **62**: 386.

Stegelmeier BL. Pyrrolizidine alkaloid-containing toxic plants (Senecio, Crotalaria, Cynoglossum, Amsinckia, Heliotropium, and Echium spp.). *Vet Clin North Am Food Anim Pract* 2011; **27**: 419–28.

Seow LJ, Beh HK, Umar MI, Sadikun A, Asmawi MZ. Anti-inflammatory and antioxidant activities of the methanol extract of *Gynura segetum* leaf. *Int Immunopharmacol* 2014; **23**: 186–191.

Shahraki MR, Ahmadimoghadm M, Shahraki AR. The antinociceptive effects of hydroalcoholic extract of *Borago officinalis* flower in male rats using formalin test. *Basic Clin Neurosci* 2015; **6**: 285–90.

Shen CC, Syu WJ, Li SY, Lin CH, Lee GH, Sun CM. Antimicrobial activities of naphthazarins from *Arnebia euchroma*. *J Nat Prod* 2002; **65**: 1857–62.

Shylesh BS, Padikkala J. Antioxidant and anti-inflammatory activity of *Emilia sonchifolia*. *Fitoterapia* 1999; **70**: 275–278.

Shylesh BS, Padikkala J. *In vitro* cytotoxic and antitumor property of *Emilia sonchifolia* (L.) DC in mice. *J Ethnopharmacol* 2000; **73**: 495–500.

Shylesh BS, Nair AKS, Subramoniam A. Induction of cell-specific apoptosis and protection from Dalton's lymphoma challenge in mice by an active fraction from *Emilia sonchifolia*. *Indian J Pharmacol* 2005; **37**: 232–237.

Singh M, Kamal YT, Rabea P, Ahmad S. *In vitro* antioxidant activity and HPTLC analysis of *Borago officinalis* Linn. *Indian J Pharm Edu Res* 2013; **47**: 24–30.

Smith LW, Culvenor CCJ. Plant sources of hepatotoxic pyrrolizidine alkaloids. *J Nat Prod* 1981; **44**: 129–15.

Subramanian SS, Nagarajan S. Flavonoids of the seeds of *Crotalaria retusa* and *Crotalaria striata*. *Curr Sci* 1969; **38**: 65.

Suna QH, Yangb JJ, Weia XH, Xua H, Chou GX. Two new pyrrolizidine alkaloids from *Crotalaria albida*. *Phytochemistry* 2013; **6**: 449–452.

Tinker RB, Lauter WM. Constituents of *Crotalaria spectabilis* Roth. *Econ Bot* 1956; **10**: 254–57.

Tomimori K, Nakama S, Kimura R, Tamaki K, Ishikawa C, Mori N. Antitumor activity and macrophage nitric oxide producing action of medicinal herb, *Crassocephalum crepidioides*. BMC *Complement Altern Med* 2012; **12**: 78. Published online 2012 Jun 21. doi: 10.1186/1472-6882-12-78.

Tundis R, Loizzo MR, Statti GA, Passalacqua NG, Peruzzi L, Menichini F. Pyrrolizidine alkaloid profiles of the *Senecio cineraria* group (Asteraceae). *Z Naturforsch* C 2007; **62**: 467–72.

Wiedenfeld H, Narantuya S, Altanchimeg D, Roeder E. Pyrrolizidine Alkaloids in *Senecio nemorensis* L. from Mongolia. *Scientia Pharmaceutica* 2000; **68**: 207–211.

Wiedenfeld H, Andrade-Cetto A. Pyrrolizidine alkaloids from *Ageratum houstonianum* Mill. *Phytochemistry* 2001; **57**: 1269–71.

Wiedenfeld H, Altanchimeg D, Gantur A, Narantuya S. Toxic pyrrolyzidine alkaloids from three mongolian plants. A possible risk for cattle poisoning. *J Nat Toxins* 2002; **11**: 187–92.

Woerdenbag HJ. *Eupatorium cannabinum* L. A review emphasizing the sesquiterpene lactones and their biological activity. *Pharm Weekbl* 1986; **8**: 245–51.

Wu CC, Lii CK, Liu KL, Chen PY, Hsieh SL. Anti-inflammatory activity of *Gynura bicolor* ether extract through inhibits nuclear factor kappa B activation. *J Trad Compl Med* 2013; **3**: 48–52.

Xie WD, Li X, Row KH. A new pyrrolizidine Alkaloid from *Senecio vulgaris. Bull Korean Chem Soc* 2010; **31**: 2715–19.

Yua CH, Tangb WZ, Penga C, Suna T, Liuc B, Lia M, Xiea XF, Zhangd H. Diuretic, anti-inflammatory, and analgesic activities of the ethanol extract from *Cynoglossum lanceolatum. J Ethnopharmacol* 2012; **139**: 149–154.

Zheng ZX, Tang XW, Xue CY, Zhang Y. Hypoglycemic effect of ethanol extract from *Gynura bicolor* DC. in healthy mice. *J Clin Rehabil Tissue Engineer Res* 2007; **11**: 9503–9507.

Chapter 32

The Phytocannabinoids

32.1 Introduction

Cannabis as a medicine was used before the Christian era in Asia, mainly in India. The introduction of Cannabis in Western medicine occurred in the midst of the 19th century, reaching the climax in the last decade of that century, with the availability and usage of Cannabis extracts or tinctures (Lambert 2001). In the first decades of the 20th century, the Western medical use of Cannabis significantly decreased largely due to difficulties to obtain consistent results from batches of plant material of different potencies. The identification of the chemical structure of Cannabis components and the possibility of obtaining its pure constituents were related to a significant increase in scientific interest in such plants since 1965. This interest was renewed in the 1990's with the description of cannabinoid receptors and the identification of an endogenous cannabinoid system in the brain (Zuardi 2006).

32.2 Cannabis in Traditional Chinese Medicine

It is estimated that around 6000 BCE seeds of Cannabis were used for food purpose in China (Joe et al. 1993). The first recorded use for medicinal purpose in China pharmacopoeia dates back to 2727 BCE. It was in 1500 BCE that Cannabis was cultivated in China. The *pen-ts'aoching,* the world's oldest pharmacopoeia, has described use of Cannabis in the treatment of rheumatism, malaria, diseases of the female genital tract, and constipation (Hou 1977). With regard to the treatment of constipation, study reported efficacy of a Chinese herbal proprietary medicine (Hemp Seed Pill) in the treatment of functional constipation (Whiting and Ford 2011). HuaT'o, the father of Chinese surgery used a compound derived from Cannabis along with wine for inducing anesthetsia in patients undergoing surgical procedures (Li and Lin 1974; Auvinen and Peltola 1999).

32.3 Cannabis in Indian Medicine

The Atharva Veda, written sometime around 1200–800 BCE, has mentioned Cannabis to be one of the five sacred plants. Here Cannabis refers to Bhang (dried leaves, stem, and seeds). *Anandakanda*, the ancient Indian alchemy text has described 10 types of

Cannabis users. Further, 50 other different preparations of Cannabis have also been described for the purpose of rejuvenation, aphrodisiac effect, and cure of several diseases. *Anandakanda* has given a detailed description of the toxic effects of hemp that appear in human beings in nine successive stages (Ethan 2005).

In ayurvedic works, about 51 important formulations containing Cannabis have been described and they are of different categories (Dominik et al. 2003). The Cannabis root is described to be a poison in *Susruta Samhita* (700 B.C.) and in many medieval ayurvedic works on Materia Medica it is described as an upavisa (poison of minor importance). Sushruta has indicated use of Cannabis during surgery (Wujastyk 2002).

Sharangadhara Samhita, a compendium of therapeutics (thirteenth century A.D.), has included medicaments titrated with the fresh extract of *bhang*. Authoritative *Ayurvedic* works on materia medica such as *Dhanwantarinighantu* (eighth century A.D.), *Madanapalanighantu* (1374 A.D.), and *Rajanighantu* (1450 A.D.) have described the properties, actions, and indications of both Cannabis and opium. *Bhava mishra* (fifteenth century A.D.), a contemporary of Paracelsus, has in his compendium on medicine and therapeutics, *Bhava prakasha*, described the properties, actions, indications, and formulations of Cannabis (Dwarakanath 1965; Chaturvedi et al. 1981). *Bhava prakasha* described Cannabis as antiphlegmatic, digestive, bile affecting, pungent, and astringent, prescribing it to stimulate the appetite, improve digestion, and better the voice.

32.4 Cannabis in Traditional Arabic and Persian Medicine

Arab physicians knew and used its anti-emetic, anti-epileptic, anti-inflammatory, analgesic antipyretic, and diuretic properties (Lozano 1997). Dioscorides (first century A.D.) had described both Cannabis and opium, and he is seen to have made use of them for therapeutic purposes. Galien (138–201 A.D.) and Rhazes (865–925 A.D.) have given detailed descriptions of these drugs, their actions, therapeutics, and uses.

Zoroaster wrote an multi-volume ancient Persian religious text known as The Zoroastrian Zend-Avesta between 559 BCE and 379 CE. In this work, Zoroaster has mentioned *Bhanga* as a valuable narcotic. In addition, Bhanga has also been described as an abortifacient. The credit of introducing Cannabia in Europe goes to Scythian tribes around 500 BCE. As per estimates, it was between 500 BCE-1000 BCE, Cannabis spread through our Northern Europe.

Authoritative Arabic and Persian medical works such as (1) *Firdousul-Hikmat* and (2) *Mujardat Quanan* have not only described the properties of these drugs, but have also included a number of formulations containing them. It would appear that potions containing Cannabis and linctus containing opium were popular in Arabia, Persia, and Muslim India.

32.5 Cannabis in Herbal Materia Medica and Western Pharmacopoeia

Dioscorides (77 CE) described properties of Cannabis in 'De Materia Medica'. In his third book, Dioscorides describes that the root when boiled and applied reduces pain, oedema, and inflammation in inflamed joints. The seeds cosumed in certain quantity supresses conception. The expressed juice is used in the treatment of earaches (Osbaldeston 2000).

Although there is evidence of Cannabis use in Europe from the thirteenth century, after Marco Polo returned from his journey to the east in 1297, its medical use became more popular in the nineteenth century, when the British physician William B. O'Shaugnessy brought back an account of the remarkable effects of this plant from India. Even Queen Victoria is said to have sipped marijuana tea prescribed by her court physician to treat menstrual cramps (Frazzetto et al. 2003).

Cannabis has been used for centuries for both the symptomatic and prophylactic treatment of migraine. It was highly esteemed as a headache remedy by the most prominent physicians of the age between 1874 and 1942, remaining part of the Western pharmacopoeia for this indication even into the mid-twentieth century (Russo 1998).

Squire in *Companion to the Latest Edition of the British Pharmacopoeia* has described Cannabis as sedative, anodyne, and hypnotic (Squire 1899). In *Merck's 1899 Manual of the Materia Medica*, a preparation made with Cannabis extracts named *Cannabine Tannate Merck* is recommended as a hypno-sedative in the treatment of hysteria, delirium, and nervous insomnia.

Cushny (1906) has defineds Cannabis as a hypnotic in *A Textbook of Pharmacology and Therapeutics*. The 1907 and 1930 editions of *Merck Indexes* have defined Cannabis as having "hypnotic property" of Cannabis. *Bruce and Dilling's Materia Medica and Therapeutics* also maintains the hypnotic property of Cannabis (Dilling 1933). *The British Pharmaceutical Codex 1934* has again mentioned Cannabis and called it a "sedative or hypnotic" (Pharmaceutical Society of Great Britain 1934).

Cannabis is listed as a hypnotic and a sedative in the 1940 edition of the *Merck Manual of Therapeutics and Materia Medica*, and the pharmaceutical trade journal *Ciba Symposia* in 1946 (Robinson 1946) points out that Cannabis "is sometimes employed as a hypnotic in those cases where opium, because of long-continued use, has lost its efficiency." *Merck Index* (O'Neil et al. 2001) continued to list drowsiness as an effect of Cannabis smoking/inhalation.

32.6 Government Reports

32.6.1 Indian Hemp Drugs Commission Report

Indian Hemp Drugs Commission Report was initiaed by the British in the year of 1894. It noted sleep to be major effect of Cannabis (Kaplan 1969; Kalant 1972). In some persons, Cannabis was reported to produce peculiar loss of sensation of space and time. This is followed by a stage of deep sleep. The intoxication lasts about three hours, when sleep supervenes. The Commission also noted slight narcotic effects of hemp more or less complete (Mikuriya 1968).

32.6.2 La Guardia Committee Report

Fifty years later, the *La Guardia Committee Report*, after in-depth investigation of smoked Cannabis in the United States reported prolonged drowsiness as major effect (Mayor's Committee on Marihuana 1944).

32.6.3 The Wootton Report

In order to investigate the effects of Cannabis, the British in the year of 1968 published the *Wootton Report*. The report emphasized the calming and relaxing effects (U.K. Home Office 1968).

32.6.4 The Le Dain Commission

In order to investigate the effects of Cannabis, Canada in the year of 1972 appointed the *Le Dain Commission*. The *Commission* published on the calming effect of Cannabis (Le Dain 1970).

32.6.5 The Shafer Commission

Nearly simultaneously, president Richard Nixon commissioned a report to study Cannabis abuse in the U.S., commonly referred to as the *Shafer Commission*. The *Shafer Commission* like the *Wootton Report* and *The Le Dain Commission, reported* relaxation as major effect of Cannabis (Shafer 1972).

32.6.6 Marihuana Tax Act of 1937 (MTA)

The Marihuana Tax Act of 1937 was a US Act that placed a tax on the sale of Cannabis. The act was drafted by Harry Jacob Anslinger, the first commissioner of the U.S. Treasury Department's Federal Bureau of Narcotics (FBN).

32.7 Cannabis and the American Herbal Pharmacopoeia

The American Herbal Pharmacopoeia has recently published the two-part "Cannabis monograph". The monograph has been written and reviewed by the world's leading experts. It tends to brings together scientific data and issues long-awaited standards for the plant's identity, purity, quality, and botanical properties. The monograph gives doctors who want to prescribe Cannabis therapy a full scientific understanding of the plant, its constituent components, and its biologic effects.

32.8 Botany of Cannabis

32.8.1 Cannabis sativa L.

Distribution: *C. sativa* is widespread throughout North America. It is further found globally, actually growing in the wild in Northern India. It is also found in Southern Siberia, and probably in China.

Botany: *C. sativa* is an annual, the erect stems growing from 3 to 10 feet or more high, very slightly branched, having greyish-green hairs. The leaves are palmate, with five to seven leaflets (three on the upper leaves), numerous, on long thin petioles with acute stipules at the base, linear-lanceolate, tapering at both ends, the margins sharply serrate, smooth and dark green on the upper surface, lighter and downy on the under one. The small flowers are unisexual, the male having five almost separate, downy, pale yellowish segments, and the female a single, hairy, glandular, five-veined leaf enclosing the ovary in a sheath. The ovary is smooth, one-celled, with one hanging ovule and two long, hairy thread-like stigmas extending beyond the flower for more than its own length. The fruit is small, smooth, light brownish-grey in colour, and completely filled by the seed.

32.8.2 Cannabis indica Lam

Distribution: India, Afghanistan, Bangladesh, and Pakistan.

Botany: The typical example of *C. indica* is a more compact, thick-stemmed bush than its cousins, usually reaching a height of less than two metres. The foliage is generally a dark shade of green, some examples appearing to have almost blue or green-black leaves. These leaves are composed of short, wide blades. Indica strains tend to produce more side-branches and denser overall growth than Sativas, resulting in wider, bushier plants. Indica flowers form in thick clusters around the nodes of the female plant (the points at which pairs of leaves grow from the stem and branches). They usually weigh more than Sativa flowers of similar size, as they are more solid 2.3: *C. sativa* ssp. *Ruderalis*.

32.8.3 Cannabis ruderalis Janisch. (Wild hemp)

Distribution: Central Russia.

Botany: *C. ruderalis* grows to a height of only 60 cm. It has few branches and small leaves. The inflorescences are small and form on the end of the stalk.

32.9 Chemistry of Cannabis

32.9.1 Resin

Cannabin stands for a biologically active resin extracted from Indian hemp. Smith (1846) reported about presence of cannabin in *Cannabis indica*. Resin is composed of cannabinol, pseudo-cannabinol, and cannabinin. Nowadays, resin is considered to be residual THC and plant matter. The *cannabin* is the alcoholic or resinous extract employed in medicine.

32.9.2 The volatile oil

Personne (1857) separated the volatile oil in cannabene and *cannabene hydride.* Cannabene is a colourless fluid having intoxicating properties similar to Cannabis.

32.9.3 Alkaloids

Cannabinine: Cannabinine was isolated by Siebold and Bradbury (1881), in a very small quantity. The extraction methodology was similar to that of nicotine.

Tatano-cannabinine: Matthew Hay (1883) isolated tatano-cannabinine in crystalline form having tetanic property. Jahns in 1889 proposed tatano-cannabinine identical with choline.

Cannabamines A–D: There structure is not yet confirmed (El Feraly and Turner, 1975).
Choline, trigonelline, muscarine, and an unidentified betain: These have been reported along with cannabamines.

Cannabisativine (Fig. 32.1): An ethanol extract of the root of a Mexican variant of *C. sativa* afforded, after partitioning and chromatography yielded, the new spermidine alkaloid Cannabisativine (Turner et al. 1976).

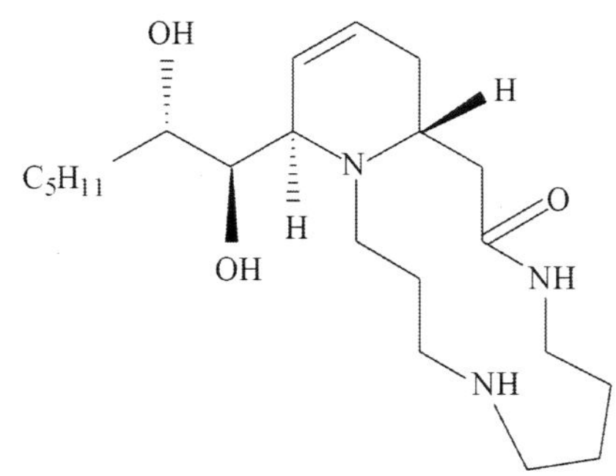

Fig. 32.1: Structure of Cannabisativine.

Anhydrocannabisativine (Fig. 32.2): Ethanol extracts of the leaves and roots of a Mexican variant of *C. sativa* afforded the new spermidine alkaloid, anhydrocannabisativine (Elsohly et al. 1978).

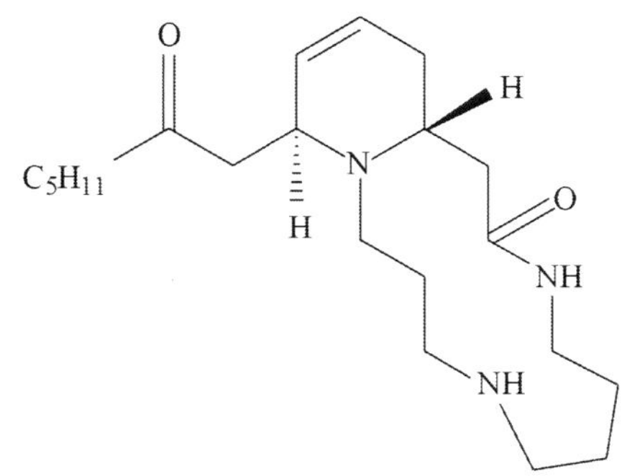

Fig. 32.2: Structure of anhydrocannabisativine.

32.10 Cannabinoids

32.10.1 Introduction

Cannabinoids are diverse chemical compounds that were first discovered in 1964. Thay were originally defined as a group of C21 compounds uniquely produced by the Cannabis plant. 113 cannabinoids have been isolated from the flower, leaf, and stem of the Cannabis plant. Since they exist naturally in the Cannabis plant, they are also known as phytocannabinoids. They have been classified as natural cannabinoids, herbal cannabinoids, and classical cannabinoids also. Tetrahydrocannabinol is the most important cannabinoid (Argurell et al. 1984; Appendino et al. 2011). Cannabichromene is found in other plants also.

32.10.2 Classification

Cannabinoids, a class of meroterpenoids derived from the alkylation of an olivetol-like alkyl resorcinol with a monoterpene unit, are the most typical constituents of *Cannabis*. This class includes over a hundred members belonging to several structural types, mainly differing by the constitution of their terpenoid moiety (Makriyannis and Rapaka 1987; Appendino and Taglialatela-Scafat 2013). Each of the 113 cannabinoid compounds falls within one of six different subclasses. These six cannabinoid subclasses include:

- Tetrahydrocannabinols (THCs)
- Cannabierols (CBGs)
- Cannabinodiols and cannabinols (CBDL and CBNs)
- Cannabidiols (CBDs)
- Cannabichromenes (CBCs)
- Miscellaneous cannabinoids—cannabitriol (CBT), cannabicyclol (CBL), cannabielsoin (CBE), and others.

Another way of classification is as follows:

- Phytocannabinoids occur uniquely in the Cannabis plant.
- Endogenous cannabinoids are produced in the body of humans and other animals.
- Synthetic cannabinoids are similar compounds produced in a laboratory.

32.10.3 Important cannabinoids

Tetrahydrocannabinol (THC)

Tetrahydrocannabinol (Fig. 32.3) is more precisely known as (−)-*trans*-Δ^9-tetrahydrocannabinol. It is the chief psychoactive constituent of Cannabis and hasish. Tetrahydrocannabinol is found in the resin secreted by glands of the Cannabis plant. The buds are often preferred because of their higher tetrahydrocannabinol content. The credit of isolation of tetrahydrocannabinol goes to Raphael Mechoulam, an Isreal based Scientist.

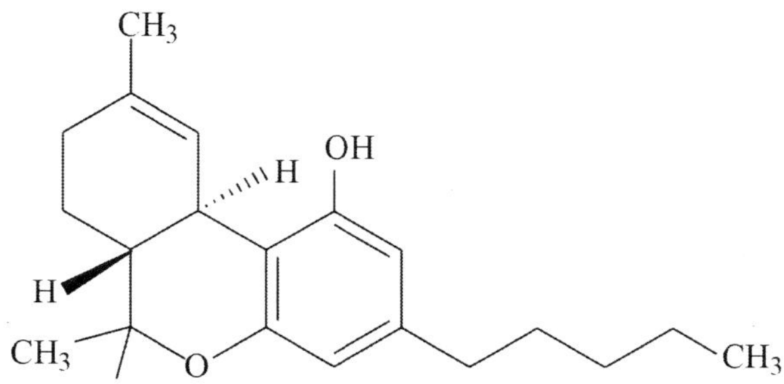

Fig. 32.3: Structure of tetrahydrocannabinol.

The role of THC in Cannabis seems to protect the plant from herbivores or pathogens. Initially, tetrahydrocannabinol was included in Schedule I of the 1971 Convention of Psychotropic Substances. Recently, the World Health Organisation (WHO) reclassified tetrahydrocannabinol in the less stringent Schedule III.

Cannabidiol (CBD)

Cannabidiol (Fig. 32.4) accounts for 40% of the Cannabis extract.

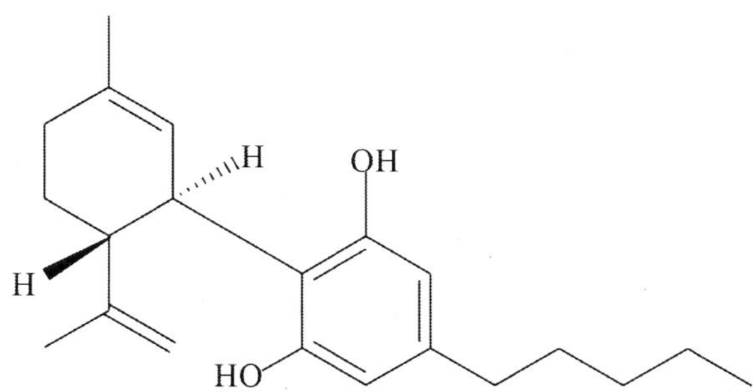

Fig. 32.4: Structure of cannabidiol.

Cannabinol (CBN)

Cannabinol (Fig. 32.5) is a degradation product of tetrahydrocannabinol.

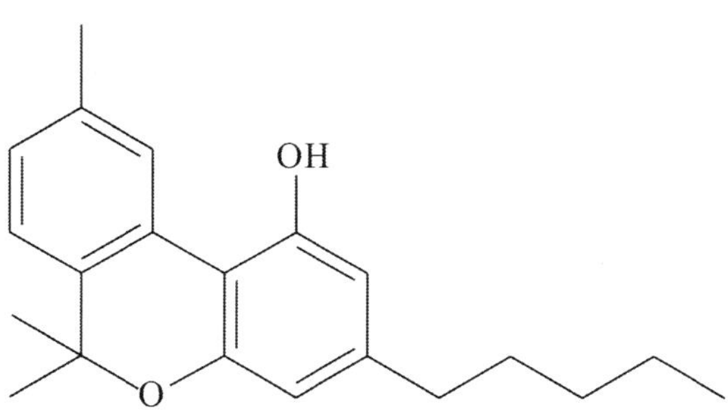

Fig. 32.5: Structure of cannabinol.

Cannabichromene (CBC)

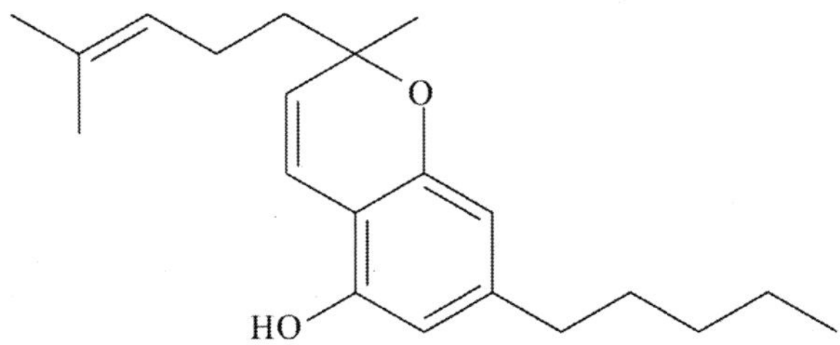

Fig. 32.6: Structure of cannabichromene.

Cannabigerol (CBG)

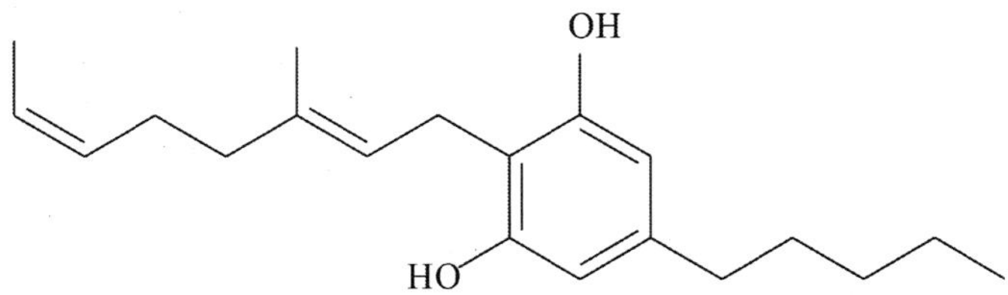

Fig. 32.7: Structure of cannabigerol.

Cannabidiolic acid (CBDA)

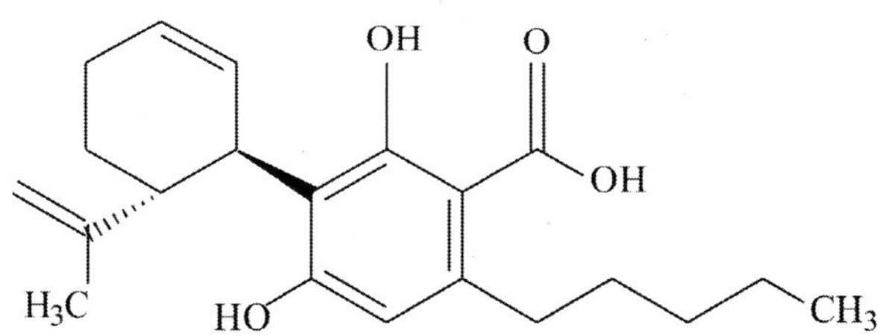

Fig. 32.8: Structure of cannabidiolic acid.

Tetrahydrocannabivarin (THCV)

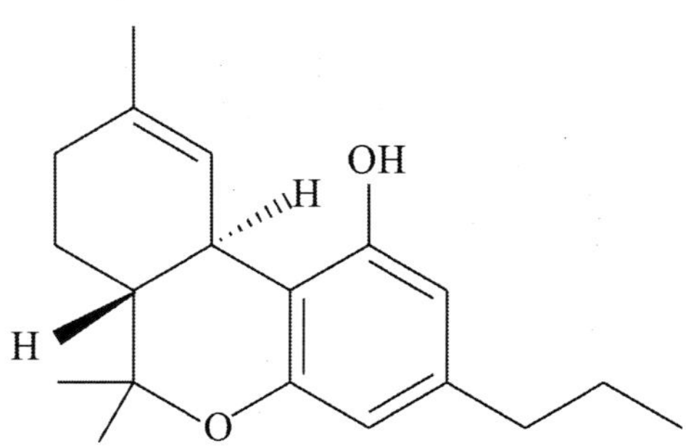

Fig. 32.9: Structure of tetrahydrocannabivarin.

Cannabivarin (CV)

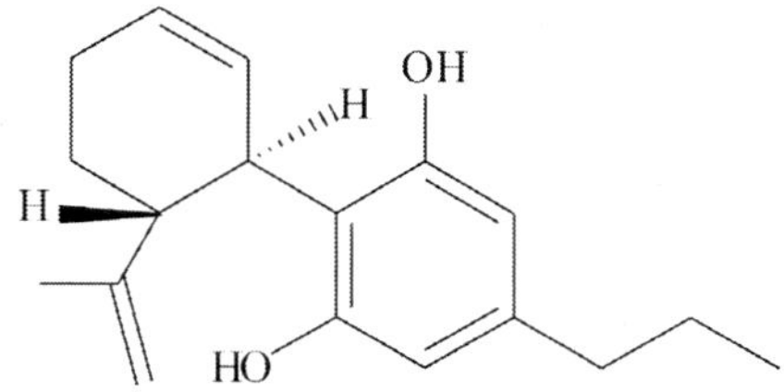

Fig. 32.10: Structure of cannabivarin.

32.11 Other Herbal Cannabinomimetics

32.11.1 Artemisia absinthum L. (Asteraceae)

As per data published in Nature, *Artemisia absinthum* (Absinthe) has been speculated to activate the CB1 cannabinoid receptor (Nature 253: 365-356; 1975). Thujone (the active constituent in Absinthe oil) (Fig. 32.11) product displaced [3H]CP55940, a cannabinoid agonist, only at concentrations above 10 microM.

> [35S]GTPgammaS binding assays revealed that thujone failed to stimulate G-proteins even at 0.1 mM. Thujone failed to inhibit forskolin-stimulated adenylate cyclase activity in N18TG2 membranes at 1 mM. Rats administered thujone exhibited different behavioral characteristics compared with rats administered a potent cannabinoid agonist, levonantradol (Fig. 32.12) (Meschler and Howlett 1999).

32.11.2 Desmodium canum Schinz & Thell. (Fabaceae)

Three isoflavanones with cannabinoid-like moieties have been isolated (Botta et al. 2003).

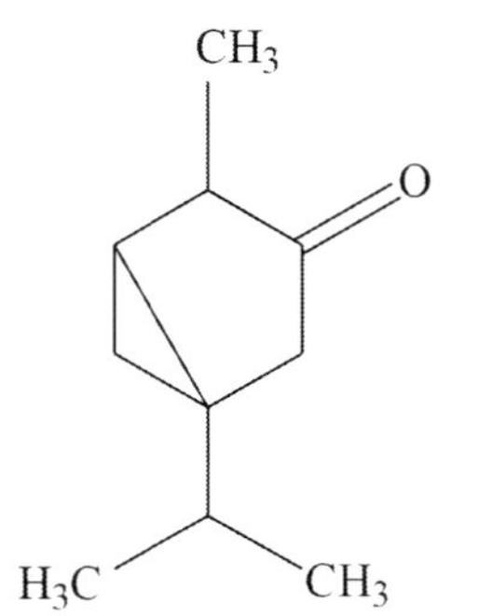

Fig. 32.11: Structure of thujone.

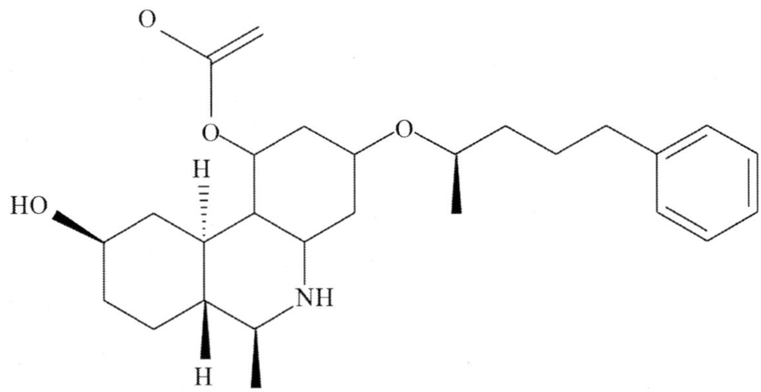

Fig. 32.12: Structure of levonantradol.

32.11.3 *Echinacea angustifolia* L. (Asteraceae)

Alkamides have structural similarity with anandamide. Alkamides exhibited selective affinity especially to CB2 receptors and can therefore be considered as CB ligands (Woelkart et al. 2005). Alkamides (structural similarity with anandamide). Dodeca-2E,4E,8Z,10Z-tetraenoic acid isobutylamide, dodeca-2E,4E-dienoic acid isobutylamide bind to the CB2 receptor more strongly than the endogenous cannabinoids (Raduner et al. 2006).

32.11.4 *Echinacea purpurea* L. (Asteraceae)

The bioactivity of isolated compounds including three alkamides and nitidanindiisovalerianate (Fig. 32.13) from *E. purpurea* were studied in [^{35}S]GTPγS-binding experiments performed on rat brain membrane preparations. Both partial and inverse agonist compounds for cannabinoid (CB1) receptors were identified among the metabolites, characterized by weak to moderate interactions with the G-protein signaling mechanisms. However, upon coadministration with arachidonyl-2'-chloroethylamide (Fig. 32.14), a number of them proved capable of inhibiting the stimulation of the pure agonist, thereby demonstrating cannabinoid receptor antagonist properties (Hohmann et al. 2011).

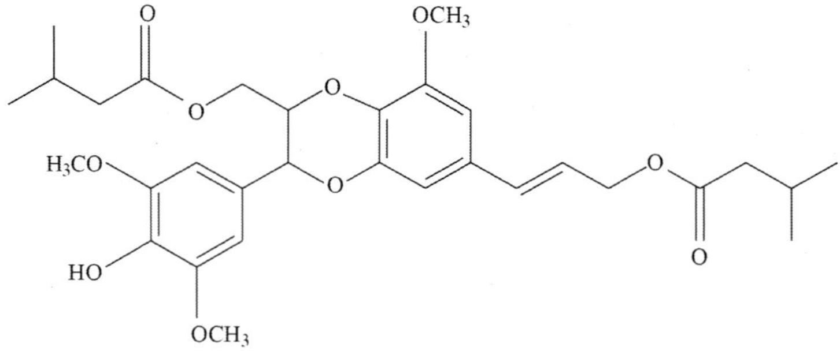

Fig. 32.13: Structure of nitidanindiisovalerianate.

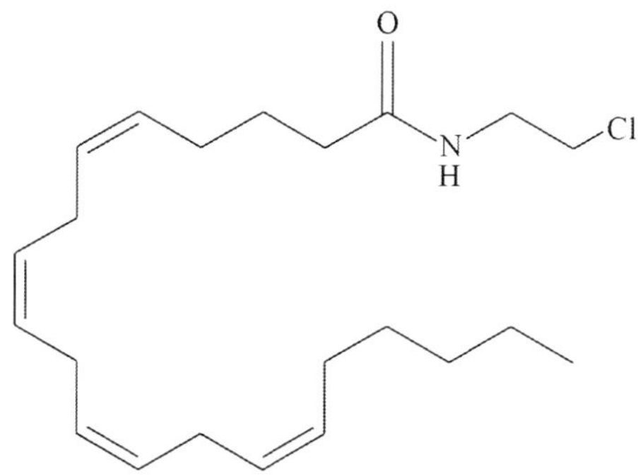

Fig. 32.14: Structure of arachidonyl-2'-chloroethylamide.

32.11.5 *Helichrysum umbraculigerum* Less. (Asteraceae)

Cannabigerol has been reported.

32.11.6 *Radula marginata* Taylor

Cannabinoid type bibenzyl compounds: perrottetinenic acid (Fig. 32.15), perrottetinene (Fig. 32.16), and isoperrottetin A (Toyota et al. 2002). The chemical structure of perrottetinene resembles that of THC. It is thought that perrottetinene may also be an active cannabinoid agonist although detailed pharmacological investigation of the compound has yet to be reported.

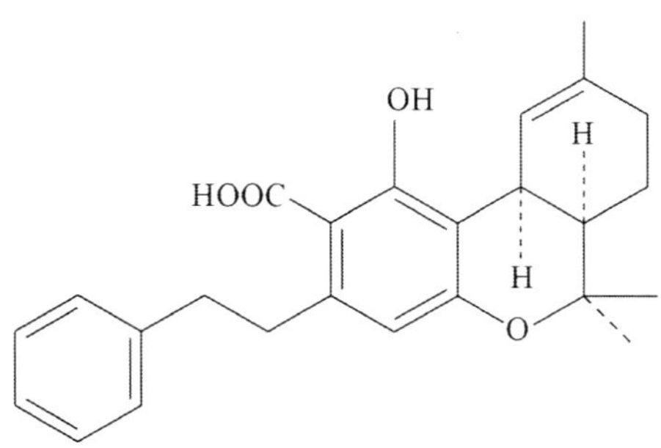

Fig. 32.15: Structure of perrottetinenic acid.

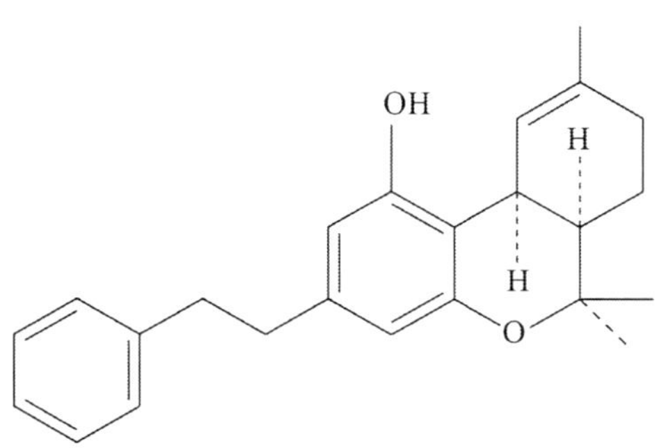

Fig. 32.16: Structure of perrottetinene.

32.11.7 *Rutamarin from Ruta graveolens*

Rutamarin (Fig. 32.17) in *Ruta graveolens* has micromolar affinity for CB_2 (Rollinger et al. 2009).

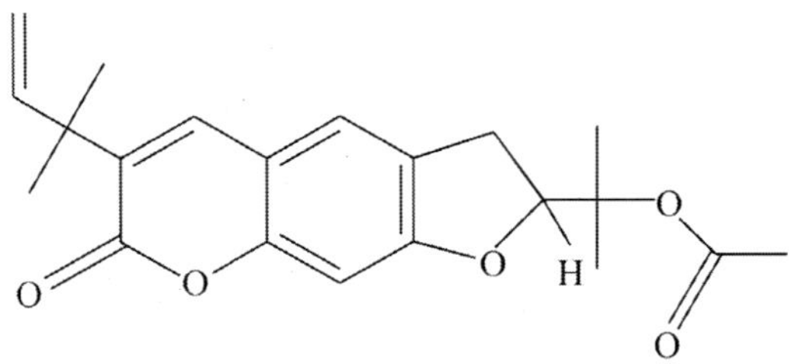

Fig. 32.17: Structure of rutamarin.

32.11.8 *(E)-β-caryophyllene*

(*E*)-β-caryophyllene [(*E*)-BCP] (Fig. 32.18) selectively binds to the CB_2 receptor ($K_i = 155 \pm 4$ nM) and is a functional CB_2 agonist. (*E*)-BCP is a major component in *Cannabis*. Upon binding to the CB_2 receptor, (*E*)-BCP inhibits adenylate cylcase, leads to intracellular calcium transients, and weakly activates the mitogen-activated kinases Erk1/2 and p38 in primary human monocytes.

(*E*)-BCP (500 nM) has inhibitory lipopolysaccharide (LPS)-induced proinflammatory cytokine expression in peripheral blood and attenuates LPS-stimulated Erk1/2 and JNK1/2 phosphorylation in monocytes. Furthermore, peroral (*E*)-BCP at 5 mg/kg strongly reduces the carrageenan-induced inflammatory response in wild-type mice but not in mice lacking CB_2 receptors, providing evidence that it exerts cannabimimetic effects *in vivo* (Gertsch et al. 2008).

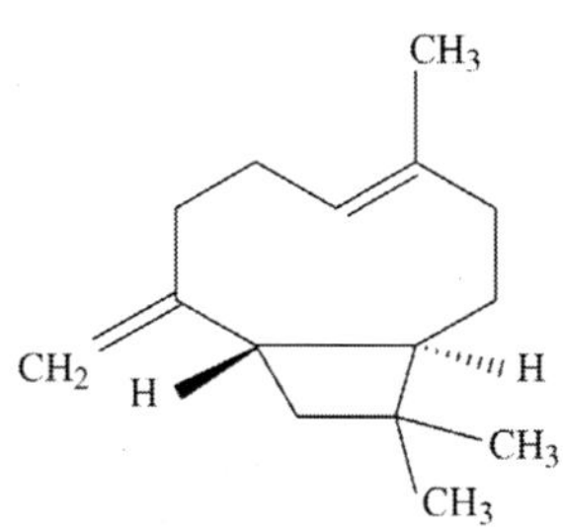

Fig. 32.18: Structure of β-caryophyllene.

32.11.9 *Falcarinol (panaxynol, carotatoxin) from* Seseli praecox

Falcarinol (Fig. 32.19) is a fatty alcohol found in members of Apiaceae including carrots, parsley, and celery, and in *Panax ginseng*. Falcarinol was isolated from *Seseli praecox*, a plant endemic to Sardinia. Falcarinol exhibited binding affinity to both human CB receptors but selectively alkylates the anandamide binding site in the CB(1) receptor (K(i) = 594 nM), acting as covalent inverse agonist in CB(1) receptor-transfected CHO cells.

In human HaCaT keratinocytes falcarinol resulted in increased expression of the pro-allergic chemokines IL-8 and CCL2/MCP-1 in a CB(1) receptor-dependent manner. Moreover, falcarinol inhibited the effects of anandamide (Fig. 32.20) on TNF-alpha stimulated keratinocytes. *In vivo*, falcarinol strongly aggravated histamine-induced oedema reactions in skin prick tests.

The findings suggest anti-allergic effects of anandamide and that falcarinol-associated dermatitis is due to antagonism of the CB(1) receptor in keratinocytes, leading to increased chemokine expression and aggravation of histamine action (Leonti et al. 2010).

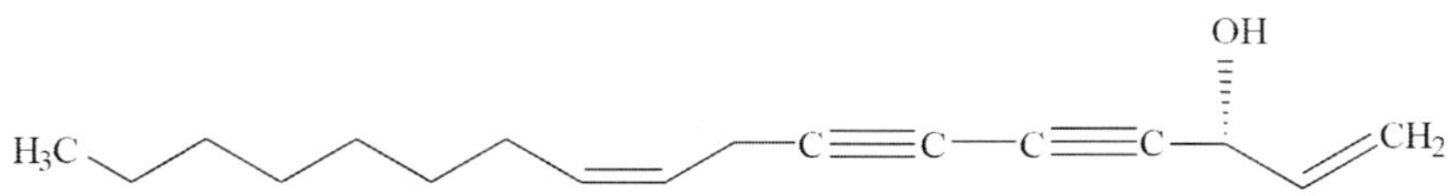

Fig. 32.19: Structure of (R)-(−)-Falcarinol.

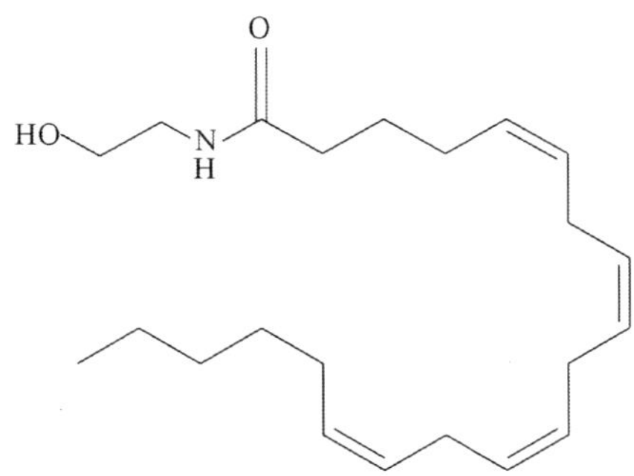

Fig. 32.20: Structure of anandamide (*N*-arachidonoylethanolamine).

32.11.10 Lyngbya majuscula Harvey ex Gomont (Oscillatoriaceae)

Cyclopropyl-containing matabolites (grenadadiene, debromogrenadiene, and grenadamide) were isolated from *L. majuscula* collected from Grenada. Grenadamide exhibited modest brine shrimp toxicity (LD50 = 5 microg/mL) and cannabinoid receptor binding activity (Ki = 4.7 microM) (Sitachitta and Gerwick 1998).

NMR-guided fractionation of two independent collections of the marine cyanobacteria *Lyngbya majuscula* obtained from Papua New Guinea and

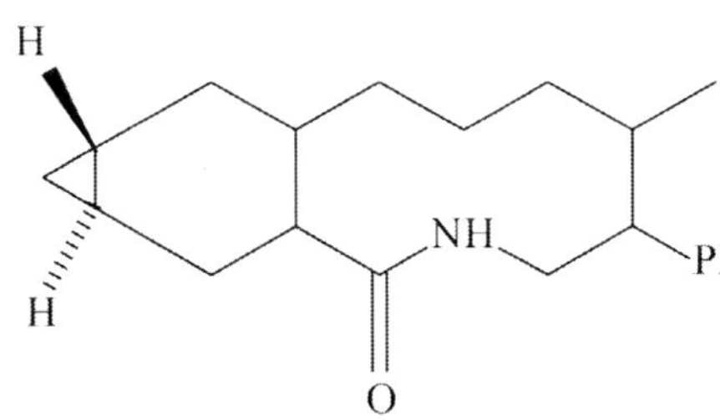

Fig. 32.21: Structure of grenadamide.

Oscillatoria sp. collected in Panama led to the isolation of the new lipids serinolamide A (Fig. 32.22) and propenediester. Serinolamide A exhibited a moderate agonist effect and selectivity for the CB1 cannabinoid receptor (Ki = 1.3 μM, > 5-fold) (Gutierrez et al. 2011).

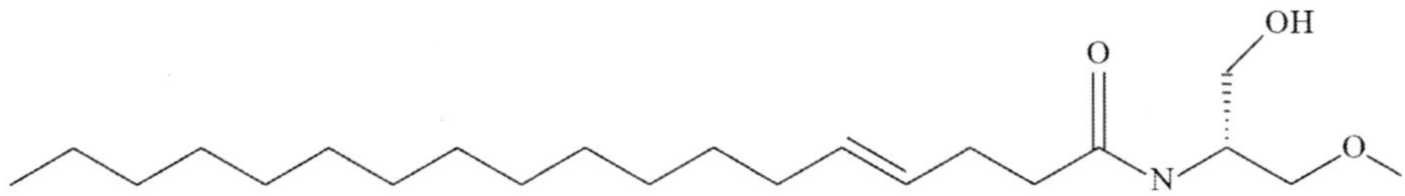

Fig. 32.22: Structure of serinolamide A.

Further Readings

Appendino G, Chianese G, Taglialatela-Scafati O. Cannabinoids: occurrence and medicinal chemistry. *Curr Med Chem* 2011; **18**: 1085–99.

Appendino G, Taglialatela-Scafat O. Cannabinoids: chemistry and medicine. *Nat Prod* 2013; **18**: 3415–3435.

Argurell S, Dewey WL, Wilette RE (eds.). The Cannabinoids: Chemical, Pharmacologic and Therapeutic Aspects. Orlando: Academic Press, 1984.

Auvinen A, Peltola J. On oenodotes and oenotherapy-wine in medicine. *Duodecim* 1999; **115**(23): 2623–32.

Botta B, Gacs-Baitz E, Vinciguerra V, Delle Monache G. Three isoflavanones with cannabinoid-like moieties from Desmodium canum. *Phytochemistry* 2003; **64**: 599–602.

Cannabis. Merck's Index. An Encyclopedia for the Chemist, Pharmacist and Physician Rahway, NJ: Merck & Co., Inc. 1930; 4th ed., p. 147.

Cannabis. *The* Merck Manual of Therapeutics and Material Medica. A Source of Ready Reference for the Physician 1940; 7th ed., p. 1356.

Chaturvedi GN, Tiwari SK, Rai NP. Medicinal use of opium and cannabis in medieval India. *Indian J Hist Sci* 1981; **16**: 31–5.

Cushny A. Cannabis indica. *In*: A Textbook of Pharmacology and Therapeutics or the Actions of Drugs in Health and Disease 1906; 4th ed., pp. 232–234.

Dilling WJ. Cannabis indica. In Bruce and Dilling's Materia Medica and Therapeutics an Introduction to the Rational Treatment of Disease 1933; 14th Rev. ed., p. 383.

Dominik W, Gupta RD, Reynolds L, Brady CM. History of medicinal use of Cannabis in ancient India. *J Urol* 2003; **169**: 253.

Dwarakanath C. Use of opium and Cannabis in the traditional systems of medicine in India. *Bull Narcotics* 1965; 17, 1, January-March, W.H.O., Geneva.

El Feraly FS, Turner CE. Alkaloids of *Cannabis sativa* leaves. *Phytochemistry* 1975; **14**: 2304.

Elsohly MA, Turner CE, Phoebe CH, Knapp JE, Schiff PL, Slatkin DJ. Anhydrocannabisativine, a new alkaloid from *Cannabis sativa* L. *J Pharm Sci* 1974; **67**: 124.

Ethan R. Cannabis in India: ancient lore and modern medicine. R. Mechoulam (ed.). 2005; 1–22.

Frazzetto G. Does marijuana have a future in pharmacopoeia? *EMBO Rep* 2003; **4**: 651–653.

Gertsch J, Leonti M, Raduner S, Racz I, Chen JZ, Xie XQ, Altmann KH, Karsak M, Zimmer A. Beta-caryophyllene is a dietary cannabinoid. *Proc Natl Acad Sci USA* 2008; **105**: 9099–104.

Gutierrez M, Pereira AR, Debonsi HM, Ligresti A, Di Marzo V Garwick WH. Cannabinomimetic lipid from a marine cyanobacterium. *J Nat Prod* 2011; **74**: 2313–2317.

Hohmann J, Rédei D, Forgo P, Szabó P, Freund TF, Haller J, Bojnik E, Benyhe S. Alkamides and a neolignan from Echinacea purpurea roots and the interaction of alkamides with G-protein-coupled cannabinoid receptors. *Phytochemistry* 2011; **72**: 1848–53.

Hou JP. The development of Chinese herbal medicine and the Pen-ts'ao. *Comp Med East West* 1977; **5**: 117–22.

Iorno I. The therapeutic uses of *Cannabis sativa* (L.) in Arabic medicine. *J Cannabis Ther* 2001; **1**: 63–70.

Joe Z, Stark H, Seligman J, Levy R, Werker E, Breuer A, Mechoulam R. Early medical use of Cannabis. *Nature* 1993; **363**: 215.

Joyce CBR, Curry SH. The Botany and Chemistry of Cannabis. Churchill: London, 1970.

Kalant OJ. Report on the Indian hemp drugs commission 1893–94; a critical review. *Int J Addict* 1972; **7**: 177–96.

Kaplan J. Marijuana-Report of the Indian Hemp Drugs Commission, 1893–1894. Thomas Jefferson Publishing Co., Silver Spring, MD, 1969.

Lambert DM. Medical use of cannabis through history. *J Pharm Belg* 2001; **56**: 111–8.

Le Dain, G. Interim Report of the Commission of Inquiry into the Non-medical Use of Drugs. Ottawa, ON: Information Canada 1970.

Leonti M, Casu L, Raduner S, Cottiglia F, Floris C, Altmann KH, Gertsch J. Falcarinol is a covalent cannabinoid CB1 receptor antagonist and induces pro-allergic effects in skin. *Biochem Pharmacol* 2010; **79**: 1815–26.

Li HL, Lin H. An archaeological and historical account of cannabis in China. *Econ Bot* 1974; **28**: 437–47.

Lozano I. Therapeutic use of *Cannibis sativa* L. in Arab medicine. *Asclepio* 1997; **49**: 199–208.

Makriyannis A, Rapaka RS. The medicinal chemistry of cannabinoids: an overview. *NIDA Res Monogr* 1987; **79**: 204–10.

Mayor's Committee on Marihuana. The Marihuana Problem in the City of New York: Sociological, Medical, Psychological and Pharmacological Studies. Lancaster, PA: The Jaques Cattell Press 1994.

Mechoulam R. The pharmacohistory of Cannabis sativa. pp. 1–19. *In*: R. Mechoulam ed., 1986.

Mechoulam R, Hanuš L. A historical overview of chemical research on cannabinoids. *Chem Phys Lipids* 2000; **108**: 1–13.

Meschler JP, Howlett AC. Thujone exhibits low affinity for cannabinoid receptors but fails to evoke cannabimimetic responses. *Pharmacol Biochem Behav* 1999; **62**: 473–480.

Mikuriya TH. Physical, mental, and moral effects of marijuana: the Indian Hemp Drugs Commission Report. *Int J Addict* 1968; **3**: 2.

O'Neil, MJ et al. (eds.). Cannabis. *In*: The Merck index. An Encyclopedia of Chemicals, Drugs, and Biologicals. Whitehouse Station, NJ: Merck & Co., Inc. 2001; 13th ed., p. 292.

Osbaldeston TA. De Materia Medica: Being an Herbal with Many Other Medicinal Materials. Johannesburg: Ibidis Press, 2000.

Pharmaceutical Society of Great Britain. Cannabis. *In*: The British Pharmaceutical Codex, 1934: An Imperial Dispensatory for the Use of Medical Practitioners and Pharmacists 1934; p. 270.

Raduner S, Majewska A, Chen J-Z, Xie X-Q, Hamon J, Faller B, Altmann K-H, Gertsc J. Alkylamides from echinacea are a new class of cannabinomimetics. *Biol Chem* 2006; **281**: 14192–14206.

Robinson R. The Great Book of Hemp: The Complete Guide to the Environmental, Commercial, and Medicinal Uses of the World's Most Extraordinary Plant. Rochester, VT: Park Street Press 1996.

Rollinger JM, Schuster D, Danzl B, Schwaiger S, Markt P, Schmidtke M, Gertsch J, Raduner S, Wolber G, Langer T, Stuppner H. *In silico* target fishing for rationalized ligand discovery exemplified on constituents of *Ruta graveolens*. *Planta Med* 2009; **75**: 195–204.

Russo E. Cannabis for migraine treatment: the once and future prescription? An historical and scientific review. *Pain* 1998; **76**: 3–8.

Shafer RP. Marihuana—A Signal of Misunderstanding. The Official Report of the National Commission on Marihuana and Drug Abuse. New York, NY: Signet. 1972.

Sitachitta N, Gerwick WH. Grenadadiene and grenadamide, cyclopropyl-containing fatty acid metabolites from the marine cyanobacterium *Lyngbya majuscula*. *J Nat Prod* 1998; **61**: 681–684.

Squire PW. *Cannabis indica*. Indian hemp. *In*: Companion to the Latest Edition of the British Pharmacopoeia 1899; 17th ed., pp. 179–181.

Toyota M, Shimamura T, Ishii H, Renner M, Braggins J, Asakawa Y. New nibenzyl Cannabinoid from the New Zealand Liverwort *Radula marginata*. *Chem Pharm Bull* 2002; **50**: 1390–1392.

Turner CE, Hsu MH, Knapp JE, Schiff PL, Slatkin DJ. Isolation of cannabisativine, an alkaloid, from *Cannabis sativa* L. root. *J Pharm Sci* 1976; **65**: 1084–5.

U.K. Home Office. Cannabis: Report by the Advisory Committee on Drug Dependence. Home Office, Her Majesty's Stationery Office,

Whiting RL, Ford AC. Efficacy of traditional chinese medicine in functional constipation. *Am J Gastroenterol* 2011; **106**(5): 1003; author reply 1003–4.

Woelkart K, Xu W, Pei Y, Makriyannis A, Picone RP, Bauer R. The endocannabinoid system as a target for alkamides from *Echinacea angustifolia* roots. *Planta Med* 2005; **71**: 701–5.

Wujastyk D. Cannabis in traditional Indian herbal medicine. *In*: A. Salema (ed.). Wellcome Library: London 2002; pp. 45–73.

Zuardi AW. History of cannabis as a medicine: a review. *Rev Bras Psiquiatr* 2006; **28**: 153–7.

Chapter 33

Leonotis leonurus (L.) R. Br. (Lamiaceae)-Wild Hemp

33.1 Common name: Lion's Ear, Lion's Tail, Wild Dagga, Dacha, Daggha (Africa), Wild Hemp, Minaret Flower, Flor de Mundo, Mota (Mexico), and wild dagga (this name links the plant with cannabis).

33.2 Habitat: Native to South Africa and South America.

33.3 Botany: *L. leonurus* is a robust shrub which grows up to 2–3 m tall and 1.5 m wide. Stems are velvety and woody at the base. The leaves are long, narrow, rough above, velvety below, with serrated edges. The wild dagga flowers profusely in autumn with its characteristic bright orange flowers carried in compact clusters in whorls along the flower stalk. Apricot and creamy white flowered forms are also found.

33.4 Phytochemistry: Alkaloid: leonurine (Hayashi, 1962) and diterpenoids: leonurenones A–C, 14α-hydroxy-9α, 13α-epoxylabd-5(6)-en-7-on-16, 15-olide and 13ξ-hydroxylabd-5(6), 8(9)-dien-7-on-16, 15-olide (He et al. 2012; Narukawa et al. 2015), and luteolin 7-*O*-β-glucoside and luteolin.

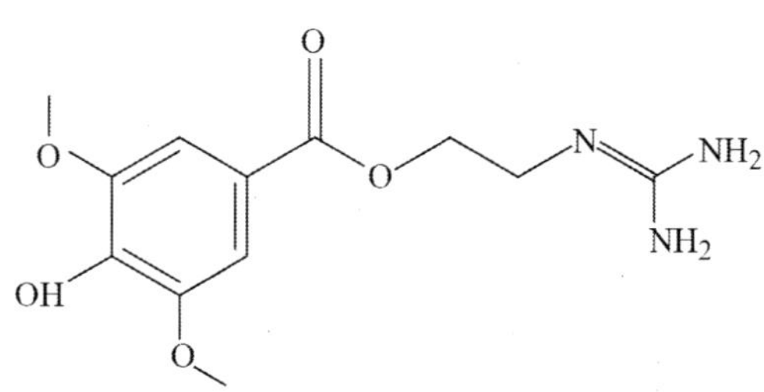

Fig. 33.1: Structure of leonurine.

33.5 Therapeutics: *L. leonurus* is used in the treatment of pyrexia, haemorrhoids, eczema, skin rashes, boils, itching, muscular cramps, headache, epilepsy, chest infections, constipation, and spider and snake bites. The plant has a mild hallucinogenic effect. The effect is seen in buds or leaves that are dried and smoked. The dried leaves and flowers are smoked to relieve epilepsy (Nsuala et al. 2015).

In some cultures, people smoke *L. leonurus* with cannabis or as a substitute for marijuana. In Mexico, *L. leonurus* is known as *flor de mundo* and *mota* and is used as a substitute for Cannabis. Some communities of South Africa use *L. leonurus* in the treatment of arthritis and inflammation and type-2 diabetes mellitus.

33.6 Pharmacology of *L. leonurus*

Antinociceptive, anti-inflammatory, and *antidiabetic*: *L. leonurus* leaf aqueous extract in dose of 50–800 mg/kg, intra-peritoneally resulted in a dose-dependent and significant ($p < 0.05$–0.001) antinociceptive effects against thermally and chemically induced nociceptive pain stimuli in mice. *L. leonurus* leaf aqueous extract in the same dose significantly ($p < 0.05$–0.001) inhibited fresh egg albumin-induced paw edema, and caused significant ($p < 0.05$–0.001) hypoglycemic effects in rats (Ojewole 2005).

Anti-epileptic: Water extract of *L. leonurus* in the doses of 200 and 400 mg/kg resulted in delayed pentylenetetrazole (90 mg/kg)-induced tonic seizures. Similarly, water extract of *L. leonurus* in same doses significantly ($p < 0.05$; Student's t-test) delayed the onset of tonic seizures produced by picrotoxin (8 mg/kg) and N-methyl-DL-aspartic acid (400 mg/kg) (Bienvenu et al. 2002).

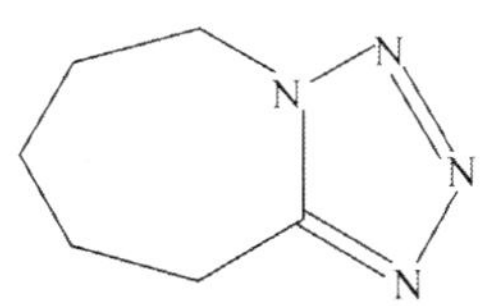

Fig. 33.2: Structure of pentylenetetrazole.

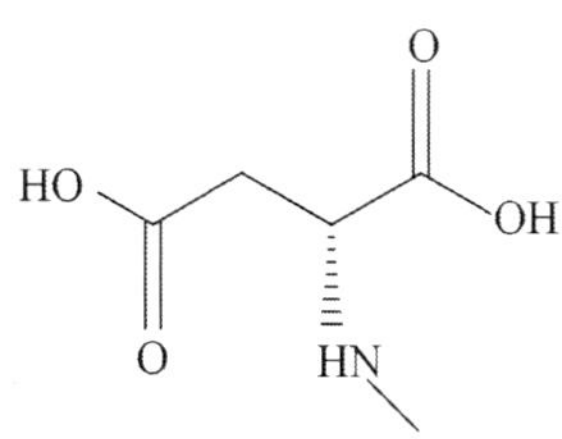

Fig. 33.3: Structure of N-methyl-DL-aspartic acid.

33.7 Pharmacology of Leonurine

Neuroprotective: An investigatory study was undertaken for evaluating therapeutic effect of leonurine on ischemic stroke. Middle cerebral artery occlusion was selected as the model of study. The animals were pretreated with leonurine orally for seven days and the surgery was performed.

In *in vivo* experiments, the alkaloid pretreatment resulted in reduction of infarct volume, improved neurological deficit in stroke groups, and increased activities of antioxidant enzymes. It further and decreased levels of malondialdehyde (the lipid peroxidation marker) (Fig. 33.4).

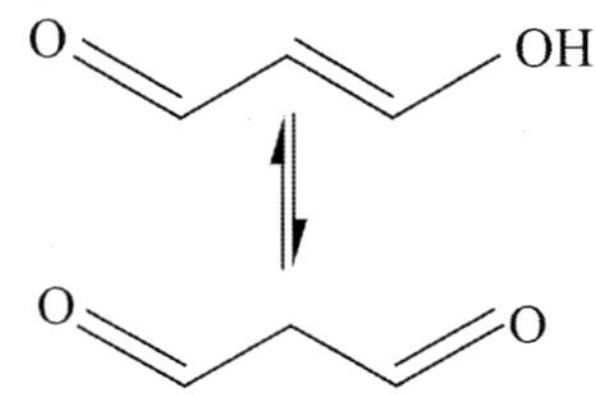

Fig. 33.4: Structure of malondialdehyde.

Cardioprotective: In this study, cardioprotective effects of leonurine were studied. The alkaloid pretreatment concentration-dependently attenuated lipopolysaccharide-induced mRNA expression of intercellular adhesion molecule-1 (ICAM-1), vascular cell adhesion molecule-1 (VCAM-1), E-selectin, and monocyte chemoattractant protein-1.

Lipopolysaccharide-mediated expression/release of ICAM-1, VCAM-1, and cyclooxygenase-2, and tumor necrosis factor-α was reduced by leonurine. The alkaloid exerted anti-inflammatory response through inhibition of reactive oxygen species and NF-κB signaling pathways (Liua et al. 2012).

Anti-tumour: A study investigated effects of leonurineon human non-small cell lung cancer H292 cells. After treatment with different concentrations of leonurine (0, 10, 25, and 50 μmol/L) for 6, 12, 24, 48, and 72 h, the cell viability was assessed by the MTT assay. The alkaloid significantly inhibited the proliferation of H292 cells in a time- and dose-dependent manner, and induced G0/G1 cell-cycle arrest (Mao et al. 2015).

Nephroprotective: A study examined the effect of leonurine on lipopolysaccharide-induced acute kidney injury in mice. The results suggest that the alkaloid suppress NF-κB activation and inhibited pro-inflammatory cytokine production via decreasing cellular reactive oxygen species production. Leonurine reduces kidney injury and protects renal functions from lipopolysaccharide-induced kidney injury (Xu et al. 2014).

Further Readings

Agnihotri VK, ElSohly HN, Smillie TJ, Khan IA, Walker LA. Constituents of *Leonotis leonurus* flowering tops. *Phytochem Lett* 2009; **2**: 103–105.

Bienvenu E, Amabeoku GJ, Eagles PK, Scott G, Springfield EP. Anticonvulsant activity of aqueous extract of *Leonotis leonurus*. *Phytomedicine* 2002; **9**: 217–223.

Hayashi Y. Studies on the ingredients of *Leonurus sibiricus* L. II. Structure of leonurine (2) Yakugaku Zasshi 1962; **82**: 1025–1027.

He F, Lindqvist C, Hardinga WW. Leonurenones A–C: Labdane diterpenes from *Leonotis leonurus*. *Phytochemistry* 2012; **83**: 168–172.

Liua XH, Pana LL, Yanga HB, Gonga QH, Zhua YZ. Leonurine attenuates lipopolysaccharide-induced inflammatory responses in human endothelial cells: involvement of reactive oxygen species and NF-κB pathways. *European J Pharmacol* 2012; **680**: 108–114.

Loh KP, Qi J, Tan BK, Liu XH, Wei BG, Zhu YZ. Leonurine protects middle cerebral artery occluded rats through antioxidant effect and regulation of mitochondrial function. *Stroke* 2010; **41**: 2661–8.

Mao F, Zhang L, Cai MH, Guo H, Yuan HH. Leonurine hydrochloride induces apoptosis of H292 lung cancer cell by a mitochondria-dependent pathway. *Pharm Biol* 2015; **53**: 1684–90.

Narukawa Y, Komori M, Niimura A, Noguchi H, Kiuchi F. Two new diterpenoids from *Leonotis leonurus* R. *Br J Nat Med* 2015; **69**: 130–4.

Nsuala BN, Enslin G, Viljoen A. "Wild cannabis": a review of the traditional use and phytochemistry of *Leonotis leonurus*. *J Ethnopharmacol* 2015; **174**: 520–39.

Ojewole JA. Antinociceptive, anti-inflammatory and antidiabetic effects of *Leonotis leonurus* (L.) R. BR. [Lamiaceae] leaf aqueous extract in mice and rats. *Meth Find Exp Clin Pharmacol* 2005; **27**: 257–64.

Xu D, Chen M, Ren X, Ren X, Wu Y. Leonurine ameliorates LPS-induced acute kidney injury via suppressing ROS-mediated NF-κB signaling pathway. *Fitoterapia* 2014; **97**: 148–55.

ANNEXURES

ANNEXURE 1: REGULATION OF HERBAL MEDICINE IN NETHERLANDS

Authorisation of Traditional Herbal Medicine

For a traditional herbal medicine the following procedures are available:

* National procedure
* Decentralised procedure
* Mutual recognition procedures

The decentralised and mutual recognition procedures can only be completed if:

(a) A Community herbal monograph has been drawn up, or

(b) The herbal medicine consists of herbal substances, herbal preparations or combinations thereof, which are mentioned in the list drawn up by EMA's Committee for Herbal Medicinal Products.

Bringing traditional herbal medicine to the market

Up to now only a few herbal preparations have been authorized as medicinal products in the Netherlands. Most herbal preparations are not registered. Medicines with recognized efficacy are generally used in medical practice; however, registration issues are major barriers in enhancing the accountability of the herbal medicines.

The EU member states have introduced a range of different procedures and provisions to keep these herbal preparations on the market. These differences have various negative consequences: they impede the harmonisation of the EU market in herbs, and might be detrimental to the protection of public health since the necessary guarantees of quality, safety and efficacy are not always currently available. That is why European pharmaceutical legislation was changed on 1 November 2005 by the introduction of a simplified marketing authorisation procedure for traditional herbs.

Applicants can ask The Medicines Evaluation Board (MEB) for advice on dossier requirements prior to submitting a dossier. They then submit a dossier containing all the necessary information for assessment. The Botanicals department of the MEB assesses the application on the basis of criteria laid down in the Dutch Medicines Act

(Geneesmiddelenwet) and establishes the conditions under which the product can be allowed onto the Dutch market. Safety and quality are the key criteria. Efficacy is assessed on the basis of a simplified procedure (see under general information). Once the MEB has given a positive assessment of the medicinal product, the manufacturer receives marketing authorisation. The medicinal product is then added to the Register of Medicinal Products and given a marketing authorisation number.

The "Summary of Product Characteristics" or product information is part of the marketing authorisation. This is the scientific text which contains all the key data about the product. Package leaflets are based on this text. Manufacturers submit a draft for these texts but the final version is drawn up by the MEB.

The MEB also decides on the legal status of the medicinal product, indicating whether it can only be obtained with a GP's prescription or can be purchased without a prescription from a drugstore, for example. Applicants have the choice between two forms of marketing authorisation for a medicinal product: a national marketing authorisation and a European marketing authorisation.

After marketing authorisation has been granted, any adverse events can be recorded. Pharmacovigilance is the process by which unwanted side-effects are detected, evaluated and wherever possible prevented. The MEB is responsible for pharmacovigilance with regard to authorised medicinal products.

Dossier requirement for traditional herbal medicine products

Applications for traditional herbal medicinal products must be supplied in NTA format. More information about this can be found on the European Commission website.

Directive 2001/83/EC does not contain any specific requirements for the pharmaceutical quality of traditional herbal medicinal products. Consequently, the normal quality requirements apply to traditional herbal medicinal products. Information about the quality requirements can be found in the European pharmacopoeia and on the EMA website.

Safety

Safety has to be demonstrated with a bibliography of data relating to safety (literature) and an experts' report. The report must show that the literature data is applicable. Additionally, data from experimental studies may be provided. Criteria for evidence of safety can be found in the **HMPC guideline on 'Non-clinical documentation for herbal medicinal products in applications for marketing authorisation (bibliographical and mixed applications) and in applications for simplified registration'**. If the herbal substances or herbal preparations are included in the Community list and if the preparation plus recommended doses for the product to be authorised match the product on the list, then a bibliography does not have to be provided. A reference to the list will then be sufficient.

Contra-indications

Medicinal products are granted marketing authorisation on the basis of the safety-efficacy ratio. As the efficacy of traditional herbal medicinal products is not demonstrated in the usual way, the MEB takes a cautious approach to accepting contra-indications or adverse events for these products. This means that contra-indications which are accepted for other OTC medicinal products of proven efficacy are not necessary acceptable for traditional herbal medicinal products without further consideration.

Evidence of efficacy

The efficacy of traditional herbal medicinal products does not have to be supported using clinical tests. The pharmacological effects and efficacy must be shown to be plausible on the basis of long-standing use and experience. It should be noted that under Directive 2001/83/EC, simplified registration is not available for products for which there is sufficient scientific evidence of efficacy.

In the case of traditional herbal medicinal products that appear on the Community list, a reference to this list is sufficient. If an established Community monograph for traditional use exists for the product, additional data may be requested to support the product's efficacy. To support its traditional use, applicants may refer to the section in the established Community monograph dealing with traditional use.

Efficacy can be shown by, inter alia:

- data from experience
- manuals
- results of pharmacological studies
- case studies

Indications

Traditional herbal medicinal products may only be recommended for use with conditions that do not require a doctor to establish the diagnosis, prescribe treatment, or monitor the patient. In other words: only OTC (over-the-counter) indications are permissible.

ANNEXURE 2: REGULATION OF HERBAL MEDICINE IN IRELAND

Scope

This guideline concerns the traditional herbal medicinal products registration scheme, under which applications may be made for the granting of certificate of traditional-use registration to relevant herbal medicines. It aims to provide information and guidance on the documents and particulars required to make such an application.

Introduction

The Directive on Traditional Herbal Medicinal Products (2004/24/EC) as published in the Official Journal of the European Union (Ref: OJ L 136, 30.04.2004, p. 85) has now been transposed into Irish law by the Department of Health and Children. The Medicinal Products (Control of Placing on the Market) Regulations 2007 (S.I. No. 540 of 2007) were implemented on 23 July 2007 by the Minister for Health and Children.

This legislation is designed to provide an appropriate legal framework for placing traditional herbal medicinal products on the market within the European Community. It introduces a simplified registration procedure that gives traditional herbal medicinal products recognition and enhanced status, while ensuring protection of public health.

The Department of Health and Children designated the HPRA as the competent authority for implementation of this legislation and on this basis the HPRA has established the Traditional Herbal Medicinal Products Registration Scheme. Under this registration scheme an applicant can apply for a certificate of traditional-use registration for their traditional herbal medicinal product. A registration will be called a traditional-use registration and will be allocated a TR number.

The following guidance is for applicants to the traditional herbal medicinal products registration scheme on the format and content of applications. Applicants also should ensure that they are familiar with the relevant EU legislation and guidelines published for human medicines including:

- Directive 2001/83/EC, as amended by 2004/24/EC and 2004/27/EC available on the European Commission website.
- EUDRALEX Volume 2—Pharmaceutical Legislation : Notice to Applicants available on the European Commission website.

Scientific guidelines for human medicinal products published by the EMA and available on their website.

Format and Content of the Dossier

The format of the dossier is based on the Common Technical Document (CTD). The CTD is an internationally agreed structure and format for an application dossier and is the format currently used for marketing authorisation applications. General guidance on the compilation of dossiers in CTD format is given by the European Commission and the Committee on Herbal Medicinal Products at the EMA has also prepared guidance in submitting an application for a traditional herbal medicinal product in this format. These documents should be consulted, available on the European Commission website and the EMA website:

- Notice to Applicants-Volume 2B, incorporating the Common Technical Document (CTD) (June 2006).
- Committee on Herbal Medicinal Products (HMPC)-Guideline on the use of the CTD format in the preparation of a registration application for traditional herbal medicinal products-EMEA/HMPC/71049/2007.

Application form

An application form must be submitted as part of Module 1 of the application dossier. The application form for a certificate of traditional-use registration is the EU application form required for all medicinal product applications, available on the European Commission website:

- Application Form: Module 1.2 Application form February 2007 – pdf document.
- Application Form: Module 1.2 Application form February 2007 – word document.

Product Information

A Summary of Product Characteristics (SPC) is required as part of the product information in Module 1 of the application. The SPC includes the name of the product, strength, pharmaceutical form, quantity of active ingredients, posology, method of administration, indications, contraindications, excipients, shelf life and any special warnings and precautions for use, etc. According to Article 16c of Directive 2004/24/EC, section 5 of the SPC, which relates to pharmacodynamic, pharmacokinetic and pre-clinical data, is not required for traditional herbal medicinal products.

In addition to the Notice to Applicants SPC guideline, the Committee on Herbal Medicinal Products has published guidance on the quantitative and qualitative declaration of the active substance in section 2 of the SPC for herbal medicinal products, available on the European Commission website

- Guideline on Summary of Product Characteristic, Revision 1, October 2005.
- Guideline on Declaration of Herbal Substances and Herbal Preparations in Herbal Medicinal Products/Traditional Herbal Medicinal Products in the SPC.

The proposed product labelling and package leaflet must be submitted as part of the product information in Module 1. The proposed label information and the user package leaflet should be in English and meet the requirements of Articles 54 to 65 of Directive 2001/83/EC, as amended. It will be necessary to submit a mock-up of the label and package leaflet for each product and strength. Article 56a of Directive 2001/83/EC as amended, requires certain information on the packaging and package leaflet to be in Braille for the blind and partially sighted. Please see information on these requirements on the 'Publications and Forms' section of www.hpra.ie.

In addition to this, Article 16(g)(2) of Directive 2004/24/EC requires the labelling and user package leaflet of any relevant product to contain a statement to the effect that:

...this is a traditional herbal medicinal product for use in specified indication(s) exclusively based upon long-standing use; and the user should consult a doctor or a qualified health care practitioner if the symptoms persist during the use of the medicinal product or if adverse effects not mentioned in the package leaflet occur.

The following guidance should also be consulted, available on the European Commission website and the EMA website:

- Guidance concerning the Braille requirements for labelling and the package leaflet (Article 56a of Directive 2001/83/EC amended by Directive 2001/27/EC).
- Volume 3B Guidelines-Medicinal Product for human use-Safety, environment and information-Excipients in the label and package leaflet of medicinal products for human use, July 2003.
- A guideline on the readability of the label and package leaflet of medicinal products for human use.

Quality

The quality aspect of a medicinal product is independent of traditional use and so the normal quality requirements applicable to all authorised medicinal products, also apply to traditional herbal medicinal products for human use.

In addition to the EU quality guidance on medicinal products for human use, specific guidance on the quality requirements for herbal medicinal products is available on the EMA website. Applicants should be familiar with all the relevant available guidance on quality when considering the quality aspects of their product.

The quality data are submitted in Module 3 of the dossier. A pharmaceutical expert is required to provide a Quality Overall Summary in Module 2.3 of the application.

Compliance with Good Manufacturing Practice (GMP) is required and there is also a requirement to hold a manufacturer's authorisation or a wholesaler's authorisation where appropriate. For further information on obtaining a manufacturer's or wholesaler's authorisation please see the 'Publications and Forms' section of www.hpra.ie.

Safety

According to Article 16c 1(d) of Directive 2004/24/EC, a bibliographic review of safety data, together with an expert report, must be submitted with each application. This review must be up-to-date, comprehensive and objective. It is in the interest of the applicant to ensure the expert compiling these reports has appropriate qualifications and experience. The HPRA, where justified, may request more data in order to assess the safety of the product. The applicant is reminded that products, including their indications must be intended and designed for use without the intervention of a medical practitioner for diagnosis, prescription or monitoring of treatment.

The bibliographic review of safety, together with the expert report, is submitted in Module 2.4 of the dossier and the supporting safety literature is submitted in Module 4.

Traditional use

A traditional herbal medicinal product, as defined in Article 16a 1 of Directive 2004/24/EC must be:

- intended and designed for use without the intervention of a medical practitioner for diagnosis, prescription or monitoring of treatment.
- taken orally, for external use or inhalation.
- administered exclusively at a specified strength and dose.
- on the market for a 'period of traditional use'.

To demonstrate 'traditional use', the applicant will need to prove that the traditional herbal medicinal product or a 'corresponding product has been in medicinal use for at least 30 years at the time of application. At least 15 years of this period must have been within the European Community. The efficacy of the product must be *'plausible on the basis of long-standing use and experience.'*

In accordance with Directive 2004/24/EC, a corresponding product refers to a product that has the same active ingredient (irrespective of excipients used), the same indication(s) for use, contains the same strength and dose, and has the same or similar route of administration.

Applicants are required to produce bibliographic or expert evidence documenting the traditional use of the product for the proposed indication. There is a wide range of possible sources which, taken together if necessary, can be used to provide the required evidence. The following are examples of the types of bibliographical and/or expert evidence which may be used:

- Information from handbooks of medicine, pharmacy, pharmacology, pharmacognosy, phytotherapy, herbal medicine, etc.
- Official expert committee reports or monographs from learned societies, such as WHO, Commission E, ESCOP and national formularies/compendia, etc.
- A monograph in the Ph. Eur. or an official national pharmacopoeia will be accepted as a general proof of medicinal use during the years the monograph has been valid. It may also provide relevant information on strength/type of extract.
- Product-related documentation, such as post-marketing studies, product information leaflets, sales catalogues, sales statistics, etc.

The bibliographic or expert evidence of traditional-use overview should be submitted in Module 2.1 of the dossier and the supporting evidence of traditional use should be submitted in Module 5.

European List

In accordance with Directive 2004/24/EC, a list of traditional herbal medicinal 'substances/preparations or combinations thereof' will be established by the European Commission on the basis of scientific advice provided by the Committee on Herbal Medicinal Products. This list will state the therapeutic indication, specified strength, dose, route of administration and any relevant safety information relating to each substance/preparation or combination. An applicant seeking to register a product containing a substance/preparation or combination on the list (in the form and for the indications specified on the list) can refer to this list rather than have to demonstrate traditional use and safety. The applicant must, however, still demonstrate quality.

The absence of a substance/preparation or combination from the positive list will not prevent a successful traditional-use registration, subject to full quality, safety and traditional-use requirements being met.

Community Herbal Monographs

The Committee on Herbal Medicinal Products also has responsibility for the development of Community herbal monographs. These are documents based on the format of a summary of product characteristics under either Article 10a for 'well-established' medicinal products (i.e., full marketing authorisations) or Article 16a(1) for traditional herbal medicinal products. Community herbal monographs may be used to support an application to the registration scheme.

Herbal monographs and list entries published by the HMPC are available from the EMA website.

Pharmacovigilance Requirements for Traditional Herbal Medicinal Products

In accordance with Article 16g of Directive 2004/24/EC, the pharmacovigilance requirements described in Articles 101–108 of Directive 2001/83 EC as amended, also apply in respect of traditional herbal medicinal products.

It is important to note that revised legislation providing for strengthened and increased harmonisation of pharmacovigilance obligations will come into force in July 2012. These revisions will introduce extensive changes to all aspects of pharmacovigilance including routine activities, such as adverse reaction reporting arrangements and requirements for periodic safety update reports (PSUR), as well as for the evaluation of new and emerging safety concerns. For further information, please see: Directive 2010/84/EU and Regulation (EU) No. 1235/2010, available on the European Commission website.

Submission of Periodic Safety updates Reports

At present, once a medicinal product is authorised in the EU, even if it is not marketed, the marketing authorisation holder is required to submit PSURs. PSURs are normally required to be prepared and submitted if requested by a national Competent Authority, at 6-monthly intervals for the first two years following the medicinal product's authorisation in the EU, annually for the following 2 years, and thereafter at 3-yearly intervals.

There may, however, be exceptions where the cycle may be re-started, or an exemption to the requirement for 6-monthly and annual PSURs is granted.

Where an amendment is proposed, the applicant should submit, as part of the application for a marketing authorisation, a reasoned request for the amendment which, if granted, becomes part of the conditions of authorisation. If an amendment is applied for after authorisation, such an application should follow the procedures for a type II variation. Refer to Volume 9A of The Rules Governing.

Medicinal Products in the European Union for further details regarding the current requirements.

The revised pharmacovigilance legislation introduces a number of new provisions, including changes to the current requirements for PSURs. In accordance with the provisions of this legislation (specifically, Article 107b (3) of Directive 2010/84/EU), it is intended to provide a general exemption for the requirement for PSUR submission for Traditional Herbal Medicinal Products, unless:

- The authorisation provides for the submission of PSURs as a condition.
- PSURs are requested by a Competent Authority on the basis of the grounds defined in legislation.
- The active substance is included on the list of Union Reference Dates (URDs) and the requirement for submission of a PSUR according to the harmonised frequency is indicated on the list in accordance with Competent Authority consultation.

Supplementary legislation and guidance is currently being developed to facilitate implementation of the new requirements and additional information regarding these developments will be highlighted at EU and national level, with public consultation related to good vigilance practice activities, expected early in 2012. The HPRA advises checking in regularly with the EMA website as well as our own to keep up to date on this evolving area, particularly in relation to the publication of the list of URDs.

ANNEXURE 3: REGULATION OF HERBAL MEDICINE IN BRAZIL

Introduction

In recent decades, efforts in Brazil have been undertaken to stimulate studies of medicinal plants, to promote training of qualified personnel on research, and to introduce herbal alternatives into basic health care. However, only in 2006 did the Brazilian Government define policies, considered a milestone for the industry of medicinal plants and herbal medicines at the time, in the form of the PNPIC (Ministerial GM/MS 971/06) and PNPMF (Decree 5,813/06). These policies contain among their guidelines, fostering of research on medicinal plants and development of herbal medicines of quality, safety and efficacy standards that must be made available to the population while prioritizing the protection of biodiversity and promoting greater access to safe and effective treatments.

The Brazilian Health Surveillance System, coordinated by ANVISA, is responsible for assessing the herbal medicine industry in Brazil, while State and Municipal assessments are performed by their respective local governments. Therefore, these two policies govern surveillance of marketing, dispensing, handling and distribution of raw materials of plant origin, as well as the registration and supervision of herbal medicine production.

Herbal Medicine Registration in Brazil

The registration of pharmaceuticals follows parameters so as to demonstrate the quality, effectiveness and safety of the product. The registration dossier consists of a documentary part, a technical report containing data on production and quality control, and a safety and efficacy report.

Herbal medicines can be registered only by pharmaceutical industries previously authorized by ANVISA to manufacture these products. The company must also confirm that it is able to manufacture pharmaceuticals in accordance with the Good Manufacturing Practices and Control protocol (BPFC). In order to achieve this certification, the company must follow internationally standardized norms, internalized in Brazil by the newly published RDC 17/10. For herbal medicine production, industries must prove they have employed a qualified pharmacist who will be the technically responsibly officer for the product.

Safety and Efficacy Information

Requirements for proof of safety and efficacy of herbal medicines have not essentially changed since RDC 48/04. Companies must submit one of the following alternatives: preclinical and clinical safety and efficacy trials must be carried out, as required to register any other medicine in Brazil; these tests may be exempted if the plant is included in the "list of simplified registration of herbal medicines" or if there is sufficient data in the literature validating its safety and efficacy. This validation is a score that can be obtained from a list of standard technical-scientific literature and books, all compiled in IN 05/10. The fourth alternative is attesting the traditional use of the product.

IN 05/10 encompasses the "list of bibliographic references for assessment of safety and efficacy of herbal medicines" and presents 35 reference books that are classified in the form of "points". Information on efficacy and safety can also be extracted from monographs and articles on plant species published in indexed journals.

The selection of the books was made through assessing technical-scientific references and those selected are divided into three groups (A, B and C), according to the theoretical reference and the monograph model for medicinal plants found in each book. Each reference must inform the herbal derivative used, the plant part this derivative is from, how it is extracted, its recommended posology and therapeutic indication in order to prove its safety and efficacy.

The simplified registration species list, IN 05/08, comprises 36 plant species. Companies must follow all parameters that are specified in the list mentioned, which are: plant organ, chemical standard/marker, plant derivative, therapeutic indications/ actions, daily dose, route of administration and restrictions of use. This list was developed considering the various scientific papers published on selected species for simplified registration. Thus, if the plant is present in this list, it is not necessary to submit any additional safety and efficacy data.

The safety and efficacy data can also be confirmed through the observation of traditional use, ethnopharmacology or ethno-oriented studies and use, technical-scientific documentation, or other publications that show a period equal to 20 years or more of proven efficacy and safety of the product. The proposed period of use for the registered medicinal product, in this case, must be occasional or short.

The use of medicinal plants, mostly in the form of infusions and decoctions, supports the ethnopharmacological studies, and the companies, when filing for registration of their herbal products, should follow the traditional use patterns as closely as possible. Such caution is necessary to ensure the similarity of the chemical profile of the registered medicinal product to its traditional use.

Only when the product safety and efficacy cannot be accomplished by one of the three mentioned ways above are preclinical and clinical trials required for a specific product. For herbal medicine products, there is a guide to conducting preclinical toxicological trials (RE 90/04) (Brasil, 2004b). This guide sets out the minimum acceptable criteria for conducting acute, sub-chronic and chronic toxicological studies. Additional reproductive toxicity tests and toxicology studies must be conducted for topical products (skin sensitivity; skin and eye irritability). The studies shall be conducted using standardized herbal medicines product samples or the herbal derivatives from which they are originated.

Once preclinical tests are finalized, clinical studies are necessary to confirm the proposed actions and safety of the drug. In addition to these studies, it is also important to ascertain pharmacokinetic and pharmacodynamic effects and occurrence of possible adverse events. There are Brazilian Health Council norms (CNS) for this purpose: Resolution 196/96, which regulates any clinical studies conducted in the country, and Resolution 251/97 (Brasil, 1997), specific for drugs and medicines. In order to complement the specific requirements for the registration of medicinal products, ANVISA also published RDC 39 in 2008 as a guide to the conduction of clinical trials.

Good Clinical Practice guidelines shall be applied to all stages of clinical trials to ensure that the requirements of quality control and ethics are covered (WHO, 2008). ANVISA's guide on Good Clinical Practice determines that, before undergoing any research, the study proposals must be approved by ANVISA then followed by a Special Announcement issuance.

The World Health Organization guide on clinical research states that the information on the plant species' traditional usage that is going to be clinically tested should be taken into consideration, and, in this case, the chemical composition and form of manufacturing should be similar to the traditional wording used. The parameters related to the quality of the product to be tested must also be observed. In the case of herbal medicines, the guarantee of the identity of the plant species, its standardization and the presence of potential contaminants and adulterations must be observed (OMS, 2008).

Registering medicinal products requires companies to prove, in addition to the safety and efficacy of the product, that they have a pharmacovigilance system that is able to detect adverse effects or harm caused by the use of their products. RDC 04/09 presents guidelines on pharmacovigilance to support the regulated sector.

Quality Control Information

It is necessary to conduct various tests in order to demonstrate product's quality. Such tests verify, for example, the identity of plant species and the absence or presence of contaminants within acceptable limits. Control is performed in all production stages: herbal drugs, herbal derivatives, and herbal medicines.

Contamination of medicinal plants by fungi may give rise to mycotoxins, e.g., aflatoxins, which are carcinogenic (Bugno, 2006). Ways of avoiding this type of contamination include constant monitoring and evaluation of these substances in raw plant materials. There is no specific regulation for aflatoxin limits in herbal medicines in Brazil. The WHO has determined the analysis of aflatoxin types B1, B2, G1, and G2. The presence of any of these in raw medicinal plant material is regarded as highly dangerous.

Waste material through ash tests; bacteria and fungi; other plant species, or even parts of the plant species other than the one indicated as a place of greater concentration of actives. The plant species gathering or cropping site and whether methods of eliminating contaminants were used must be informed followed by the search for possible residues.

For herbal derivatives, extraction methods and the detection of solvent residues should be reported. Physicochemical tests are requested including: organoleptic characterization, solid residues, pH, alcoholic content, density (for liquid extracts); moisture loss due to drying, apparent solubility and density (for dry extracts); refractive index, optical rotation (for essential oils); ester, iodine (for fixed oils) and acidity indexes.

The qualitative and quantitative analysis of markers via chromatographic techniques and spectrometric must be presented. RDC 14/09 lays down specific requisites based on quality assurance, requiring reproducibility of herbal medicines. Marker quantitative control can be replaced by the therapeutic activity's biological control.

The company that applies for the herbal medicine registration should preferably use the methodologies described in pharmacopeias recognized by ANVISA (RDC 37/2009) for quality control of herbal materials, excipients and herbal medicine. Otherwise it is necessary to validate the methodologies used. To do so, the method must present the parameters stated by RE 899/03; despite having been developed for synthetic drugs, they are also applicable to herbal medicines. Bioanalytical method parameters must be used.

Once the herbal medicine registration is granted, any changes made to the product should be communicated to ANVISA, accompanied by proof of the product's continued quality, safety and efficacy. Companies must follow the procedures specified in the "Guide to making changes and additions to post-registered herbal medicines", RE 91/04.

Other Applicable Regulations

Other regulations can be applied to herbal medicines, including: restriction of sale criteria, i.e., whether or not the medicinal product will be sold as a prescription medicine; a stability study guide, good manufacturing practices, quality control and storage outsourcing, medicinal products advertising norms, standardization of package leaflet, packaging models and wordings, and good manufacturing practices for raw material.

RDC 138/03 sets down the medicinal products that do not require prescription prior to sale, except for parenterally administered products. All therapeutic indications that are not described in this list are sold under prescription.

RE 01/05 provides for the information that must be present in all stability study reports for each pharmaceutical form, and also covers stability study monitoring and photostability. The stability study of medicines should be carried out in both accelerated mode as well as long lasting mode. The accelerated study is designed to accelerate the chemical degradation and/or physical changes of a pharmaceutical product in forced storage conditions. The long-term study is designed for the verification of physical, chemical, biological, and microbiological characteristics of a pharmaceutical product during (and optionally after) the expected expiry date. The results are used for the establishment or confirmation of the product's expiry date and recommendations for the appropriate storage conditions for the product at the marketing point and by the user.

RDC 47/09 lays down rules for patients and health professionals concerning leaflets. In parallel, package leaflets for herbal medicines have been standardized. There are leaflets standardized for 18 plant species in order to ensure that standardized product information is available to the population.

RDC 71/09 states medicine labeling parameters. It was published in order to improve the form and content of the labels of all registered medicinal products marketed in Brazil. It also introduced the obligation of manufacturing labels to appear in Braille.

Labels must contain alerts on after-preparation conservation of the medicinal products and inform that its expiry date is reduced once the package is open. The inclusion of identification and security mechanisms to enable tracking of the product from manufacturing to dispensing are also included in the labels.

ANNEXURE 4: REGULATION OF HERBAL MEDICINE IN JAPAN

Regulation of Herbal Medicines in Japan

In Japan, two overlapping types of traditional herbal medicines coexisted for centuries. The first one was the traditional Japanese and Chinese medicine. These medical systems were damaged by the first Medical Care Law in 1874 that proclaimed the abrogation of traditional Japanese medicine. The second type of herbal medicine used in Japan originated in Europe and south-east Asia and became popular after the law in 1874 was announced. Some of those products are still used today as prescription drugs. Although the renaissance of the traditional medicines has been on the rise since approximately 1960, the confusion and decline of the traditional Japanese medicines was further strengthened by introduction of dietary supplements. Regulation of herbal medicines, except 'Kampo' formulas is the same as the approval for both prescription and OTC drugs. Typical characteristics of the Japanese herbal medicines is the existence of Japanese traditional medicines, 'Kampo' formulas and combinations of the traditional medicines with vitamins and pharmaceuticals. Regulation of quality standards of those herbal products was established in Japanese Pharmacopoeia for more than 90% of them.

Kampo medicines are the main traditional herbal medicines in Japan and are classified as pharmaceuticals. They are based on ancient Chinese medicine and have evolved to the Japanese original style over a long period of time. Ethical Kampo formulations are prescribed in general practice by physician under the National Health Insurance reimbursement system. Over-the-counter (OTC) Kampo formulations can be purchased and used for self-medication in primary health care settings. Kampo medicines have a substantial role in the Japanese healthcare system.

In the early 1970s, "The Internal Assignments on the Review for Approval of OTC Kampo Products", known as "210 OTC Kampo Formulae", was published by the Ministry of Health and Welfare (currently the Ministry of Health, Labour and Welfare). In 2008, "210 OTC Kampo Formulae" was revised and presented as "The Approval Standards for OTC Kampo Products" and now 294 Kampo formulae are listed in the standards. These products have had wide spread usage in Japan. Crude drugs and Kampo extracts have been listed in The Japanese Pharmacopoeia. Both The Approval Standards and The Quality Standards play a key role in regulation of Kampo products. "Application Guideline for Western Traditional Herbal Medicines as OTC Drugs" was published in 2007. Other ethnopharmaceuticals mostly from Europe could be approved as OTC drugs in Japan.

Application of Herbal Drugs to Health Care

Kampo, derived from traditional Chinese medicine, has been adopted in Japan for centuries, and the demand for herbal drugs is increasing. At present herbal drugs are utilized in pharmaceutical forms such as granules of the extracts. A special commission has evaluated and selected traditional prescriptions for their efficacy and safety by clinical experience. The Kampo preparations are also accepted by the national health care insurance. About 80% of the plants used are imported. The Japanese Pharmacopoeia reports 116 herbal drugs, the majority of Chinese origin, under specifications established and reviewed by the Pharmacopoeia Committee. In Japan, high quality research, which has developed during the last century, has partly ascertained the active principles in the herbal drugs and pharmacological tests have also been adopted, although limitations exist in the modern pharmacological methods.

Toxicological Considerations of Kampo Medicines in Clinical Use

Kampo medicines, produced by combining multiple crude drugs, almost all of plant origin but with some of animal or mineral origin, contain great many substances. Since Kampo medicine results from the combination of many substances, their effect is a combination of the various interactions of the constituent substances. It has been demonstrated that several potential side effects such as allergic reactions, diarrhea and vomiting may be experienced when administering Kampo medicine. In addition, it has been reported that Kampo medicine may have antagonistic or synergistic interactions with western drugs or with some foods such as grapefruit juice.

ANNEXURE 5: REGULATION OF HERBAL MEDICINE IN CANADA

Introduction

All natural health products (NHPs) sold in Canada are subject to the *Natural Health Products Regulations*, which came into force on January 1, 2004. The Regulations help give Canadians access to a wide range of natural health products that are safe, effective and of high quality.

NHPs are defined in the Regulations to include traditional medicines, homeopathic medicines, vitamins and minerals, herbal remedies, probiotics, amino acids, plant isolates and essential fatty acids (like Omega-3). Health Canada, in conjunction with the new NHP Expert Advisory Committee, examine the status of bulk herbs for legislative purposes.

About the Regulations

The Natural Health Products Regulations were created after many consultations with Canadian consumers, academics, health care practitioners and industry stakeholders. They address Canadians' concerns about NHP availability and safety, as well as the House of Commons Standing Committee on Health's 53 recommendations on the regulation of natural health products (NHPs) in Canada.

To be legally sold in Canada, all natural health products must have a product licence, and the Canadian sites that manufacture, package, label and import these products must have site licences.

To get product and site licences, specific labelling and packaging requirements must be met, good manufacturing practices must be followed, and proper safety and efficacy evidence must be provided.

Licensing Requirements

The licensing requirements of the Natural Health Products Regulations apply to any person or company that manufactures, packages, labels and/or imports NHPs for commercial sale in Canada. They do not apply to health care practitioners who compound products on an individual basis for their patients, or to retailers of NHPs.

Product licensing

All natural health products must have a product licence before they can be sold in Canada. To get a licence, applicants must give detailed information about the product to Health Canada, including: medicinal ingredients, source, dose, potency, non-medicinal ingredients and recommended use(s).

Once Health Canada has assessed a product and decided it is safe, effective and of high quality, it issues a product licence along with an eight-digit Natural Product Number (NPN) or Homeopathic Medicine Number (DIN-HM), which must appear on the label. This number lets you know that the product has been reviewed and approved by Health Canada.

Because Health Canada has not yet evaluated all natural health products currently on the market, products with exemption numbers can also legally be sold in Canada. The exemption number will be listed on the product label in the form EN-XXXXXX.

These products have not been fully evaluated by Health Canada, but have gone through an initial assessment to make sure that information supporting their safety, quality and efficacy has been provided, and that specific safety criteria have been met. This will allow Canadians access to the full range of NHPs they are used to while Health Canada continues to fully assess each product.

Evidence requirements for safety and efficacy

The safety and efficacy of NHPs and their health claims must be supported by proper evidence so that consumers and Health Canada know the products are indeed safe and effective. Evidence may include clinical trial data or references to published studies, journals, pharmacopoeias and traditional resources. The type and amount of supporting evidence required depends on the proposed health claim of the product and its overall risks.

Labelling

All NHPs must meet specific labelling requirements, to help you make safe and informed choices about the NHPs you choose to use. Information required on NHP labels includes: product name product licence number quantity of product in the bottle complete list of medicinal and non-medicinal ingredients recommended use (including purpose or health claim, route of administration and dose) any cautionary statements, warnings, contra-indications and possible adverse reactions associated with the product any special storage conditions.

Site licensing

All Canadian manufacturers, packagers, labellers, and importers of natural health products must have site licenses. To get a licence, sites must maintain proper distribution records, have proper procedures for product recalls and for the handling, storage and delivery of their products, and demonstrate that they meet good manufacturing practice requirements.

Good manufacturing practices

Good Manufacturing Practices make sure proper standards and practices for the testing, manufacture, storage, handling and distribution of natural health products are met. Good Manufacturing Practices for NHPs cover: product specifications premises equipment personnel sanitation program operations quality assurance stability records sterile products lot or batch samples recall reporting.

Good Manufacturing Practices are meant to ensure safe and high quality products while giving manufacturers, packagers, labellers, importers and distributors the flexibility to implement quality systems appropriate for their product lines and businesses.

Adverse Reaction Reporting

The Natural Health Products Regulations require product licence holders to monitor all adverse reactions related to their product. License holders must report serious adverse reactions to Health Canada.

Reporting side effects is important because it helps Health Canada identify rare or serious adverse reactions, make changes in product safety information, issue public warnings and advisories, and/or remove unsafe products from the Canadian market.

Clinical Trials

A clinical trial is when natural health products are tested using human subjects. Clinical trials are intended:

- To discover or verify the product's effects
- To identify any adverse events that are related to its use
- To study its absorption, distribution, metabolism and excretion
- To test its safety or efficacy

ANNEXURE 6: REGULATION OF HERBAL MEDICINE IN AUSTRALIA

What Complementary Medicines are?

In Australia, medicinal products containing such ingredients as herbs, vitamins, minerals, nutritional supplements, homoeopathic and certain aromatherapy preparations are referred to as 'complementary medicines' and are regulated as medicines under the Therapeutic Goods Act 1989.

A complementary medicine is defined in the Therapeutic Goods Regulations 1990 (link is external) as a therapeutic good consisting principally of one or more designated active ingredients mentioned in Schedule 14 of the Regulations, each of which has a clearly established identity and traditional use.

How Complementary Medicines are Regulated in Australia?

Australia has a risk-based approach with a two-tiered system for the regulation of all medicines, including complementary medicines:

- Lower risk medicines can be listed on the Australian Register of Therapeutic Goods (ARTG).
- Higher risk medicines must be registered on the ARTG.

Some complementary medicines are exempt from the requirement to be included on the ARTG, such as certain preparations of homoeopathic medicines. The Australian Regulatory Guidelines for Complementary Medicines (ARGCM) provides detail on the regulation of complementary medicines and assist sponsors to meet their legislative obligations.

TGA post market regulatory activity of complementary medicines

TGA post-market regulatory activities relate to the monitoring of the continuing safety, quality and efficacy of listed, registered and included therapeutic goods once they are on the market. Information on the TGA's approach to managing compliance risk is available at: TGA regulatory framework. The TGA Manufacturing Quality Branch inspects manufacturers on an ongoing basis for compliance with good manufacturing practice.

Adverse events to complementary medicines

Sometimes medicines, including complementary medicines, have unexpected and undesirable effects. The TGA has a strong pharmacovigilance program, which involves the assessment of adverse events that are reported to the TGA by consumers, health professionals, the pharmaceutical industry, international medicines regulators or by the medical and scientific experts on TGA advisory committees.

Sponsors of medicines are required to report to the TGA suspected adverse reactions for their medicines that they are aware of. Guidance for sponsors is provided in 'Australian requirements and recommendations for pharmacovigilance responsibilities of sponsors of medicines'.

Advertising of complementary medicines

The marketing and advertising of therapeutic goods, including complementary medicines, is to be conducted in a manner that promotes the quality use of the product, is socially responsible and does not mislead or deceive the consumer. The advertising of therapeutic goods in Australia is subject to the advertising requirements of the

Therapeutic Goods Act, which adopts the Therapeutic Goods Advertising Code (TGAC) and the supporting Regulations, the Trade Practices Act 1975 and other relevant laws.

Purchasing complementary medicines over the Internet

Products available on international websites are not regulated by the TGA. The TGA advises that consumers do not order medicines, including dietary supplements and herbal preparations, over the Internet unless you know exactly what is in the preparation and have checked the legal requirements for importation and use in Australia. For more information refer to: Buying medicines and medical devices over the Internet.

Australian regulatory guidelines for complementary medicines (ARGCM)

Part A: General guidance on complementary medicine regulation in Australia

Part A provides an overview of the regulatory framework for complementary medicines in Australia. The guidance is provided for sponsors, healthcare professionals and the general public.

Information is provided on:

- What complementary medicines are
- Legislation applicable to complementary medicines
- The different types of complementary medicines and ingredients
- Approved terminology
- Exempt medicines
- Practitioner medicines and exemptions
- Medicine presentation
- Changes to information in the Australian Register of Therapeutic Goods (ARTG) for complementary medicines
- Post market regulatory activity
- Complementary medicine interface issues
- Appeal mechanisms for decisions made under the *Therapeutic Goods Act 1989*.

Part B: Listed complementary medicines

Part B provides guidance on the regulatory framework for ‘low risk’ listed complementary medicines. The guidance is mainly directed at sponsors and manufacturers of listed medicines.

Information is provided on:

- Ingredients and indications permitted for use in listed medicines

- Legislative requirements for listed medicines
- Quality of listed medicines
- Guidance on how to list a medicine on the ARTG.

Part C: New complementary medicine substance evaluation

This guidance is for applicants proposing new substances for use as an ingredient in listed medicines. You can submit an application for evaluation for suitability for use in listed medicines for:

- A new complementary medicine substance not currently a permitted ingredient; or
- A proposed new role or a change to a regulatory requirement of use for a permitted ingredient, for example: a proposal for an ingredient permitted for use as an excipient to be used as an active ingredient; or change to the permitted level of use; or change the permitted route of administration.

A request for evaluation of a new complementary medicine substance is considered under Regulation 16GA of the Therapeutic Goods Regulations 1990 (the Regulations). There is an associated fee.

Part D: Registered complementary medicines

This guidance is provided for an 'applicant'—the person who submits an application for a new registered complementary medicine. The applicant may or may not become the sponsor of the new registered medicine if it is approved. This guidance applies to proposed registered medicines that are eligible for evaluation by the TGA's Office of Complementary Medicines—refer to Route of evaluation for complementary medicines.

Overview of registered complementary medicines

Registered medicines are considered to be of higher risk than listed medicines based on their ingredients and/or therapeutic indications they carry. Medicines must be registered on the Australian Register of Therapeutic Goods (ARTG), where they:

- Do not solely comprise ingredients permitted for use in listed medicines or
- Contain an ingredient or ingredient component that is subject to the conditions of a Schedule or relevant appendix to the Poisons Standard (link is external); or
- Are required to be sterile; or
- Have indications that are not indications permitted for use in listed medicines.

Prior to being approved for entry on the ARTG, registered medicines are subject to critical assessment by the TGA to determine whether the proposed medicine meets the requirements for quality, safety and efficacy.

How a new registered complementary medicine is evaluated?

A request for a new registered complementary medicine substance is made under Section 23 of the Therapeutic Goods Act 1989 (link is external) (the Act). There are associated application and evaluation fees.

In determining if a medicine can be approved for registration, consideration is given to:

- Whether the quality, safety and efficacy of the medicine for the purposes for which it is to be used have been satisfactorily established;
- The presentation of the medicine is acceptable; and
- The medicine complies with all applicable legislative requirements (under section 25 of the Act).

Application phases for a new registered complementary medicine

The application phases for a new registered complementary medicine are as follows:

- Phase 1: Pre-submission meeting (recommended)
- Phase 2: Submission of application and payment of application fee
- Phase 3: Screening of application, determination and receipt of evaluation fee
- Phase 4: Evaluation
- Phase 5: Decision
- Phase 6: Implementation.

Information required in an application for a new registered medicine

An application for a new registered complementary medicine must include a comprehensive dossier of relevant safety, quality and efficacy data. It is preferred that the data is presented in an electronic format that is able to be electronically searched and copied. If provided in hard copy two complete sets must be provided. The dossier must be sequentially numbered (either the whole dossier or within each module/part).

All documents, including references, must be in English and legible. If original documentation is in another language, it should be translated to English by a certified translator and both the English version and the original document should be provided. Non-English documents without certified translations and non-certified translations will not be considered as valid data.

It is recommended that the data is presented in a manner consistent with the European Medicines Agency (EMA)—Common technical document (CTD). Although the CTD format is not a mandatory requirement for a new registered complementary medicine application, presentation in this manner will expedite evaluation. The CTD is divided into five modules and is an internationally agreed set of specifications for a submission dossier:

- Module 1: Administrative information and prescribing information for Australia
- Module 2: Summaries of quality, safety and clinical data
- Module 3: Quality
- Module 4: Nonclinical data
- Module 5: Clinical data.

Overview and summaries of quality, safety and efficacy data in an application for a new registered complementary medicine

Name

Provide the proposed name of the medicine.

Dosage of medicine

State the dosage form, dosage range, frequency and duration of use and the pack size/s.

Composition of the medicine

List the ingredients in the medicine formulation.

Route of administration for the medicine

Provide the proposed route of administration, for example: oral, topical.

Container type for medicine

Describe the container type/s, for example: 'PET bottle with child-resistant closure' or 'Blister pack in carton'.

Proposed therapeutic use of medicine

Provide the proposed therapeutic indications and state the target population.

History of use of medicine

Provide a summary of human exposure data, dietary, traditional and commercial use in Australia and internationally.

Details of the number of people estimated to have been exposed to the medicine since the start of supply should be provided and categorised, as appropriate, by indication, dosage and route of administration, treatment duration and geographical location.

Traditional use

Where evidence of traditional use is provided in the dossier, it must be demonstrated that the proposed medicine is consistent with the traditional preparation and the traditional use (including dose, route of administration and duration of use).

Traditional use cannot fully establish the safety and efficacy of a proposed new registered complementary medicine. However, traditional, long-term and safe therapeutic use may be taken into account in evaluating the safety of a medicine.

Information on quality required for an application for a new registered complementary medicine

One should present the data on quality in an application for evaluation of a new registered complementary medicine in a manner consistent with the European Medicines Agency (EMA) Common Technical Document (CTD) module 3: ICH M4Q Common Technical Document for the registration of pharmaceuticals for human use—Quality. While presentation of data in the CTD format is not mandatory, it is encouraged. Quality issues relating to the active ingredient/s and the finished product should be addressed. A list of the scientific guidelines on quality matters that have been adopted in Australia is available on the TGA website.

Information on safety and efficacy required for a new registered complementary medicine

If an ingredient or medicine is well described and appropriately referenced in reputable texts or publications (for example: Martindale—The Complete Drug Reference) the TGA will consider these sources in the assessment of safety and efficacy where these are provided in the application. Indications, dosage and route of administration must be consistent with the reference provided. The ARGOM Appendix 1: Guidelines on efficacy and safety aspects of OTC applications provide guidance for applicants choosing to submit a literature-based submission.

For other new medicines that are not well described in literature, nonclinical and clinical data will be required to support the safety and efficacy of the medicine. Safety and efficacy data should be presented as 'nonclinical' and 'clinical' data modules.

Data that demonstrate the safety of the medicine include information on history and pattern of use, biological activity, toxicology, clinical data and reports of adverse reactions. The overall safety of the medicine is dependent upon its formulation, its intended therapeutic purpose, dosage, method or route of administration, duration of use, the target patient group (such as children or the elderly) and the potential for interaction with other medication/s.

Safety may be established by detailed reference to the published literature and/or the submission of original study data. Where there is sufficient evidence based on human experience to support safety, the absence of extensive nonclinical investigations may be justifiable. Note that anecdotal or limited clinical reports of efficacy alone are not considered evidence of efficacy and safety.

Nonclinical Data

Pharmacology

Primary pharmacodynamics: in vitro and in vivo

Studies on primary pharmacodynamics should be provided and evaluated.

Secondary pharmacodynamics: in vitro and in vivo

Studies on secondary pharmacodynamics should be provided by organ system, where appropriate, and evaluated.

Safety pharmacology

Safety pharmacology studies should be provided and evaluated. In some cases, secondary pharmacodynamic studies can contribute to the safety evaluation when they assess potential adverse effects in humans.

Pharmacodynamic drug interactions

Where they have been performed, pharmacodynamic drug interactions should be provided.

Pharmacokinetics

Analytical methods and validation reports

Provide the methods of analysis for biological samples, including the detection and quantification limits of analytical procedures.

Absorption

Provide data on the extent and rate of absorption (*in vivo* and *in vitro* studies) and kinetic parameters, bioequivalence and/or bioavailability.

Distribution

Where available, provide data tissue distribution studies, protein binding and distribution in blood cells and placental transfer studies.

Metabolism

Where available, provide data on

- Chemical structures and quantities of metabolites in biological samples
- Possible metabolic pathways
- Pre-systemic metabolism
- *In vitro* metabolism including P450 studies
- Enzyme induction and inhibition

Excretion

Where available provide data on routes and extent of excretion and excretion in breast milk.

Pharmacokinetic drug interactions (nonclinical)

If they have been performed, provide nonclinical pharmacokinetic drug interaction studies (*in vitro* and *in vivo*).

Provide details of any contraindications or interactions with conventional and non conventional medicines.

Other pharmacokinetic studies

If studies have been performed in nonclinical models of disease they should be provided and evaluated.

Toxicology

Single dose toxicity

The single dose data should be provided in order of species, by route and evaluated.

Repeat dose toxicity

Studies should be provided in order of species, by route and by duration and evaluated.

Genotoxicity: in vitro and in vivo

Where available, *in vitro* and *in vivo* mammalian and non-mammalian cell system genotoxicity studies should be provided and evaluated.

Carcinogenicity: long term studies and short or medium term studies

Where available, carcinogenicity studies should be provided and evaluated.

Reproductive, developmental toxicity:

Where available, provide and evaluate studies on:

- Fertility and early embryonic development
- Embryo-foetal development
- Prenatal and postnatal development
- Studies in offspring.

Local tolerance

If local tolerance studies have been performed, these should be provided and evaluated.

Other toxicity studies

Provide any other studies such as: antigenicity, immunotoxicity, mechanistic studies, dependence, metabolites and impurities.

Clinical Data

Clinical data should preferably be presented as specified in Modules 2.5 Clinical Overview, 2.7 Clinical Summary and Module 5 Clinical Study Reports of the CTD format. The clinical overview provides a critical analysis of the clinical data in the dossier while the clinical summary is provides a detailed, factual summarisation of the clinical information.

Proposed 'Approved Herbal Name' (AHN) application form

The form is used to propose a botanical name for a herb (an 'Approved Herbal Name' (AHN)). This application will be assessed by the TGA's Herbal Ingredient Names Committee (HINC).

Proposed 'Herbal Component Name' (HCN) application form

The form is used for the name of a component comprising either a single chemical constituent or a particular group of chemical constituents found in herbal ingredients, where the ingredient is standardised to the constituent or group of constituents (a 'Herbal Component Name' (HCN)).

Proposed 'Herbal Substance Name' (AHS) application form

The form is used to apply for a name for a herbal material that is fully characterised in a monograph or pharmacopoeia (a 'Herbal Substance Name' (AHS)).

Registered medicine variation form (complementary medicines)

It is used to vary the particulars of complementary medicines which are already registered in the ARTG.

Notification of selective non-disclosure of Active Herbal Extract details

'Notification of selective non-disclosure of Active Herbal Extract details' form should be used to request the TGA to keep certain details regarding an 'Active Herbal Extract' (e.g., solvent type, concentration extract ratio and extract excipient details) confidential from sponsors of therapeutic goods. Please note that information such as herb species, plant part, basic preparation details (e.g., concentrated extract), ratio of equivalent dry weight to extract amount, and component percentage (if relevant) may not be confidential from the sponsor.

ANNEXURE 7: REGULATION OF HERBAL MEDICINE IN GERMANY

Commission E

The German Commission E is a scientific advisory board of the "BundesinstitutfürArzneimittel und Medizinprodukte" (the German equivalent of the Food and Drug Administration (FDA)), formed in 1978. The commission gives scientific expertise for the approval of substances and products previously used in traditional, folk and herbal medicine. The American Botanical Council based in Austin, Texas, published an English translation of the Commission E monograph text.

Germany's Commission E Monographs

Promoters have touted Germany's Commission E as the most accurate information in the world on herbal remedies; and some have made it appear as if Commission E was the equivalent of the U.S. FDA. However, neither of these descriptions is valid. Commission E obtains clinical reports from practitioners, which have little validity in a country with a strong tradition of romantic vitalism. In Germany, 7 out of 10 general practitioners employ "alternative" methods. It is true that Commission E has been responsible for the removal of over 100 herbal drugs from the marketplace, but a large dubious medicine industry still thrives there, with 90,000 of the 126,000 medicines on the market in Germany are homeopathic and natural medicines.

Market Importance of Herbal Medicines

Herbal remedies represent an important share of the German pharmaceutical market. According to an InstitutfürMedizinischeStatistik (IMS) report, presented during an ESCOP Symposium in Brussels in October 1990, the German herbal medicines market was worth US$ 1.7 billion (incl. VAT) in 1989, which was equal to 10% of the total pharmaceutical market in Germany.

A representative study carried out by the Allensbach Institute among the German population in June 1989 confirmed that a large number of people use natural medicines. The study showed that 58% of the population had taken such remedies, 44% of them within the previous year. It could also be shown that over the years the number of younger people using natural medicines had increased significantly. According to the study report, natural medicines were generally considered to be more harmless than chemical drugs. A majority among the German population (85%) believed that the experience of physicians, practitioners, and patients should be accepted as a proof for the efficacy of natural medicines.

Herbal medicines are distributed through over-the-counter sales in pharmacies and other distribution channels and on medical prescription through pharmacies. They are, in principle, reimbursable by the health insurance system unless special criteria for their exclusion apply, for example, specified indications such as common cold or laxatives, or substances, with a negative assessment by Commission E. Except for a few preparations, herbal medicines are not prescription-bound but can be prescribed by physicians or practitioners for reimbursement.

The total turnover of non-prescription-bound herbal medicines in pharmacies was DM 4.5 billion in 1995 (public price level), which is equal to almost 30% of the total turnover of non-prescription-bound medicines (DM 15.2 billion). Preparations sold on prescription amounted to DM 2.4 billion and those purchased through self-medication to DM 2.1 billion of the total turnover of non-prescription-bound phytomedicines. Herbal medicines can be found among the 2000 most important drugs prescribed by medical doctors and reimbursed by health insurances.

ANNEXURE 8: REGULATION OF HERBAL MEDICINE IN AFRICAN REGION

Introduction

In most parts of Africa, herbal medicines are formulated in small quantities and sold in open markets and stores without appropriate regulation by Government authorities. Most herbal products sold in public places lack scientific evidence for safety, efficacy and quality. Only 8 countries out of 34 that responded to WHO survey (WHO, 2002) had national regulations on Traditional Medicine. In response to this deplorable situation, the WHO has developed generic guidelines on various aspects of the development of herbal medicines including generic regulations and law. In a few countries, the situation is changing. This paper will present case studies on the regulatory situation on local production of herbal medicines in Ghana, Republic of Benin, Egypt, Nigeria and South Africa.

South Africa

The Bill/Law for regulating Traditional Health Practitioners (THPs) which is part of Complimentary and Alternative Medicine (CAM) was approved by parliament in 2004 (personal communication, Medicines Control Council, Department of Health, South Africa). Subsequently, a separate Bill on African Traditional Medicine was developed which was subjected to mandatory review processes by stakeholders. Presently, formulated African traditional medicines (ATMs) are regulated as nutritional supplements. Over 1,500 herbal products were received by the Medicines Control Council (MCC) of the Department of Health and listed for purposes of monitoring. The MCC would only consider an African traditional medicine for registration after appropriate clinical trials have been done with evidence of safety and efficacy.

At the moment, most CAM products are imported from India and China. However, many African traditional medicines are locally manufactured by pharmaceutical companies usually on contract basis and marketed as nutritional supplements. ATMs which target HIV/AIDS patients are marketed as immune boosters and are usually fortified with some relevant vitamins, minerals and amino acids. It has been estimated that the annual turnover of CAM in South Africa was about 2 billion rands (about $90 million) while the corresponding figure for essential conventional medicines was about 10 billion rands (about $450 million) in 2005 (personal communication, Medicines Control Council).

Republic of Benin

The Traditional Medicine policy was adopted in 2002. Subsequently, a Traditional Medicine Programme in the Ministry of Public Health with a Director was established thereby providing the administrative structure for implementing Traditional Medicine programme within the Ministry of Public Health. There are 14 Medicinal Plant Gardens located at different ecological zones in Benin. The Gardens were established by Government in collaboration with the THP Association. The Gardens are managed by THP under supervision of the Ministry of Public Health. There is limited production of herbal medicines for management of malaria, HIV/AIDS, sickle cell disorder, diabetes and hypertension. However, there is no commercial production of herbal medicines. Centre de MedecineTraditionelleExperimentalle, Takpe developed anti-malarial herbal medicine used widely in Benin and exported to Guinee Bissau, Ghana, Togo and Zimbabwe.

Egypt

Herbal medicine is officially recognized. The national registration requirements for herbal medicines are published. There are four pharmaceutical companies which are engaged in the manufacture of herbal medicines in accordance with GMP. The products are formulated into conventional dosage forms and they are generally well packaged.

Kenya

The Traditional Medicine Policy is actively being developed. The process has engaged all the stakeholders. Consequently, there are no regulations on Traditional Medicines until the policy is adopted. The complimentary and alternative medicines in the market are essentially imported from India and China.

Ghana

The Traditional and Alternative Medicine Directorate (TAMD) of the Ministry of Health is responsible for policy, institutional and regulatory aspects of traditional medicine. Traditional Medicine Practice Act 575 was adopted in 2000. The directorate has organised all the THP associations in Ghana into Ghana Federation of Traditional Medicine Associations (GHAFTRAM). Based on the National Traditional Medicine Policy, a second National Strategic Plan for the Development of Traditional and Alternative Medicine (2005–2009) was developed. Food and Drugs Board is responsible for registration of herbal medicines for sale in Ghana, monitors advertisements on herbal medicines, issues manufacturing and export licenses of herbal medicines and conducts GMP audit inspection on manufacturing premises. On the other hand, Ghana Standards Board (GSB) is responsible for setting standards for all goods locally manufactured. The Centre for Scientific Research into Plant

Medicine (CSRPM) is principally engaged in conducting research into medicinal plants and undertakes quality, safety and efficacy assessments of herbal medicines. There is a National Centre of Pharmacovigilance which routinely conducts safety monitoring of all medicines including herbal medicines. The plant raw materials are collected essentially from the wild. Herbal medicines are produced and marketed by THPs, private entrepreneurs, government institutions, NGOs or imported from India, USA, China, Korea and Egypt. Government supports CSRPM to produce herbal medicines while all other producers of herbal medicines make their own financial arrangements.

Nigeria

The Traditional Medicine policy was adopted in 2005 while the Traditional Medicine Bill was prepared, reviewed by stakeholders and approved by Federal Executive Council in 2006. The Traditional Medicine Bill is pending ratification by the National Assembly. There is a Deputy Director in the Federal Ministry of Health who is responsible for Traditional Medicine programmes. The National Agency for Food Administration and Control (NAFDAC) has developed guidelines for registration of herbal medicines. The agency has listed over 53 herbal medicines. The Nigerian Herbal Pharmacopoeia has been developed in collaboration with the World Health Organization and would be published in 2008. The National Pharmacovigilance Unit was established in 2005 under NAFDAC.

ANNEXURE 9: REGULATION OF HERBAL MEDICINE IN OTHER COUNTRIES

Albania

Herbal medicine is regulated in the "Law on Health Care" from 2009, section 20, Alternative medicine, as a therapeutic system that does not follow the general medical methods that are accepted and may not have a scientific explanation for their effectiveness. Treatments, conditions and manner of use of herbal medicine shall be determined by order of the Minister of Health. Advertisement and practice of herbal medicine treatment unlicensed by MoH are prohibited. There are no follow up regulations approved by MoH that regulates herbal medicine treatment.

Bulgaria

The use of non-pharmaceutical products of organic and mineral origin is regulated under the health law article 166.1.1 and 2. and art 167. It is understood that herbal medicine may be covered by this regulation.

Chile

In 1992 the Unidad de Medicina Tradicional was established with the aims of incorporating traditional medicine with proven efficacy into health programmes and of contributing to the establishment of their practice. Herbal products with therapeutic indications and/or dosage recommendations are considered to be drugs. Distribution of these products is restricted to pharmacies. A registration for marketing authorization is needed for herbal products, homeopathic products, and other natural products. An application for such registration consists of the complete formula, the labelling, samples of the product, and a monograph which permits identification of the formula and characteristics of the product.

Hungary

The CAM legislation from 1997 regulates Phytotherapy. Physicians and practitioners must have a licence to practise according to regulations. The non-medical practitioners have to be official registered members of the public health system.

Italy

The Italian "National Federation of the Orders of Doctors and Dentists" (FNOMCeO) has acknowledged phytotherapy treatment as a responsibility of a medical doctor or a dentist.

Korea (Republic of)

The Pharmaceutical Act of 1993 explicitly allowed pharmacists to prescribe and dispense herbal drugs.

Saudi Arabia

Registration of medicinal products by the Ministry of Health is obligatory, as is that of products, in addition to drugs, with medicinal claims or containing active ingredients having medicinal effects such as herbal preparations, health and supplementary food, medicated cosmetics, antiseptics or medical devices.

Turkey

The Turkish government has on October 27, 2014 passed a new law on Regulation of Traditional and Complementary Medicine Practice.

Further Readings for Annexures

AESGP (Association Européenne des Spécialités Pharmaceutiques Grand Public; The Association of the European Self-Medication Industry) (1998) Herbal Medicinal Products in the European Union. Study Carried out on Behalf of the European Commission, Brussels [http://pharmacos.eudra.org/F2/pharmacos/docs/doc99/Herbal%20Medecines%20EN.pdf.

Angell M, Kassirer JP. Alternative medicine—The risks of untested and unregular remedies. *New Engl J Med* 1998; **339**: 839–841.

Awang DVC. Quality control and good manufacturing practices: Safety and efficacy of commercial herbals. *Food Drug Law Inst* 1997; **52**: 341–344.

Blumenthal M, Busse WR, Goldberg A, Gruenwald J, Hall T, Riggins CW, Rister RS (eds.) (1998) The Complete German Commission E Monographs: Therapeutic Guide to Herbal Medicines. Austin, TX/Boston, MA: American Botanical Council/Integrative Medicine Communications.

Chang J. Scientific evaluation of traditional Chinese medicine under DSHEA: A conundrum. *J Altern Complem Med* 1999; **5**: 181–189.

Cho B-H. The politics of herbal drugs in Korea. *Soc Sci Med* 2000; **51**: 505–509.

EMEA (European Agency for the Evaluation of Medicinal Products). Working Party on Herbal Medicinal Products: Position paper on the risks associated with the use of herbal products containing Aristolochia species (EMEA/HMPWP/23/00), London, 2000.

ESCOP (European Scientific Cooperative on Phytotherapy). ESCOP Monographs on the Medicinal Uses of Plant Drugs, Exeter, UK, 1999.

European Commission. Council Directive 65/65/EEC of 26 January 1965 on the approximation of provisions laid down by Law, Regulation or Administrative Action relation to proprietary medicinal products. *Off J* 1965; **22**: 369–373.

European Commission. Council Directive 92/28/EEC of 31 March 1992 on the advertising of medicinal products for human use. *Off J* 1992a; **L113**: 13–18.

European Commission. Council Directive 92/25/EEC of 31 March 1992 on the wholesale distribution of medicinal products for human use. *Off J* 1992b; **L113**: 1–4.

European Commission. Second Council Directive 75/319/EEC of 20 May 1975 on the approximation of provisions laid down by Law, Regulation or Administrative Action relating to proprietary medicinal products. *Off J* 1975; **L147**: 13–22.

Food and Drug Administration. Guidance for Industry: Botanical Drug Products, Washington DC, Center for Drug Evaluation and Research [http://www.fda.gov/cder/ guidance/index.htm], 2000.

Food and Drug Administration. Good Manufacturing Practices (GMP)/Quality System (QS) Regulation [http://www.fda.gov/cdrh/dsma/cgmphome.html], 2002.

Ikegami F, Fujii Y, Satoh T. Toxicological considerations of Kampo medicines in clinical use. *Toxicology* 2004; **198**: 221–8.

Keller K. Legal requirements for the use of phytopharmaceutical drugs in the Federal Republic of Germany. *J Ethnopharmacol* 1991; **32**: 225–229.

Maegawa H, Nakamura T, Saito K. Regulation of traditional herbal medicinal products in Japan. *J Ethnopharmacol* 2014; **158** Pt B: 511–5.

Mei N, Guo L, Fu PP, Fuscoe JC, Luan Y, Chen T. Metabolism, genotoxicity, and carcinogenicity of comfrey. *J Toxicol Environ Health B Crit Rev* 2010; **13**: 509–26.

Ministry of Agriculture, Fisheries and Food (MAFF). Herb Legislation, London, Herb Society, 1998.

Natori S. Application of herbal drugs to health care in Japan. *J Ethnopharmacol* 1980; **2**: 65–70.

Oberlies NH, Kim NC, Brine DR, Collins BJ, Handy RW, Sparacino CM, Wani MC, Wall ME. Analysis of herbal teas made from the leaves of comfrey (*Symphytum officinale*): reduction of N-oxides results in order of magnitude increases in the measurable concentration of pyrrolizidine alkaloids. *Public Health Nutr* 2004; **7**: 919–24.

Saito H. Regulation of herbal medicines in Japan. *Pharmacol Res* 2000; **41**: 515–9.

Stickel F, Seitz HK. The efficacy and safety of comfrey. *Public Health Nutr* 2000; **3**: 501–8.

Therapeutic Goods Administration. Medicines Regulation and the TGA (December 1999), Woden, ACT, Australia, 1999.

WHO. Annex II. Guidelines for the Assessment of Herbal Medicines (WHO Technical Report Series No. 863), Geneva, 1996.

WHO. WHO Monographs on Selected Medicinal Plants, Vol. 1, Geneva, 1999.

Zhang X. Regulatory Situation of Herbal Medicines. A Worldwide Review (WHO/trm/98.1), Geneva, World Health Organization, 1998.

Index

T

V

W

X

Y

Z